LONDON MATHEMATICAL SOCIETY LECTURE NOTE

Managing Editor: Professor N.J. Hitchin, Mathematical Insti
University of Oxford, 24–29 St Giles, Oxford OX1 3LB, Un

The titles below are available from booksellers, or, in case of
at www.cambridge.org.

London Mathematical Society Lecture Note Series. 299

Kleinian Groups and Hyperbolic 3-Manifolds

Proceedings of the Warwick Workshop,
September 11–14, 2001

Edited by

Y. Komori
Osaka City University, Japan

V. Markovic
University of Warwick

C. Series
University of Warwick

CAMBRIDGE
UNIVERSITY PRESS

PUBLISHED BY THE PRESS SYNDICATE OF THE UNIVERSITY OF CAMBRIDGE
The Pitt Building, Trumpington Street, Cambridge, United Kingdom

CAMBRIDGE UNIVERSITY PRESS
The Edinburgh Building, Cambridge CB2 2RU, UK
40 West 20th Street, New York, NY 10011–4211, USA
477 Williamstown Road, Port Melbourne, VIC 3207, Australia
Ruiz de Alarcón 13, 28014 Madrid, Spain
Dock House, The Waterfront, Cape Town 8001, South Africa

http://www.cambridge.org

First published 2003

Printed in the United Kingdom at the University Press, Cambridge

Typeface Times 10/13 pt.   *System* LATEX   [TB]

A catalogue record for this book is available from the British Library

ISBN 0 521 54013 5 paperback

Contents

Part III   Related topics

Preface

This volume forms the proceedings of the workshop *Kleinian Groups and Hyperbolic 3-Manifolds* which was held at the Mathematics Institute, University of Warwick, 11–15 September 2001. Almost 80 people took part, many travelling large distances to come.

The workshop was organised around six expository lectures by Yair Minsky on the combinatorial part of his programme to extend his results on Thurston's ending lamination conjecture for once punctured tori to general surfaces. Not long after the workshop, a complete proof of the conjecture was announced by Brock, Canary and Minsky. This is undoubtedly one of the most important developments in the subject in the last decade, paving the way for a complete understanding of the internal geometry of hyperbolic 3-manifolds, and involving deep understanding of the fascinating links between this geometry and the combinatorics of the curve complex. Minsky's lectures, reproduced in this volume, give an invaluable overview.

As was clear from the talks at the conference, hyperbolic geometry is currently developing with astonishing rapidity. We hope that the expositions here will provide a useful guide. The volume is divided into three parts. Part I contains Minsky's lectures together with other articles on the geometry of hyperbolic 3-manifolds. The paper by Hodgson and Kerckhoff is an exposition of their recent work on cone manifolds. This is key to many recent developments, in particular Brock and Bromberg's proof, outlined here, of the long standing Bers' density conjecture for incompressible ends. Otal's result on unknottedness is an important ingredient. Other exciting developments concern the extension of known results from incompressible to compressible boundaries, based on a recent breakthrough by Kleineidam and Souto who here explain and develop some of their advances.

Part II revolves around several articles which revisit Troels Jørgensen's famous but hitherto unpublished paper *On pairs of once-punctured tori*. We are delighted to have Jørgensen's permission to include his paper here. Part III contains three expository papers on closely related topics, notably Epstein, Marden and Markovic's counterexample to Thurston's $K = 2$ conjecture.

We would like to thank the London Mathematical Society for funding the conference. The organisers were the first and last editors, Makoto Sakuma and Young Eun Choi. The editors owe a considerable debt to David Sanders whose TEXpertise rapidly got the articles for this volume into their present coherent shape. We would also like to thank Nicholas Wickens for assisting with the editorial work and Yasushi Yamashita for putting Jørgensen's article into TEX.

Finally, it will not have escaped notice that our conference began on September 11, 2001. Inevitably the terrible events of that day overshadowed the conference and occupied the thoughts of all the participants. We were very thankful that everyone arrived safely and the conference was able to proceed as planned.

Yohei Komori, Vladimir Markovic and Caroline Series
Mathematics Institute, University of Warwick, February 2003

Part I

Hyperbolic 3-manifolds

Kleinian Groups and Hyperbolic 3-Manifolds   Y. Komori, V. Markovic & C. Series (Eds.)
Lond. Math. Soc. Lec. Notes **299**, 3–40   Cambridge Univ. Press, 2003

Combinatorial and geometrical aspects of hyperbolic 3-manifolds

Yair N. Minsky[1]

1. Introduction

This is the edited and revised form of handwritten notes that were distributed with the lectures that I gave at the conference in Warwick on September 11–15 of 2001.[2] The goal of the lectures was to expose some recent work [Min03] on the structure of ends of hyperbolic 3-manifolds, which is part of a program to solve Thurston's Ending Lamination Conjecture (the conclusion of the program, which is joint work with J. Brock and R. Canary, will appear in [BCM]). In the interests of simplicity and the ability to get to the heart of the matter, the notes are quite informal in their treatment of background material, and the main results are often stated in special cases, with detailed examples taking the place of proofs. Thus it is hoped that the reader will be able to extract the main ideas with a minimal investment of effort, and in the event he or she is still interested, can obtain the details in [Min03], which will appear later on.

I would like to thank the organizers of the conference for inviting me and giving me the opportunity to talk for what must have seemed like a very long time.

1.1. Object of Study

If the interior N of a compact 3-manifold $\overline{N}$ admits a complete infinite-volume hyperbolic structure, then there is a multidimensional *deformation space* of such structures. The study of this space goes back to Poincaré and Klein, but the modern theory began with Ahlfors-Bers in the 1960's and received the perspective that we will focus on from Thurston and others in the late 70's. The deformation theory depends deeply on an understanding of the geometry of the *ends* of N (in the sense of Freudenthal [Fre42]), which one can think of as small neighborhoods of the boundary components of $\overline{N}$.

[1]Based on work partially supported by NSF grant DMS-9971596.

[2]The terrible events in New York that coincided with the beginning of this conference overshadow its subject matter in significance, and yet those same events demand of us to continue with our ordinary work.

The interior of the deformation space, as studied by Ahlfors, Bers [AB60, Ber60, Ber70b], Kra [Kra72], Marden [MMa79, Mar74], Maskit [Mak75], and Sullivan [Sul85], can be parametrized using the Teichmüller space of $\partial \overline{N}$ – that is, by choosing a "conformal structure at infinity" for each (non-toroidal) boundary component of $\overline{N}$. (See also [KS93] and [BO01] for other approaches to the study of the interior). The boundary contains manifolds with parabolic cusps [Mak70, McM91], and more generally, with *geometrically infinite ends* [Ber70a, Gre66, Thu79]. The Teichmüller parameter is replaced by Thurston's *ending laminations* for such ends. Thurston conjectured [Thu82] that these invariants are sufficient to determine the geometry of N uniquely – this is known as the Ending Lamination Conjecture (see also [Abi88] for a survey).

In these notes we will consider the special case of *Kleinian surface groups*, for which $\pi_1(N)$ is isomorphic to $\pi_1(S)$ for a surface S. This case suffices for describing the ends of general N, provided $\partial \overline{N}$ is incompressible. (In the compressible case the deeper question of Marden's *tameness conjecture* comes in, and this is beyond the scope of our discussion. See Marden [Mar74] and Canary [Can93b].)

We will show how the the ending laminations, together with the combinatorial structure of the set of simple closed curves on a surface, allows us to build a *Lipschitz model* for the geometric structure of N, which in particular describes the thick-thin decomposition of N. These results, which are proven in detail in [Min03], will later be followed by *bilipschitz* estimates in Brock–Canary–Minsky [BCM], and these will suffice to prove Thurston's conjecture in the case of incompressible boundary.

1.2. Kleinian surface groups

From now on, let S be an oriented compact surface with $\chi(S) < 0$, and let

$$\rho : \pi_1(S) \to \mathrm{PSL}_2(\mathbb{C})$$

be a discrete, faithful representation. If $\partial S \neq \emptyset$ we require $\rho(\gamma)$ to be parabolic for γ representing any boundary component. This is known as a (marked) Kleinian surface group. We name the quotient 3-manifold

$$N = N_\rho = \mathbb{H}^3 / \rho(\pi_1(S)).$$

Periodic manifolds   Before discussing the general situation let us consider a well-known and especially tractable example.

Let $\varphi : S \to S$ be a pseudo-Anosov homeomorphism (this means that φ leaves no finite set of non-boundary curves invariant up to isotopy). The mapping torus of φ is

$$M_\varphi = S \times \mathbb{R} / \langle (x,t) \mapsto (\varphi(x), t+1) \rangle,$$

a surface bundle over S^1 with fibre S and monodromy φ. Thurston [Thu86b] showed, as part of his hyperbolization theorem, that $int(M_\varphi)$ admits a hyperbolic structure which we'll call N_φ (see also Otal [Ota96] and McMullen [McM96]). Let $N \cong int(S) \times \mathbb{R}$ be the infinite cyclic cover of N_φ, "unwrapping" the circle direction (Figure 1). After identifying S with some lift of the fibre, we obtain an isomorphism $\rho : \pi_1(S) \to \pi_1(N) \subset \mathrm{PSL}_2(\mathbb{C})$, which is a Kleinian surface group.

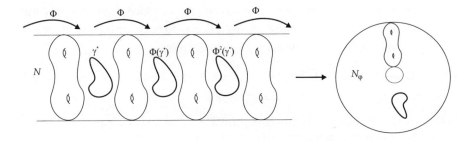

Figure 1: N covers the surface bundle N_φ.

The deck translation $\Phi : N \to N$ of the covering induces $\Phi_* = \varphi_* : \pi_1(S) \to \pi_1(S)$. We next consider the action of φ on the space of *projective measured laminations* $\mathscr{PML}(S)$ (see [FLP79, Bon01], and Lecture 3). For every simple closed curve γ in S, the sequences $[\varphi^n(\gamma)]$ and $[\varphi^{-n}(\gamma)]$ converge to two distinct points v_+ and v_- in $\mathscr{PML}(S)$. After isotopy, φ can be represented on S by a map that preserves the leaves of both v_+ and v_-, stretching the former and contracting the latter.

We can see $v_\pm$ directly in the asymptotic geometry of N: For a curve γ in S, let γ^* be its geodesic representative in N. Now consider $\Phi^n(\gamma^*)$ – these are all geodesics of the same length, marching off to infinity in both directions as $n \to \pm\infty$, and note that $\Phi^n(\gamma^*) = \varphi^n(\gamma)^*$. So, we have a sequence of simple curves in S, converging to v_+ as $n \to \infty$, whose geodesic representatives "exit the $+$ end" of N (similarly as $n \to \infty$ they converge to v_- and the geodesics exit the other end).

The laminations $v_\pm$ are the ending laminations of ρ in this case. To understand the general case we will need to develop a bit of terminology, and recall the work of Thurston and Bonahon.

Ends   Let N_0 denote N minus its cusps (each cusp is an open solid torus, whose boundary in N is a properly embedded open annulus). The relative version of Scott's core theorem (see McCullough [McC86], Kulkarni–Shalen [KSh89] and Scott [Sco73b]) gives us a compact submanifold K in N_0, homeomorphic to $S \times [0,1]$, which meets each cusp boundary in an annulus (including the annuli $\partial S \times [0,1]$). The components of $N_0 \setminus K$ are in one to one correspondence with the topological ends of N_0, and are called neighborhoods of the ends (see Bonahon [Bon86]).

N also has a *convex core* C_N, which is the smallest closed convex submanifold whose inclusion is a homotopy equivalence. Each end neighborhood either meets C_N in a bounded set, in which case the end is called *geometrically finite*, or is contained in C_N, in which case the end is *geometrically infinite*.

From now on, let us assume that N has *no extra cusps*, which means that the cusps correspond only to the components of ∂S. In particular N_0 has exactly two ends, which we label $+$ and $-$ according to an appropriate convention.

Simply degenerate ends   In [Thu79], Thurston made the following definition, which can be motivated by the surface bundle example:

Definition 1.1. An end of N is **simply degenerate** if there exists a sequence of simple closed curves α_i in S such that α_i^* exit the end.

Here "exiting the end" means that the geodesics are eventually contained in an arbitrarily small neighborhood of the end, and in particular outside any compact set. Note that a geometrically finite end cannot be simply degenerate, since all closed geodesics are contained in the convex hull.

Thurston then established this theorem (stated in the case without extra cusps):

Theorem 1.2. [Thu79]   *If an end e of N is simply degenerate then there exists a unique lamination ν_e in S such that for any sequence of simple closed curves α_i in S,*

$$\alpha_i \to \nu_e \iff \alpha_i^* \text{ exit the end } e.$$

A sequence $\alpha_i \to \nu_e$ can be chosen so that the lengths $\ell_N(\alpha_i^*) \le L_0$, where L_0 depends only on S.

Furthermore, ν_e fills S – its complement consists of ideal polygons and once-punctured ideal polygons.

(We are being cagey here about just what kind of lamination ν_e is, and what convergence $\alpha_i \to \nu_e$ means. See Lecture 3 for more details.)

Thurston also proved that simply degenerate ends are *tame*, meaning that they have neighborhoods homeomorphic to $S \times (0, \infty)$, and that manifolds obtained as limits of quasifuchsian manifolds have ends that are geometrically finite or simply degenerate. Bonahon completed the picture with his "tameness theorem",

Theorem 1.3. [Bon86] *The ends of N are either geometrically finite or simply degenerate.*

In particular N_0 is homeomorphic to $S \times \mathbb{R}$, and ending laminations are well-defined for each geometrically infinite end.

Geometrically finite ends are the ones treated by Ahlfors, Bers and their coworkers, and their analysis requires a discussion of quasiconformal mappings and Teichmüller theory (see [Ber60, Ber70b, Sul86] for more). In order to simplify our exposition we will limit ourselves, for the remainder of these notes, to Kleinian surface groups ρ with no extra cusps, and with no geometrically finite ends. In particular the convex hull of N_ρ is all of N_ρ, and there are two ending laminations, ν_+ and ν_-. This is called the *doubly degenerate* case.

1.3. Models and bounds

Our goal now is to recover geometric information about N_ρ from the asymptotic data encoded in $\nu_\pm$. The following natural questions arise, for example:

- Thurston's Theorem 1.2 guarantees the existence of a sequence $\alpha_i \to \nu_+$ whose geodesic representatives have bounded lengths $\ell_N(\alpha_i^*)$. How can we determine, from ν_+, which sequences have this property?

- The case of the cyclic cover of a surface bundle is not typical: because it covers a compact manifold (except for cusps), it has "bounded geometry". That is,

$$\inf_\beta \ell_N(\beta) > 0$$

  where β varies over closed geodesics. The bounded geometry case is considerably easier to understand. In particular the Ending Lamination Conjecture in this category (without cusps) was proven in [Min93, Min94].

  Can we tell from $\nu_\pm$ alone whether N has bounded geometry?

- If N doesn't have bounded geometry, there are arbitrarily short closed geodesics in N, cach onc encased in a *Margulis tube*, which is a standard collar neighborhood. Such examples were shown to exist by Thurston [Thu86b] and Bonahon-Otal [BO88], and to be generic in an appropriate sense by McMullen [McM91].

  In the unbounded geometry case, can we tell *which* curves in N are short? How are they arranged in N?

We will describe the construction of a "model manifold" M_ν for N, which can be used to answer these questions. M_ν is constructed combinatorially from $\nu_\pm$, and contains for example solid tori that correspond to the Margulis tubes of short curves in N. M_ν comes equipped with a map

$$f : M_\nu \to N$$

which takes the solid tori to the Margulis tubes, is proper, Lipschitz in the complement of the solid tori, and preserves the end structure. This will be the content of the Lipschitz Model Theorem, which will be stated precisely in Lecture 6.

Note that if f is *bilipschitz* then the Ending Lamination Conjecture follows: If N_1, N_2 have the same invariants $\nu_\pm$ then the same model M_ν would admit bilipschitz maps $f_1 : M_\nu \to N_1$ and $f_2 : M_\nu \to N_2$, and $f_2 \circ f_1^{-1} : N_1 \to N_2$ would be a bilipschitz homeomorphism. By Sullivan's Rigidity Theorem [Sul81a], N_1 and N_2 would be isometric.

1.4. Plan

Here is a rough outline of the remaining lectures:

§2 **Hierarchies and model manifolds:** We will show how to build M_ν starting with a geodesic in the *complex of curves* $\mathscr{C}(S)$. The main tool is the *hierarchy of geodesics* developed in Masur-Minsky [MM00]. Much of the discussion will take place in the special case of the 5-holed sphere $S_{0,5}$, where the definitions and arguments are considerably simplified.

§3 **From ending laminations to model manifold:** Using a theorem of Klarreich we will relate ending laminations to *points at infinity* for $\mathscr{C}(S)$, and this will allow us to associate to a pair of ending laminations a geodesic in $\mathscr{C}(S)$, and its associated hierarchy and model manifold.

§4 **The quasiconvexity argument:** We then begin to explore the linkage between geometry of the 3-manifold N_ρ and the curve complex data. We will show that the subset of $\mathscr{C}(S)$ consisting of curves with bounded length in N is *quasiconvex*. The main tool here is an argument using pleated surfaces and Thurston's Uniform Injectivity Theorem.

§5 **Quasiconvexity and projection bounds:** In this lecture we will discuss the Projection Bound Theorem, a strengthening of the Quasiconvexity Theorem that shows that curves that appear in the hierarchy are combinatorially close to the bounded-length curves in N. We will also prove the Tube Penetration Theorem, which controls how deeply certain pleated surfaces can enter into Margulis tubes.

§6 **A priori length bounds and model map:** Applying the Projection Bound Theorem and the Tube Penetration Theorem, we will establish a uniform bound on the lengths of all curves that appear in the hierarchy.

We will then state the Lipschitz Model Theorem, whose proof uses the a priori bound and a few additional geometric arguments. As consequences we will obtain some final statements on the structure of the set of short curves in N.

2. Curve complex and model manifold

In this lecture we will introduce the complex of curves $\mathscr{C}(S)$ and demonstrate how a geodesic in $\mathscr{C}(S)$ leads us to construct a "model manifold". For simplicity we will mostly work with $S = S_{0,5}$, the sphere with 5 holes. (In general let $S_{g,n}$ be the surface with genus g and n boundary components).

2.1. The complex of curves

$\mathscr{C}(S)$ will be a simplicial complex whose vertices are homotopy classes of simple, essential, unoriented closed curves ("Essential" means homotopically nontrivial, and not homotopic to the boundary). Barring the exceptions below, we define the k-simplices to be unordered $k+1$-tuples $[v_0 \dots v_k]$ such that $\{v_i\}$ can be realized as pairwise disjoint curves. This definition was given by Harvey [Hav81].

Exceptions: If $S = S_{0,4}$, $S_{1,0}$ or $S_{1,1}$ then this definition gives no edges. Instead we allow edges $[vw]$ whenever v and w can be realized with

$$\#v \cap w = \begin{cases} 1 & S_{1,0}, S_{1,1} \\ 2 & S_{0,4}. \end{cases}$$

(see Figure 2). In this case $\mathscr{C}(S)$ is the *Farey graph* in the plane: a vertex is indexed by the slope p/q of its lift to the planar $\mathbb{Z}^2$ cover of S, so the vertex set is $\widehat{\mathbb{Q}} = \mathbb{Q} \cup \infty$. Two vertices p/q, r/s are joined by an edge if $|ps - qr| = 1$ (see e.g. Series [Ser85a] or [Min99]).

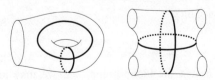

Figure 2: Adjacent vertices in $\mathscr{C}(S_{1,1})$ and $\mathscr{C}(S_{0,4})$

For $S_{0,0}, S_{0,1}, S_{0,2}, S_{0,3}$: $\mathscr{C}(S)$ is empty. (For the annulus $S_{0,2}$ there is another useful construction which we will return to later.)

Let $\mathscr{C}_k(S)$ denote the k-skeleton of $\mathscr{C}(S)$. We will concentrate on $\mathscr{C}_0$ and $\mathscr{C}_1$.

We endow $\mathscr{C}(S)$ with the metric that makes every simplex regular Euclidean of sidelength 1. Thus $\mathscr{C}_1(S)$ is a graph with unit-length edges. Consider a geodesic in $\mathscr{C}_1(S)$ – it is a sequence of vertices $\{v_i\}$ connected by edges (Figure 3), and in particular: v_i, v_{i+1} are disjoint (in the non-exceptional cases), v_i and v_{i+2} intersect but are disjoint from v_{i+1}, and v_i and v_{i+3} *fill* the surface: their union intersects every essential curve. It is harder to characterize topologically the relation between v_i and v_j for $j > i + 3$.

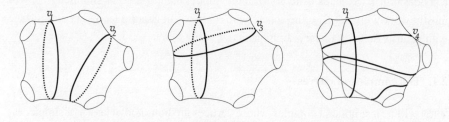

Figure 3: $\{v_1, v_2, v_3, v_4\}$ are the vertices of a geodesic in $\mathscr{C}(S_{0,5})$.

2.2. Model construction

Let $S = S_{0,5}$ – this case is considerably simpler than the general case, while preserving many of the main features.

Starting with a *bi-infinite geodesic* g in $\mathscr{C}_1(S)$ (more about the existence of such geodesics later), we will construct a manifold $M_g \cong S \times \mathbb{R}$, equipped with a piecewise-Riemannian metric. M_g is made of "standard blocks", all isometric, and "tubes", or solid tori of the form (annulus) × (interval).

Hierarchy   We begin by "thickening" g in the following sense: Any vertex $v \in \mathscr{C}_0(S)$ divides S into two components, one $S_{0,3}$ and one $S_{0,4}$. Let W_v denote the second of these. If v_i is a vertex of g then

$$v_{i-1}, v_{i+1} \in \mathscr{C}_0(W_{v_i}).$$

The complex $\mathscr{C}(W_{v_i})$ is just the Farey graph, and we may join v_{i-1} to v_{i+1} by a geodesic in that graph. Name this geodesic h_i, and represent it schematically as in Figure 4.

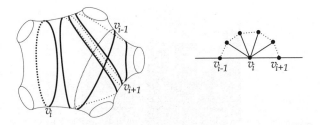

Figure 4: The local configuration at a vertex v_i of g yields a "wheel" in the link of v_i. Note, edges of h_i are not edges of $\mathscr{C}(S)$; call them "rim" edges. The other edges are called "spokes".

We repeat this at every vertex. The resulting system is called a *hierarchy of geodesics*. (In general surfaces, considerable complications arise. Geodesics must satisfy a technical condition called "tightness", and the hierarchy has more levels. This is joint work with Masur [MM99, MM00].)

Note that the construction is not uniquely dependent on g – geodesics are not always unique in the Farey graph, so there are arbitrary choices for each h_i. However what we have to say will work regardless of how the choices are made.

Blocks   To each rim edge e we associate a "block" $B(e)$, and then glue these together to form the model manifold. e is an edge of $\mathscr{C}(W_v)$ for some vertex v – denote $W_e \equiv W_v$ for convenience. Let e^-, e^+ be its vertices, ordered from left to right. Let C_+ and C_- be open collar neighborhoods of e^+ and e^-, respectively. We define

$$B(e) = W_e \times [-1, 1] - (C_+ \times (1/2, 1] \cup C_- \times [-1, -1/2)).$$

Thus we have removed solid-torus "trenches" from the top and bottom of the product $W_e \times [-1, 1]$. Figure 5 depicts this as a gluing construction.

The boundary $\partial B(e)$ divides into four *3-holed spheres*,

$$\partial_\pm B(e) \equiv (W_e - C_\pm) \times \pm 1$$

and some *annuli*. Schematically, we depict this structure in Figure 6.

Gluing   Take the disjoint union of all the blocks arising from the hierarchy over g, and glue them along 3-holed sphere, where possible. That is, if $Y \times \{1\}$ appears in $\partial_+ B(e_1)$ and $Y \times \{-1\}$ appears in $\partial_- B(e_2)$, identify them using the identity map in Y.

(A technicality we are eliding is that subsurfaces are determined only up to isotopy; one can select one representative for each isotopy class in a fairly nice and consistent way.)

There are three types of gluings that can occur:

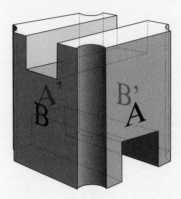

Figure 5: Construct a block $B(e)$ by doubling this object along A, A', B and B'. The curved vertical faces become $\partial W_e \times [-1, 1]$.

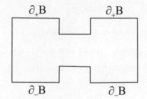

Figure 6: Schematic diagram of the different pieces of the boundary of a block.

1. Both edges occur in the link of the same vertex v; $W_{e_1} = W_{e_2} = W_v$, and $e_1^+ = e_2^-$ (Figure 7). $B(e_1)$ and $B(e_2)$ are glued along $W_v \setminus C_{e_1^+}$, which is composed of three-holed spheres Y_1 and Y_2.

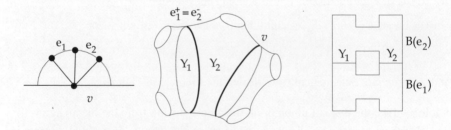

Figure 7

2. $e_1 \subset \mathscr{C}(W_u)$ and $e_2 \subset \mathscr{C}(W_v)$, where u and v are two succesive vertices (Figure 8). Now $e_1^+ = v$ and $e_2^- = u$, and the gluing is along $Y_2 = W_u \cap W_v$, which separates $S_{0,5}$.

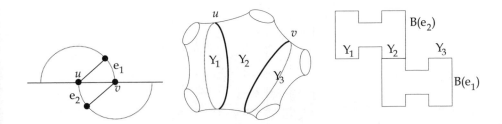

Figure 8

3. $e_1 \subset \mathscr{C}(W_u)$ and $e_2 \subset \mathscr{C}(W_w)$, where u, v, w are three successive vertices (Figure 9). In this case $e_1^+ = e_2^- = v$, and the gluing is along Y_1 which is isotopic to $W_u \cap W_w$ and does not separate $S_{0,5}$. Note that the intersection pattern of u and w is typically more complicated than pictured, as $d_{W_v}(u, w) \gg 1$.

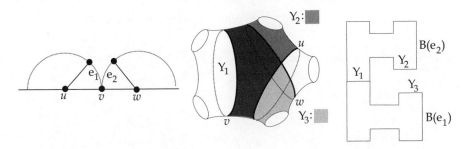

Figure 9

When we fit all the blocks together, the result can be embedded in $S \times \mathbb{R}$, in such a way that any level surface $Y \times \{t\}$ in a block is mapped to $Y \times \{s\}$ in $S \times \mathbb{R}$ by a map that is the identity on the first factor. Call such a map "straight". Note that the blocks can be stretched *vertically* in different ways.

In the gaps between blocks we find solid tori of the form

$$C \times (s, t)$$

where C is one of our collar neighborhoods of a vertex in the hierarchy. Call these the *tubes* of the model.

We should of course verify these claims about the gluing operations. Two things to check are:

1. All the vertices in the hierarchy are distinct (and hence all the tubes are homotopically distinct in $S \times \mathbb{R}$).

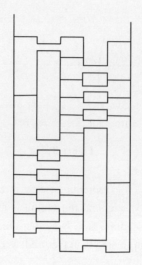

Figure 10: A schematic of the embedding of the blocks in $S \times \mathbb{R}$

2. The gluings we have shown are the only ones.

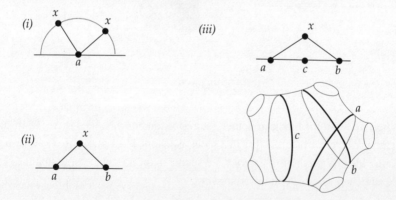

Figure 11: The three ways that a vertex can appear twice

To verify (1), suppose that a vertex x appears in two places in the hierarchy. That is, x is in the wheels (links) of vertices a and b in g. The triangle inequality in $\mathscr{C}(S)$ implies that $d(a,b) \leq 2$, and since g is a geodesic this leaves three possibilities, as in Figure 11.

In case *(i)*, $a = b$. This is not possible since the "rim" path is a geodesic in $\mathscr{C}(W_a)$.

In case *(ii)*, a, b and x make a triangle in $\mathscr{C}(S)$, but $\mathscr{C}(S)$ has no triangles for $S = S_{0,5}$.

In case *(iii)*, *a* and *b* "fill" the 4-holed sphere W_c bounded by *c*, so that if *x* is represented by a curve disjoint from both, it is equal to *c* or lies on the complement of W_c. That complement is a 3-holed sphere so the only possibility is that $x = c$. In other words *x* really only appears once, as a vertex of *g*.

To prove (2), we must consider how a gluing surface *Y* (a 3-holed sphere) can occur. There are several possibilities for the curves of ∂Y (Figure 12).

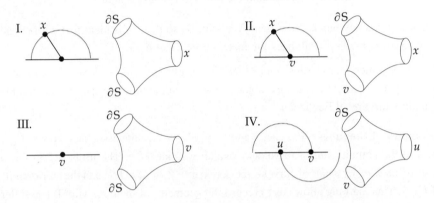

Figure 12

I. ∂Y consists of *x* and two curves of ∂S, where *x* is an interior "rim" vertex.

II. ∂Y consists of an interior rim vertex *x*, a *g* vertex *v*, and a curve of ∂S.

III. ∂Y consists of *v* and two curves of ∂S, where *v* is a vertex on *g*.

IV. ∂Y consists of two adjacent vertices *u*, *v* of *g* and a curve of ∂S.

Types I and II occur in pairs as the top and bottom surfaces of two blocks associated to adjacent rim edges meeting at the same *x*. Types III and IV occur on blocks associated to first and last edges in rim geodesics, and each one occurs in exactly two ways. It is therefore not hard to check all the possibilities and see that the gluings we described indeed produce a manifold.

The embedding of the manifold into $S \times \mathbb{R}$ can be done inductively, by "sweeping" across the hierarchy from left to right.

2.3. Geometry of the model

Fix one *standard block*: Take a copy of *W* (of type $S_{0,4}$) with two curves α, β that are neighbors in $\mathscr{C}(W)$, collars C_α, C_β, and construct a block B_0 as before out of

$W \times [-1,1]$. Give this block *some* metric with these properties:

- Symmetry of gluing surfaces: Each component of $\partial_\pm B_0$ is isometric to a *fixed* copy of $S_{0,3}$, which admits a 6-fold orientation-preserving symmetry group permuting the boundary components.

- Flat annuli: All of the annuli of $\partial B_0 \setminus \partial_\pm B_0$ are *flat* – that is, isometric to a circle cross an interval. We assume that all the circles have length 1.

Now given any block $B(e)$, identify it with B_0 so that e^- is identified with α and e^+ is identified with β. Pull back the metric from B_0 to $B(e)$.

The symmetry properties imply that all the gluings can be done by isometries (possibly after isotopy). Thus we obtain a metric on the union of the blocks, and the boundary tori are all Euclidean.

Geometry of the tubes   For each vertex v in H we have the associated "tube" $C_v \times (s,t)$ in the complement of the blocks, which we call $U(v)$. The torus $\partial U(v)$ has a *natural marking* by a pair of curves – the core curve γ_v of $C_v \times \{s\}$, and the meridian μ_v of $U(v)$. This marking allows us to record the geometry of the torus via "Teichmüller data": ∂U is a Euclidean torus in the metric inherited from the blocks, and there is a unique number

$$\omega \in \mathbb{H}^2 = \{z \in \mathbb{C} : \mathrm{Im}\, z > 0\}$$

such that $\partial U(v)$ can be identified by an orientation-preserving isometry with the quotient $\mathbb{C}/(\mathbb{Z} + \omega\mathbb{Z})$, such that $\mathbb{R}$ and $\omega\mathbb{R}$ map to the classes of γ_v and μ_v, respectively. We define $\omega_M(v) \equiv \omega$, the *vertex coefficient* of v. Note that $|\omega_M(v)|$ is the length of the meridian μ_v.

We can then extend the metric on ∂U to make U a "hyperbolic tube" as follows: Given $r > 0$ and $\lambda \in \mathbb{C}$ with $\mathrm{Re}\,\lambda > 0$, let $\mathbb{T}(\lambda, r)$ denote the quotient of an r-neighborhood of a geodesic L in $\mathbb{H}^3$ by a translation γ whose axis is L and whose complex translation distance is λ. The boundary $\partial\mathbb{T}(\lambda, r)$ is a Euclidean torus, on which there is a natural marking by a representative of γ and by a meridian. Hence we obtain a Teichmüller coefficient $\omega(\lambda, r)$ as above. It is a straightforward exercise to show that, given $\omega_M(v)$ there is a unique (λ, r) such that after identifying the markings we have $\omega(\lambda, r) = \omega_M(v)$, and the length of the γ curve in $\partial\mathbb{T}(\lambda, r)$ is 1. We then put a metric on U by identifying it with $\mathbb{T}(\lambda, r)$.

It is not hard to check that as $|\omega_M(v)| \to \infty$, the radius r of the tube goes to ∞, and the length $|\lambda|$ goes to 0.

Let M_g denote the union of blocks and tubes, with the metric we have described, and the identification with $S \times \mathbb{R}$ we have given.

3. From ending laminations to model manifold

Given a doubly degenerate Kleinian surface group $\rho : \pi_1(S) \to \mathrm{PSL}_2(\mathbb{C})$, Theorem 1.2 gives us a pair of ending laminations ν_+, ν_-. How do these determine a geodesic and a hierarchy from which we can build a model? Roughly speaking, the laminations are "endpoints at ∞" for the hierarchy.

3.1. Background

Hyperbolicity   With Masur in [MM99], we proved that

Theorem 3.1. *$\mathscr{C}(S)$ is a δ-hyperbolic metric space.*

We recall the definition, due to Cannon and Gromov [Gro87, Cnn91]: A geodesic metric space S is δ-hyperbolic if all triangles are "δ-thin". That is, given a geodesic triangle $[xy] \cup [yz] \cup [xy]$, each side is contained in a δ-neighborhood of the union of the other two.

This simple synthetic property has many important consequences, and gives X large-scale properties analogous to those of the classical hyperbolic space $\mathbb{H}^n$, and any infinite metric tree. In particular, X has a *boundary at infinity*, ∂X, defined roughly as follows: we fix a basepoint x_0 and endow X with a "contracted" metric d_0 in which $x, y \in X$ are close if

- they are close in the original metric of X, or

- they are "visually close" as seen from x_0 – that is, geodesic segments $[x_0 x]$ and $[x_0 y]$ have large initial segments $[x_0 x']$ and $[x_0 y']$ which are in δ-neighborhoods of each other (figure 13).

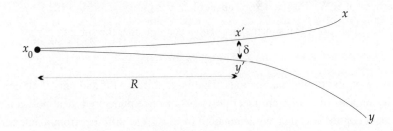

Figure 13

The completion of X in this contracted metric yields new points, which comprise ∂X. The construction does not in fact depend on the choice of x_0. See [ABC+91] for more details. The boundary of $\mathscr{C}(S)$ turns out to be a certain *lamination space*:

Laminations   Thurston introduced the space of *measured geodesic laminations* on a surface S, $\mathcal{ML}(S)$. Fixing a complete finite-area hyperbolic metric on $int(S)$, a geodesic lamination is a closed subset foliated by geodesics. A *transverse measure* on a geodesic lamination is a family of Borel measures on arcs transverse to the lamination, invariant by holonomy; that is, by sliding along the leaves. (See Figure 14).

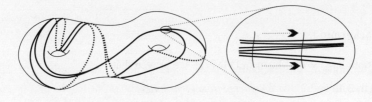

Figure 14: A geodesic lamination with closeup showing a transversal sliding along the leaves.

This space has a natural topology, which makes $\mathcal{ML}(S)$ homeomorphic to $\mathbb{R}^{6g-6+2n}$ when $S = S_{g,n}$. The choice of the hyperbolic metric is not important; all choices yield naturally homeomorphic spaces. See Bonahon [Bon01] for more. Taking the quotient of $\mathcal{ML}(S)$ (minus the empty lamination) by scaling of the measures yields the sphere $\mathcal{PML}(S)$ which was mentioned in Lecture 1. We will actually need to consider a stronger quotient, the space of "unmeasured laminations"

$$\mathcal{UML}(S) = \mathcal{ML}(S)/\text{measures}$$

Thus this is the space of all geodesic laminations which are the supports of measures, with a quotient topology obtained by forgetting the measure. This is different from from $\mathcal{PML}(S)$ because of the existence of non-uniquely ergodic laminations, and its topology is different from the topology on plain geodesic laminations obtained by Hausdorff convergence of compact subsets of S. Note that the simple closed curves, i.e. vertices of $\mathcal{C}(S)$, form a dense subset of $\mathcal{UML}(S)$.

$\mathcal{UML}(S)$ is not a Hausdorff space. However, consider the subset

$$\mathcal{EL}(S) \subset \mathcal{UML}(S)$$

consisting of all "filling" laminations. That is, $\lambda \in \mathcal{EL}(S)$ if and only if all complementary regions of λ in S are ideal polygons or once-punctured ideal polygons. An equivalent condition is that any lamination in $\mathcal{UML}(S)$ different from λ intersects it transversely. We then have (Klarreich [Kla]) that $\mathcal{EL}(S)$ is a Hausdorff space. Furthermore, elements of $\mathcal{EL}(S)$ are exactly those laminations that occur as *ending laminations* for manifolds N_ρ without extra parabolics. (In Theorem 1.2, the convergence to the ending lamination can now be understood as convergence in $\mathcal{UML}(S)$.)

Klarreich showed in [Kla] that:

Theorem 3.2. *There is a homeomorphism*

$$k : \partial \mathscr{C}(S) \to \mathscr{EL}(S)$$

such that a sequence $\beta_i \in \mathscr{C}_0(S)$ converges to $\beta \in \partial \mathscr{C}(S)$ if and only if it converges to $k(\beta)$ in the topology of $\mathscr{UML}(S)$.

Thus, ending laminations are points at infinity for $\mathscr{C}(S)$, and from now on we identify $\partial \mathscr{C}(S)$ with $\mathscr{EL}(S)$.

From lamination to hierarchy Now given a doubly degenerate $\rho : \pi_1(S) \to$ $\mathrm{PSL}_2(\mathbb{C})$, with ending laminations $\nu_\pm$, we would like to produce a bi-infinite geodesic g in $\mathscr{C}_1(S)$ whose endpoints on $\partial \mathscr{C}(S)$ are $\nu_\pm$.

If $\mathscr{C}(S)$ were *locally finite*, this would be easy: Take a sequence $\{x_i\}_{i=-\infty}^{\infty}$ in $\mathscr{C}_0(S)$ such that

$$\lim_{i \to \pm\infty} x_i = \nu_\pm$$

and note that, by hyperbolicity of $\mathscr{C}(S)$ and the definition of $\partial \mathscr{C}(S)$, the geodesic segments $[x_{-n}, x_n]$ and $[x_{-m}, x_m]$ are 2δ-fellow travelers on larger and larger segments as $n, m \to \infty$. Thus we would expect, after extracting a subsequence, to obtain a limiting geodesic with the endpoints $\nu_\pm$ at infinity.

For $\mathscr{C}(S)$, which is *not* locally finite, the convergence step is not automatic. The machinery in [MM00] gives a way of getting around this, and extracting a convergent subsequence after all. We will leave out this argument, and assume from now on that we have a geodesic with endpoints $\nu_\pm$.

Now, for $S = S_{0,5}$, we are ready to repeat the hierarchy ("wheel") construction of the previous lecture, and build from this our model manifold.

Our discussion so far yields us the following: Given a doubly-degenerate Kleinian surface group $\rho : \pi_1(S) \to \mathrm{PSL}_2(\mathbb{C})$, we obtain via Bonahon-Thurston its ending laminations $\nu_\pm \in \mathscr{EL}(S)$. Using Klarreich's theorem and the work in [MM00], we produce a geodesic g and a hierarchy H_ν, and a model manifold M_ν – all depending only on $\nu_\pm$ and not on ρ. Our next task will be to connect the geometry of M_ν to the geometry of the hyperbolic 3-manifold N_ρ.

4. The quasiconvexity argument

Our goal, in §6, is to produce a map $f : M_\nu \to N$ which is uniformly Lipschitz on each of the blocks. In particular, if v is a vertex of H_ν, it appears in some block with a *fixed length* (independent of v, since all blocks are isometric), and so its image has to have bounded length:

$$\ell_\rho(v) \leq L$$

for some uniform L. (Here $\ell_\rho(v)$ denotes the length in N of the geodesic representative of v via ρ).

To obtain a bound like this, we must exhibit some connection now between the *geometry of N_ρ* and the *combinatorics/geometry of $v_\pm$ in $\mathscr{C}(S)$*.

Recall that $v_\pm$ are by definition limits in $\mathscr{U}\mathscr{M}\mathscr{L}(S)$ of bounded-length curves: that is, there exists a sequence $\{\alpha_i\}_{i=-\infty}^{\infty}$ in $\mathscr{C}_0(S)$ with $\ell_\rho(\alpha_i) \leq L_0$ and $\lim_{i\to\pm\infty}\alpha_i = v_\pm$. The geodesic g also accumulates onto $v_\pm$ at infinity. However there seems to be no a priori reason for the α_i to be anywhere near g. Define

$$\mathscr{C}(\rho,L) = \{\alpha \in \mathscr{C}_0(S) : \ell_\rho(\alpha) \leq L.\}$$

To understand the relation of g to $\mathscr{C}(\rho,L_0)$, we will begin by proving:

Theorem 4.1 (Quasiconvexity). *For all $L \geq L_0$ there exists K, so that $\mathscr{C}(\rho,L)$ is K-quasiconvex.*

(Recall that $A \subset X$ is K-quasiconvex if for any geodesic segment γ with $\partial\gamma \subset A$, $\gamma \subset \text{Nbhd}_K(X)$.)

By hyperbolicity of $\mathscr{C}(S)$ and the definition of the boundary it is not hard to see that, since α_i converge to the endpoints of g as $i \to \pm\infty$, each finite segment G of g is, for large enough i, in a 2δ-neighborhood of $[\alpha_{-i}, \alpha_i]$.

Now $\alpha_i \in \mathscr{C}(\rho,L_0)$, so using the quasiconvexity theorem, this means that all of g is in a K'-neighborhood of $\mathscr{C}(\rho,L_0)$.

We remark that, since $\mathscr{C}$ is locally infinite, a distance bound like this is only a weak sort of control. The Projection Bound Theorem in Lecture 5 will be a considerably stronger generalization.

4.1. The bounded-curve projection

Our main tool will be a "coarsely defined map" from $\mathscr{C}(S)$ to $\mathscr{C}(\rho,L)$:

$$\Pi_{\rho,L} : \mathscr{C}(S) \to \mathscr{P}(\mathscr{C}(\rho,L))$$

where $\mathscr{P}(X)$ is the set of subsets of X. $\Pi_{\rho,L}$ (Π for short) will have the following properties:

1. Coarse Lipschitz:

$$d(x,y) \leq 1 \implies \text{diam}(\Pi(x) \cup \Pi(y)) \leq A$$

2. Coarse Idempotence:

$$x \in \mathscr{C}(\rho,L) \implies x \in \Pi(x)$$

(where A is a constant independent of ρ)

These properties imply that Π is, in a coarse sense, like a Lipschitz projection to the set $\mathscr{C}(\rho,L)$. Together with hyperbolicity of $\mathscr{C}(S)$, this has strong consequences:

Theorem 4.2. *If X is δ-hyperbolic, $Y \subseteq X$ and $\Pi : X \to \mathscr{P}(Y)$ satisfies properties (1) and (2), then Y is quasiconvex.*

The proof is similar to the proof of "stability of quasigeodesics" in Mostow's rigidity theorem. See [Min01] for more details.

4.2. Definition of $\Pi_{\rho,L}$

Pleated surfaces   A pleated surface (or pleated map) is a map $f : S \to N$, together with a hyperbolic structure σ_f on S, with the following properties:

- f takes σ_f-rectifiable paths in S to paths in N of the same length.

- There is a σ_f-geodesic lamination λ on S, all of whose leaves are mapped geodesically,

- The complementary regions of λ are mapped totally geodesically.

We call σ_f the *induced metric*, since it is determined uniquely by the map and the first condition. The minimal λ that works in the definition is called the *pleating locus* of f. Informally one can think of the map as "bent" along λ.

This definition is due to Thurston and plays an important role in the synthetic geometry of hyperbolic 3-manifolds. A standard example, which we will be making use of, is the "spun triangulation":

Begin with any set P of curves cutting S into pairs of pants, and fix a hyperbolic metric σ on $int(S)$ of finite area, so that the ends are cusps. On each component of P place one vertex, and then triangulate each pants using only arcs terminating in these vertices and in the cusps. Now "spin" this triangulation around P, by applying a sequence of Dehn twists around each component. If at each stage the triangulation is realized by geodesics in σ, then the geometric limit of the sequence will be a lamination with closed leaves P and a finite number of infinite leaves that spiral on P and/or exit the cusps. (See Figure 15)

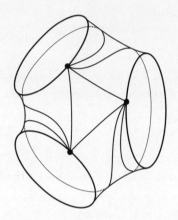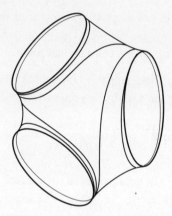

Figure 15: The triangulation and "spun" lamination on a pair of pants, when all boundary components are in P. If some are in ∂S then the leaves go out a cusp instead of spiraling.

In a similar way we can produce a pleated surface, first by mapping the curves of P to their geodesic representatives in N, and then "spinning" the images of the triangulation leaves. Finally when the leaves are in place we fill in the spaces between them with (immersed) totally geodesic ideal triangles, and obtain a surface together with induced metric. (This construction is easier to visualize equivariantly in the universal cover).

It is clear from this example that for any essential curve γ in S there is a pleated map in the homotopy class of ρ that maps γ to its geodesic representative in N. We define

$$\mathbf{pleat}_\rho(\gamma)$$

to be the set of all such pleated maps.

Now for a complete hyperbolic metric σ on $int(S)$, define

$$\mathbf{short}_L(\sigma) = \{v \in \mathscr{C}_0(S) : \ell_\sigma(v) \leq L\}.$$

We can now define:

$$\Pi_{\rho,L}(\alpha) = \bigcup_{f \in \mathbf{pleat}_\rho(\alpha)} \mathbf{short}_L(\sigma_f). \tag{4.1}$$

It is an observation originally of Bers that given S there is a number L_0 so that, for every hyperbolic metric σ on S there is a pants decomposition made up of curves of length at most L_0. We call this number the "Bers constant". Hence for $L \geq L_0$, $\Pi_{\rho,L}(\alpha)$ is always non-empty, and moreover contains a pants decomposition.

Note that if $v \in \mathscr{C}(\rho,L)$ then $v \in \Pi_{\rho,L}(v)$, since if $f \in \mathbf{pleat}_\rho(v)$, $\ell_{\sigma_f}(v) = \ell_\rho(v) \leq L$. Hence property (2) (Coarse Idempotence) is established.

Now our main claim, the Coarse Lipschitz property (1), will follow from the apparently weaker claim:

$$\mathrm{diam}_{\mathscr{C}(S)}\big(\Pi_{\rho,L}(v)\big) \leq b \qquad (4.2)$$

for a priori b (depending on L) and any simplex v.

Proof of inequality (4.2)   First note that, for any σ,

$$\mathrm{diam}_{\mathscr{C}(S)}(\mathbf{short}_L(\sigma)) \leq C(L). \qquad (4.3)$$

This is easy: If two curves have a length bound with respect to the same metric σ, their intersection number is bounded in terms of this, and a bound on the intersection number implies a bound on the $\mathscr{C}(S)$-distance by an inductive argument (see [MM99], or Hempel [Hem01]).

Thus our main point will be to show that, for some a priori constant L_1,

$$\mathbf{short}_{L_1}(\sigma_f) \cap \mathbf{short}_{L_1}(\sigma_g) \neq \emptyset \qquad (4.4)$$

for any $f,g \in \mathbf{pleat}_\rho(v)$. This would imply, together with (4.3), that $\mathrm{diam}_{\mathscr{C}(S)}(\Pi_{\rho,L_1}(v)) \leq 2C(L_1)$. In fact since $\mathbf{short}_L$ is increasing with L we can conclude that $\mathrm{diam}_{\mathscr{C}(S)}(\Pi_{\rho,L}(v)) \leq 2C(\max(L,L_1))$ for any L.

To prove inequality (4.4), let us construct a curve γ which has bounded length in both σ_f and σ_g. At first we note that $f(S)$ and $g(S)$ are only guaranteed to agree on the curve v itself, and this curve may be very long. Consider the geodesic representing v in the σ_f metric on S.

Suppose first that there are *no thin parts* on σ_f – that is that $inj(\sigma_f) > \varepsilon$ for some fixed $\varepsilon > 0$. This means that S is a closed surface, and that σ_f has no closed geodesics of length less than 2ε. In this case we can approximate v with a curve of bounded length that is composed of a segment of v concatenated with a very short "jump". That is, for $\varepsilon' < \varepsilon/2$, consider the ε'-neighborhood of a long segment s of v. If this is an embedded rectangle in S then its area is at least $2\varepsilon' l(s)$. Thus the finiteness of the area of S implies that for a certain $l(s)$, this neighborhood (which locally looks like an embedded rectangle because of the injectivity radius lower bound) will fail to embed globally, and where this happens we get a "short cut" of length $2\varepsilon'$ joining a long (but bounded) segment of v to itself (Figure 16). If we are slightly more careful we can arrange for the resulting closed curve β to be simple, and homotopically essential.

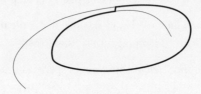

Figure 16

Now to bound the length of β with respect to σ_g requires a bound on the σ_g length of (the homotopy class rel endpoints of) the short jump part of β – the part that runs along v is already the same length in both metrics. In other words we need to prevent a certain kind of "folding" of g, as suggested by Figure 17.

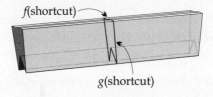

f(shortcut)

g(shortcut)

Figure 17: The thin rectangle on the top is in the image of f, whereas the image of g is folded so that the shortcut is not short in σ_g. This is what Uniform Injectivity rules out.

This is prevented by a result of Thurston called the Uniform Injectivity Theorem [Thu86a]. This theorem states, in our setting, that two leaves mapped geodesically by a pleated surface cannot line up too closely in the image unless they are already close in the domain. The two endpoints of the short cut part of β are the midpoints of long subsegments of v that are close to each other in σ_f (we can force them to be as long as we like by taking ε' small enough) and hence the same is true of their images by f, so that the leaves line up nearly parallel in the image at the endpoints of the short cut. Since $f = g$ on v, we can then apply the Uniform Injectivity Theorem to g and conclude that the endpoints of the short cut are close together in σ_g, and in fact (with a bit more care) the short cut itself is homotopic rel endpoints to an arc of length at most $\delta(\varepsilon')$.

When S has nonempty boundary, we must take a bit more care that the closed curve β is in a non-peripheral homotopy class – we may have to use two segments on v and two short cuts.

If we allow σ_f to have very short geodesics, we must consider one more case. If v does not enter any thin part of S with core length less than ε, then the previous argument applies. If v does enter the ε-thin part of a σ_f-geodesic β, then the approximation by a bounded-length curve may fail. However, in this case we see that both f and g, since they agree on v, have images that meet the ε-Margulis tube associated with β in N, and by standard properties of pleated surfaces this implies that β itself has uniformly bounded length in σ_g as well as σ_f.

Inequality (4.2) implies property (1)   Now we are ready to establish the coarse Lipschitz property (1). If $d(x, y) \leq 1$ then x and y represent disjoint curves (assume here that S is not a one-holed torus or 4-holed sphere – for those cases there is a very similar argument). Thus the simplex $[xy]$ represents a curve system on S, and $\mathbf{pleat}_\rho([xy])$ is a nonempty set of pleated surfaces. But it is clear that

$$\mathbf{pleat}_\rho([xy]) = \mathbf{pleat}_\rho(x) \cap \mathbf{pleat}_\rho(y)$$

and thus this intersection is nonempty. It follows immediately that

$$\Pi_{\rho,L}(x) \cap \Pi_{\rho,L}(y) \neq \emptyset$$

for any $L \geq L_0$. Hence the diameter bound (4.2) on $\Pi(x)$ and $\Pi(y)$ implies a bound on the union.

This concludes our sketch of the proof of the coarse Lipschitz property for $\Pi_{\rho,L}$ and hence, via Theorem 4.2, of the Quasiconvexity Theorem 4.1.

5. Quasiconvexity and projection bounds

The quasiconvexity of $\mathscr{C}(\rho, L)$ implies that the geodesic g connecting $v_\pm$ is a bounded distance from $\mathscr{C}(\rho, L)$ (as we discussed in lecture 4), and furthermore that

$$d_{\mathscr{C}(S)}(v, \Pi_{\rho,L}(v)) \leq B \tag{5.1}$$

for a priori B and all v in g. However we might now wonder *what good is such an estimate*, since $\mathscr{C}(S)$ is locally infinite?

Using a generalization of the Quasiconvexity Theorem, we will obtain a strengthening of the bound (5.1), which will then enable us in §6 to establish the A Priori Bounds Theorem and the Lipschitz Model Theorem.

5.1. Relative bounds for subsurfaces

In order to state our generalization of the projection bound (5.1), we must consider subsurfaces of S and their associated complexes.

The *arc complex* $\mathscr{A}(W)$ of a (non-annular) surface with boundary W is defined as follows: Vertices of $\mathscr{A}(S)$ are homotopy classes of either essential simple closed curves (as for $\mathscr{C}(S)$) or properly embedded arcs. In the latter case the homotopy is taken to keep the endpoints in ∂W. Simplices correspond to disjoint collections of arcs or curves. Hence $\mathscr{C}_0(W) \subset \mathscr{A}_0(W)$, and (except in the sporadic cases $S_{0,4}$ and $S_{1,1}$, which require a separate discussion) $\mathscr{C}(W) \subset \mathscr{A}(W)$.

We also note that $\mathscr{C}_0(W)$ is *cobounded* in $\mathscr{A}(W)$, that is, every point of $\mathscr{A}(W)$ is a bounded distance from $\mathscr{C}_0(W)$, and that distance in $\mathscr{A}(W)$ is estimated by distance in $\mathscr{C}(W)$. Thus the two complexes are quasi-isometric. This is easy to see with a picture (Figure 18).

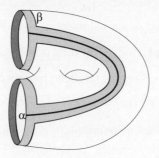

Figure 18: The regular neighborhood of an arc α and ∂W contains an essential curve β in its boundary. This gives a quasi-isometry from $\mathscr{A}_0(W)$ to $\mathscr{C}_0(W)$.

If $W \subset S$ is an essential subsurface, we obtain a map

$$\pi_W : \mathscr{A}(S) \to \mathscr{A}(W) \cup \{\emptyset\}$$

defined by taking a curve system v to the (barycenter of the) simplex formed by the essential intersections $[v \cap W]$, or to $\emptyset$ if there are no essential intersections.

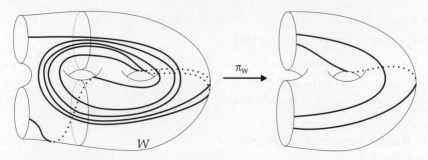

Figure 19

For annuli in S we need a different definition. Let $W \subset S$ be an essential, nonperipheral annulus, let $\widehat{W}$ be the associated annular cover of S, and $\overline{W}$ its natural compactification. (See Figure 20).

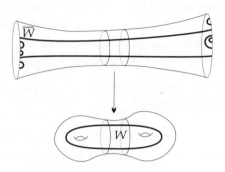

Figure 20

Let $\mathscr{A}(\overline{W})$ be as above, except that vertices are now properly embedded arcs up to homotopy with *fixed endpoints*.

Now, any $\alpha \in \mathscr{A}_0(S)$ lifts to an arc system in $\widehat{W}$, which compactify to arcs in $\overline{W}$. This system contains essential arcs in $\mathscr{A}(\overline{W})$ (those with endpoints on both boundaries) exactly if α intersects W essentially. The set of essential lifted arcs gives us $\pi_W(\alpha)$ in the annulus case.

Let

$$d_W(\alpha, \beta) = dist_{\mathscr{A}(W)}(\pi_W(\alpha), \pi_W(\beta))$$

(replacing $\mathscr{A}(W)$ with $\mathscr{A}(\overline{W})$ in the annulus case). This makes sense provided both projections are nonempty. In the annulus case, this distance measures "relative twisting" of α and β around W. We similarly define $\mathrm{diam}_W(X) = \mathrm{diam}_{\mathscr{A}(W)}(\pi_W(X))$.

We can now state the generalization of (5.1):

Projection Bound Theorem. *If v is in the hierarchy H_v and W is an essential subsurface other than a three-holed sphere, then*

$$\mathrm{diam}_W(v \cup \Pi_{\rho,L}(v)) \leq B$$

provided $\pi_W(v)$ and $\pi_W(\Pi_{\rho,L}(v))$ are nonempty, where $L \geq L_0$ and B depends on S and L.

Note that $\pi_W(\Pi_{\rho,L}(v))$ is *always* nonempty if $L \geq L_0$ and W is not an annulus, since then $\Pi_{\rho,L}(v)$ contains a pants decomposition.

The proof of this theorem is a fancier version of the argument we used for the Quasiconvexity Theorem in the previous lecture. An important ingredient is an adaptation of the "short cut" construction that yielded the curve β of bounded length on two pleated surfaces $f, g \in \mathbf{pleat}_\rho(v)$ (Figure 16). In the context of this theorem we need to make sure that β has essential intersection with the given subsurface W, and so the choice of β has to be carefully guided using Thurston's "train tracks". In addition to this, there is an inductive structure to the argument, using the hierarchy H_v.

5.2. Penetration in Margulis tubes

We can apply the Projection Bound Theorem to control the way in which a pleated surface enters a Margulis tube in N. This will then play an important role in the A Priori Bound Theorem in §6.

Let $\mathbb{T}_\varepsilon(\alpha)$ denote the ε-*Margulis tube* in N_ρ of $\rho(\alpha)$, for an element α of $\pi_1(S)$ (or a vertex α of $\mathscr{C}(S)$). This is the locus where the translation length of $\rho(\alpha)$ or some power of α is bounded by ε. If ε is less than the Margulis constant ε_0, and $0 < \ell_\rho(\alpha) < \varepsilon$, then $\mathbb{T}_\varepsilon(\alpha)$ is a solid torus, isometric to the hyperbolic tube $\mathbb{T}(r, \lambda)$ (see §2) where λ is the complex translation length of $\rho(\alpha)$ and the radius r goes to ∞ as $\frac{\ell_\rho(\alpha)}{\varepsilon} \to 0$. Our next goal is to detect the presence of Margulis tubes in N, from the structure of the hierarchy.

Tube Penetration Theorem (stated for $S = S_{0,5}$) *There exists $\varepsilon > 0$ depending on S, such that the following holds.*

Let s be a "spoke" of the hierarchy H_v. If $f \in \mathbf{pleat}_\rho(s)$, then

$$f(S) \cap \mathbb{T}_\varepsilon(\alpha) \neq \emptyset$$

only if α is one of the vertices of s.

That is, the only way for f to penetrate deeply into a tube is the "obvious" way – by pleating along the core curve of the tube.

5.3. Proof of the tube penetration theorem

We begin with this standard property of pleated surfaces (observed by Thurston in [Thu86a]): There exists $\varepsilon_1 > 0$ such that, if a pleated surface f in the homotopy class of ρ meets $\mathbb{T}_{\varepsilon_1}(\alpha)$ then $f^{-1}(\mathbb{T}_{\varepsilon_1}(\alpha))$ must be contained in an ε_0-Margulis tube in the metric σ_f – that is, only the thin part of S is mapped into the thin part of N. In particular it follows that $\ell_{\sigma_f}(\alpha) \leq \varepsilon_0$.

Now assume $\varepsilon << \varepsilon_1$. Suppose v is a vertex in g crossing α essentially, and $f \in$ **pleat**$_\rho(v)$ meets $\mathbb{T}_\varepsilon(\alpha)$. Let v^* denote the geodesic representative of v in N, which is in the image of f. Since v crosses α essentially, it must cross the ε_0-collar of α and hence by the previous paragraph v^* must meet an ε_0-neighborhood of $\mathbb{T}_\varepsilon(\alpha)$.

In either the forward or backward direction in g (suppose forward), all vertices of g after v cross α essentially. Number them $v = v_j, v_{j+1}, \ldots$. Eventually, v_i^* for some $i > j$ is outside $\mathbb{T}_{\varepsilon_1}(v)$, since $v_i \to v_+$. Let us try to see when this happens.

Lower bound: Let $f \in$ **pleat**$_\rho([v_i v_{i+1}])$. If $f(S)$ meets $\mathbb{T}_{\varepsilon_1}(\alpha)$, then both v_i^* and v_{i+1}^* cross through the ε_0-thin part of S in the metric σ_f. Any point in $v_i^* \cap \mathbb{T}_{\varepsilon_1}(\alpha)$ can be connected, via an arc in $f(S)$ of length bounded by ε_0, to $v_{i+1}^* \cap \mathbb{T}_{\varepsilon_1}(\alpha)$.

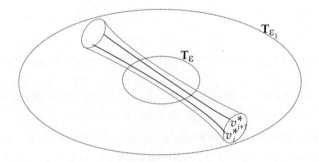

Figure 21

Suppose then that $j + Q$ is the first value of i for which v_i^* fails to meet $\mathbb{T}_{\varepsilon_1}(\alpha)$. Then applying the previous paragraph Q times we have

$$\text{dist}(\mathbb{T}_\varepsilon(\alpha), \partial \mathbb{T}_{\varepsilon_1}(\alpha)) \leq Q\varepsilon_0.$$

Since, by the collar lemmas of Brooks-Matelski [BM82] and Meyerhoff [Mey87], this distance is an increasing function of $\varepsilon_1/\varepsilon$, of the form

$$\text{dist}(\mathbb{T}_\varepsilon(\alpha), \partial \mathbb{T}_{\varepsilon_1}(\alpha)) \geq \tfrac{1}{2} \log \frac{\varepsilon_1}{\varepsilon} - C.$$

this gives us a lower bound of the form

$$Q \geq a \log \frac{\varepsilon_1}{\varepsilon} - b. \tag{5.2}$$

Upper bound: If $f \in$ **pleat**(v_i) meets $\mathbb{T}_{\varepsilon_1}(\alpha)$ then, since $\ell_{\sigma_f}(\alpha) \leq \varepsilon_0 < L_0$, we have

$$\alpha \in \Pi_{\rho, L_0}(v_i).$$

The Projection Bound Theorem then implies

$$d_{\mathscr{C}(S)}(v_i, \alpha) \le B$$

So all such v_i's lie in a ball of radius B. Since g is a geodesic, this means that

$$Q \le 2B. \tag{5.3}$$

Putting the upper and lower bounds (5.2,5.3) together, we obtain an inequality

$$\varepsilon > \varepsilon_2$$

where ε_2 depends on the previous constants. Thus let us assume now that $\varepsilon \le \varepsilon_2$. Thus if $f \in \mathbf{pleat}_\rho(v)$ meets $\mathbb{T}_\varepsilon(\alpha)$ for $v \in g$, then v and α must *not* intersect essentially. If $v = \alpha$, we are done.

If $v \ne \alpha$ then $\alpha \in \mathscr{C}(W_v)$, and we consider the spokes $\{s_j = [u_j v]\}_{j=0}^m$ around v, and try to mimic the same argument. Suppose that u_j^* meets $\mathbb{T}_\varepsilon(\alpha)$, but that $u_j \ne \alpha$. Now α can be equal to at most one of the u_i so let us assume it occurs for $i < j$ if at all. Then the last vertex u_m crosses α and, since it is just the successor of v in g, the previous argument applies to it and u_m^* must be outside $\mathbb{T}_{\varepsilon_2}(\alpha)$.

Now choose Q to be the first positive number such that u_{j+Q}^* is outside $\mathbb{T}_{\varepsilon_2}(\alpha)$. An upper bound of the form of (5.3) follows using the same argument as before, but applying the relative version of the Projection Bound Theorem,

$$\mathrm{diam}_{W_v}(u_i, \Pi_{\rho, L_0}(u_i)) \le B.$$

To obtain a lower bound of the form (5.2), but with ε_2 replacing ε_1, we need a construction to replace $\mathbf{pleat}_\rho([v_i v_{i+1}])$, since u_i and u_{i+1} do not represent disjoint curves.

Let λ_i denote the "spun" lamination, as in Lecture 4, whose closed curves are v and u_i. Let $\lambda_{i+1/2}$ be a "halfway lamination", defined as follows. $\lambda_{i+1/2}$ contains the curve v, and agrees with λ_i and λ_{i+1} on the complement of W_v. On W_v itself, $\lambda_{i+1/2}$ is as in Figure 22.

Let f_x be the pleated surface mapping λ_x geodesically, for $x = j + k/2$ ($k = 0, \ldots, 2Q$). Note that $\lambda_{i+1/2}$ has two leaves l_i and l_{i+1}, which must cross α essentially since u_i and u_{i+1} do. l_i is mapped to the same geodesic in N by $f_{i-1/2}, f_i$ and $f_{i+1/2}$ – call this geodesic l_i^*. We can now repeat the lower bound argument for Q, finding a sequence of jumps from u_j^* to u_{j+Q}^* passing through all the l_i^*. We obtain, as before, an inequality of the form

$$\varepsilon > \varepsilon_3$$

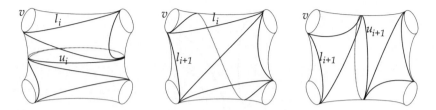

Figure 22: The related laminations λ_i, $\lambda_{i+1/2}$ and λ_{i+1}, restricted to W_v.

where ε_3 depends on the previous constants (and on ε_2). Thus if we choose $\varepsilon \leq \varepsilon_3$ we must have $\alpha = u_j$ after all, and the Tube Penetration Theorem follows.

For general S there is an inductive argument using the structure of the hierarchy, where the halfway surfaces need be used only at the last stage.

6. A priori length bounds and model map

In this last lecture we will sketch the proof of this basic bound:

A Priori Bound Theorem. *If* $\rho : \pi_1(S) \rightarrow PSL_2(\mathbb{C})$ *is a doubly degenerate Kleinian surface and v is a vertex in the associated hierarchy* $H_{v(\rho)}$, *then*

$$\ell_\rho(v) \leq B \tag{6.1}$$

where B depends only on the surface S.

We will then state the Lipschitz Model Theorem and indicate how the a priori bound is used in its proof.

Markings and elementary moves   A marking of S is a pants decomposition $\{u_i\}$ together with, for each i, a transversal curve t_i that is disjoint from $u_j, j \neq i$, and intersects u_i in the minimal possible way. For $S_{0,5}$, a marking consists of 4 curves, and t_i intersects u_i twice.

An *elementary move* $\mu \rightarrow \mu'$ taking one marking to another is one of the following operations:

Twist$_i$ performs one half-twist on t_i around u_i (Figure 24).

Flip$_i$ reverses the roles of u_i and t_i (Figure 25). Note that in this case there has to be an adjustment of the other $t_j, j \neq i$, so that they do not intersect the new u_i' which is the old t_i. There is a finite number of "simplest" ways to do this, and we just pick one.

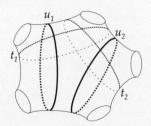

Figure 23

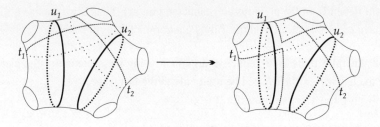

Figure 24: The move Twist$_1$

The graph whose vertices are markings and whose edges are elementary moves is connected, locally finite, and its quotient by the mapping class group of S is finite. The total length of a marking in a hyperbolic metric σ is just the sum of the lengths of the curves u_i and t_i. Note that a bound on the total length of μ in σ constrains σ to a bounded subset of Teichmüller space. We will need the following observation: If μ_0 is a marking of total length L in a hyperbolic metric σ on S, and

$$\mu_0 \to \mu_1 \to \cdots \to \mu_n$$

is a sequence of elementary moves, then μ_n has total length at most K, where K depends only on L and n.

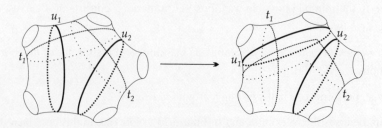

Figure 25: The move Flip$_1$

We will control elementary moves using this theorem:

Theorem 6.1 (Masur–Minsky [MM00]). *If μ and μ' are two markings and*

$$\sup_{W \subseteq S} d_W(\mu, \mu') \leq M$$

then there exists a sequence of elementary moves from μ to μ' with at most

$$CM^{\xi}$$

steps, where C and ξ depend only on S.

Here, $d_W(\mu, \mu')$ is defined as in §5, with the projection $\pi_W(\mu)$ simply being the union of $\pi_W(a)$ over components a of μ. The supremum is over all essential subsurfaces in S, including S itself. The proof of this theorem uses the hierarchy machinery discussed in §2.

6.1. Proving the a priori bounds

Again, we are working in the case that $S = S_{0,5}$. Let $s = [u_1 u_2]$ be a spoke of the hierarchy, and $f \in \mathbf{pleat}_{\rho}(s)$. Let $\varepsilon > 0$ be the constant given by the Tube Penetration Theorem.

Thick case Suppose σ_f has no geodesics of length ε or less. Then σ_f admits a marking μ of total length at most L (depending on ε).

The Projection Bound Theorem implies that

$$d_W([u_1 u_2], \mu) \leq B$$

For any W meeting u_1 or u_2 essentially, since $\mu \subset \Pi_{\rho,L}([u_1 u_2])$. This includes all W except for the annuli A_i with cores u_i ($i = 1, 2$). We therefore choose transversals t_1, t_2 for u_1, u_2, such that $d_{A_i}(t_i, \mu')$ is at most 2. Thus $\{u_1, t_1, u_2, t_2\}$ give us a marking μ' such that

$$d_W(\mu', \mu) \leq B$$

for *all* $W \subseteq S$. Applying Theorem 6.1, we bound the elementary-move distance from μ to μ', and hence obtain a bound on $\ell_{\sigma_f}(\mu')$.

This in turn bounds $\ell_{\rho}(u_i)$, which gives the a priori bound (6.1) in this case.

Thin case Suppose that σ_f does have some curve α of length less than ε. Then $f(S)$ meets $\mathbb{T}_{\varepsilon}(\alpha)$. The Tube Penetration Theorem now implies that $\alpha = u_1$ or $\alpha = u_2$. Suppose the former, without loss of generality. Thus, we repeat the argument of case (1) on the subsurface W_{α}, finding a minimal-length marking (now consisting of just one curve and its transversal) and using the Projection Bound Theorem to bound the length of u_2. Again we have the a priori bound (6.1).

6.2. Constructing the Lipschitz map

We are now ready to state, and summarize the proof of, our main theorem:

It will be convenient to define $\mathscr{U}[k]$ to be the union of tubes $\{U(v) : |\omega_M(v)| \geq k\}$, and to let $M_v[k] = M_v \setminus \mathscr{U}[k]$. Thus $M_v[0]$ is the union of blocks.

If v is a vertex of H_v, let $\mathbb{T}_{\varepsilon_0}(v)$ be the Margulis tube (if any) of the homotopy class $\rho(v)$ in N_ρ. Let $\mathbb{T}[k]$ denote the union of $\mathbb{T}_{\varepsilon_0}(v)$ over all v with $|\omega_M(v)| \geq k$.

Lipschitz Model Theorem. *There exist $K, k_0 > 0$ such that, if $\rho : \pi_1(S) \to PSL_2(\mathbb{C})$ is a doubly degenerate Kleinian surface group with end invariants $\nu(\rho)$, then there is a map*

$$F : M_v \to N_\rho$$

with the following properties:

1. *F induces ρ on π_1, is proper, and has degree 1.*

2. *F is K-Lipschitz on $M_v[k_0]$.*

3. *F maps $\mathscr{U}[k_0]$ to $\mathbb{T}[k_0]$, and $M_v[k_0]$ to $N_\rho \setminus \mathbb{T}[k_0]$.*

Note that the Lipschitz property in part (2) is with respect to the *path metric* on $M_v[k_0]$ (the distance function of M_v restricted to $M_v[k_0]$ may be smaller).

The map $F : M_v[0] \to N$ can be constructed in each block B_e individually. We first define the map on the gluing boundaries $\partial_\pm B$, which are all isometric to a fixed three-holed sphere; let Y be such a boundary component. As in Lecture 4, there is a pleated map $h : Y \to N$ in the homotopy class determined by ρ, which sends each boundary component of Y to its geodesic representative or to the corresponding cusp if it is parabolic (more accurately h is defined on $int(Y)$, and gives a metric whose completion has geodesic boundaries for the non-parabolic ends of Y and cusps for the parabolic ends).

The a priori bounds give an upper bound on the boundary lengths of this pleated surface. Thus, after excising standard collar neighborhoods of the boundaries, we obtain a surface which can be identified, with uniform bilipschitz distortion, with Y under its original model metric. Composing this identification with h, we obtain the map $F|_Y$.

Our next step is to define F on the "middle" surface of a block, which we can write as $W \times \{0\}$ with W a four-holed sphere (for general S, W can also be a one-holed torus). The "halfway surfaces" from the proof of the Tube Penetration Theorem (defined using the two vertices e^+ and e^-) provide us with a map $W \times \{0\} \to N$ and an induced hyperbolic metric. Another application of the a-priori length bounds together with Thurston's Efficiency of Pleated Surfaces [Thu86b] implies that this metric on W is within uniform bilipschitz distortion of the model metric.

We then extend to the rest of the block by a map that takes vertical lines to geodesics. A Lipschitz bound on this part of the map is then an application of the "figure-8" argument from [Min99]. (In brief, we let X be a wedge of two circles in the gluing boundary Y that generate a nonabelian subgroup of $\pi_1(Y)$ and having bounded length in both $W \times \{0\}$ and Y. The extension gives a map of $X \times [0,1]$ with geodesic tracks $\{x\} \times [0,1]$, and if the track lengths are too long then the images of the two circles in the middle are either both very short, or nearly parallel, violating discreteness either way).

Fixing k_0, the tubes U with $|\omega(U)| < k_0$ fall into some finite set of isometry types, and the map can be extended to those, again with some uniform Lipschitz bounds. If $|\omega(U)| > k_0$ then there is an *upper bound* for the corresponding vertex, $\ell_\rho(v) \leq \varepsilon$ where $\varepsilon \to 0$ as $k_0 \to \infty$. This is the main result of [Min00], which uses similar tools but is slightly different from what we have seen so far. Thus choosing k_0 appropriately, U must correspond to a Margulis tube $\mathbb{T}$ with very large radius (and short core). The map on the blocks cannot penetrate more than a bounded distance into such a tube (by the Lipschitz bounds) and so composition with an additional retraction on a collar of $\partial\mathbb{T}$ yields us a map $F : M_v[k_0] \to N \setminus \mathbb{T}[k_0]$ which takes each ∂U with $|\omega(U)| \geq k_0$ to the boundary of the corresonding $\mathbb{T}$.

We fill in the map on the remaining U and the cusps (with no Lipschitz bounds) to obtain a final map $F : M_v \to N$ with all the desired properties. The fact that F is proper essentially follows from the fact that every block contains a bounded length curve in a unique homotopy class, and thus their images cannot accumulate in a compact set. That F has degree 1 follows from the fact that blocks far toward the $+$ end of the hierarchy correspond to vertices close to v_+, and hence their images must go out the $+$ end of N.

6.3. Consequences

It is now easy to obtain the following lower bound on lengths:

Corollary 6.2. *There exists an a priori $\varepsilon > 0$ such that all curves of length less than ε in N must occur as vertices of the hierarchy.*

Proof: If $\ell_N(\gamma) < \varepsilon$ then γ has a Margulis tube $\mathbb{T}_\varepsilon(\gamma)$ in N. Since F has degree 1, this tube is in the image of F. The Lipschitz bound on $F|_{M_v[k_0]}$ keeps it out of $\mathbb{T}_\varepsilon(\gamma)$ if ε is sufficiently small. Hence $\mathbb{T}_\varepsilon(\gamma)$ is in the image of some tube $U(v)$, which means that γ corresponds to v.

(Note that this corollary does not follow directly from the Tube Penetration Lemma since that lemma assumes the existence of a pleated surface associated to a spoke which penetrates $\mathbb{T}_\varepsilon(\gamma)$. The global degree argument is necessary).

Another corollary, which requires a bit more work and uses Otal's theorem [Ota95] on the unknottedness of short curves in N, is the following, which describes the topological structure of the set of short curves in N:

Corollary 6.3. $M_v[k_0]$ *is homeomorphic to* $N \setminus \mathbb{T}[k_0]$.

For a proof of this see [BCM].

These results give us a complete description of the "short curves" in N, and in particular give a combinatorial criterion (in terms of the hierarchy and the coefficients ω_M) for when the manifold has bounded geometry. This answers the short list of questions posed in the introduction.

We can also obtain somewhat more explicit lower bounds on ρ-lengths of the vertices of the hierarchy, namely:

Corollary 6.4. *If v is a vertex of the hierarchy $H_{v(\rho)}$ then*

$$\ell_\rho(v) \geq \frac{c}{|\omega_M(v)|^2}$$

where c depends only on the surface S.

This follows from the fact that, because of the Lipschitz property of the map F, $|\omega_M(v)|$ bounds the meridian length of the corresponding Margulis tube $\mathbb{T}_{\varepsilon_0}(v)$ in N. This gives an upper bound for radius of the tube, and the collar lemmas of [BM82] and [Mey87] then give a lower bound for the core length $\ell_\rho(v)$. An upper bound for $\ell_\rho(v)$ which goes to 0 as $|\omega_M(v)| \to \infty$ also exists: it follows from the main theorems of [Min00], and was already used briefly in the proof of the Lipschitz Model Theorem.

References

[AB60]   L. Ahlfors and L. Bers (1960). Riemann's mapping theorem for variable metrics, *Ann. Math.* **72**, 385–404.

[ABC⁺91] J.M. Alonso, T. Brady, D. Cooper, V. Ferlini, M. Lustig, M. Mihalik, M. Shapiro and H. Short (1991). Notes on word hyperbolic groups. In *Group Theory from a Geometrical Viewpoint, ICTP Trieste 1990, edited by E. Ghys, A. Haefliger and A. Verjovsky, World Scientific*, 3–63.

[Abi88]   W. Abikoff (1988). Kleinian groups – geometrically finite and geometrically perverse, *Geometry of Group Representations, Contemp. Math.* **74**, Amer. Math. Soc., 11–50.

[BCM]   J. Brock, R. Canary and Y. Minsky. Classification of Kleinian surface groups II: the ending lamination conjecture, *in preparation*.

[Ber60]   L. Bers (1960). Simultaneous uniformization, *Bull. Amer. Math. Soc.* **66**, 94–97.

[Ber70a] L. Bers (1970). On boundaries of Teichmüller spaces and on Kleinian groups: I, *Ann. Math.* **91**, 570–600.

[Ber70b] L. Bers (1970). Spaces of Kleinian groups. In *Maryland conference in Several Complex Variables I, Lec. Notes Math.* **155**, Springer-Verlag, 9–34.

[BM82]   R. Brooks and J.P. Matelski (1982). Collars in Kleinian groups, *Duke Math. J.* **49**, no. 1, 163–182.

[BO88]   F. Bonahon and J.-P. Otal (1988). Varietes hyperboliques a geodesiques arbitrairement courtes, *Bull. London Math. Soc.* **20**, 255–261.

[BO01]   F. Bonahon and J.-P. Otal (2001). Laminations mesurées de plissage des variétés hyperboliques de dimension 3, *preprint*.

[Bon86]   F. Bonahon (1986). Bouts des variétés hyperboliques de dimension 3, *Ann. Math.* **124**, 71–158.

[Bon01]   F. Bonahon (2001). Geodesic laminations on surfaces. In *Laminations and Foliations in Dynamics, Geometry and Topology (Stony Brook, NY), edited by M. Lyubich, J. Milnor and Y. Minsky, Contemp. Math.* **269**, Amer. Math. Soc., 1–37.

[Can93b] R.D. Canary (1993). Ends of hyperbolic 3-manifolds, *J. Amer. Math. Soc.* **6**, 1–35.

[Cnn91]   J. Cannon (1991). The theory of negatively curved spaces and groups. In *Ergodic theory, symbolic dynamics, and hyperbolic spaces (Trieste, 1989)*, Oxford Univ. Press, 315–369.

[FLP79]   A. Fathi, F. Laudenbach and V. Poenaru (1979). Travaux de Thurston sur les surfaces, *Astérisque* **66-67**, Soc. Math. de France.

[Fre42]   H. Freudenthal (1942). Neuaufbau der Endentheorie, *Ann. Math.* **43**, 261–279.

[Gre66]   L. Greenberg (1966). Fundamental polyhedra for Kleinian groups, *Ann. Math.* **84**, 433–441.

[Gro87]   M. Gromov (1987). Hyperbolic groups. In *Essays in Group Theory, edited by S. M. Gersten, MSRI Publications* **8**, Springer-Verlag.

[Hav81]   W.J. Harvey (1981). Boundary structure of the modular group. In *Riemann Surfaces and Related Topics: Proceedings of the 1978 Stony Brook Conference, edited by I. Kra and B. Maskit, Ann. Math. Studies* **97**, Princeton.

[Hem01]   J. Hempel (2001). 3-manifolds as viewed from the curve complex, *Topology* **40**, no. 3, 631–657.

[Kla]   E. Klarreich. The boundary at infinity of the curve complex and the relative Teichmüller space, *preprint*.

[Kra72]   I. Kra (1972). On spaces of Kleinian groups, *Comment. Math. Helv.* **47**, 53–69.

[KSh89]   R.S. Kulkarni and P.B. Shalen (1989). On Ahlfors' finiteness theorem, *Adv. Math.* **76**, no. 2, 155–169.

[KS93]   L. Keen and C. Series (1993). Pleating coordinates for the Maskit embedding of the Teichmüller space of punctured tori, *Topology* **32**, 719–749.

[Mak70]   B. Maskit (1970). On boundaries of Teichmüller spaces and on Kleinian groups II, *Ann. Math.* **91**, 607–639.

[Mak75]   B. Maskit (1975). Classification of Kleinian groups. In *Proceedings of the International Congress of Mathematicians (Vancouver, 1974)* **2**, Canad. Math. Congress, 213–216.

[Mar74]   A. Marden (1974). The geometry of finitely generated Kleinian groups, *Ann. Math.* **99**, 383–462.

[McC86]   D. McCullough (1986). Compact submanifolds of 3-manifolds with boundary, *Quart. J. Math. Oxford* **37**, 299–306.

[McM91]   C. McMullen (1991). Cusps are dense, *Ann. Math.* **133**, 217–247.

[McM96] C. McMullen (1996). Renormalization and 3-manifolds which fiber over the circle, *Ann. Math. Studies* **142**, Princeton University Press.

[Mey87] R. Meyerhoff (1987). A lower bound for the volume of hyperbolic 3-manifolds, *Canad. J. Math.* **39**, 1038–1056.

[Min93] Y.N. Minsky (1993). Teichmüller geodesics and ends of hyperbolic 3-manifolds, *Topology* **32**, 625–647.

[Min94] Y.N. Minsky (1994). On rigidity, limit sets and end invariants of hyperbolic 3-manifolds, *J. Amer. Math. Soc.* **7**, 539–588.

[Min99] Y.N. Minsky (1999). The classification of punctured-torus groups, *Ann. Math.* **149**, 559–626.

[Min00] Y.N. Minsky (2000). Kleinian groups and the complex of curves, *Geometry and Topology* **4**, 117–148.

[Min01] Y.N. Minsky (2001). Bounded geometry for Kleinian groups, *Invent. Math.* **146**, 143–192.

[Min03] Y.N. Minsky (2003). Classification of Kleinian surface groups I: models and bounds, *arXiv:math.GT/0302208*.

[MM99] H.A. Masur and Y. Minsky (1999). Geometry of the complex of curves I: Hyperbolicity, *Invent. Math.* **138**, 103–149.

[MM00] H.A. Masur and Y. Minsky (2000). Geometry of the complex of curves II: Hierarchical structure, *Geom. and Funct. Anal.* **10**, 902–974.

[MMa79] A. Marden and B. Maskit (1979). On the isomorphism theorem for Kleinian groups, *Invent. Math.* **51**, 9–14.

[Ota95] J.-P. Otal (1995). Sur le nouage des géodésiques dans les variétés hyperboliques, *C. R. Acad. Sci. Paris Sèr. I Math.* **320**, 847–852.

[Ota96] J.-P. Otal (1996). *Le théorème d'hyperbolisation pour les variétés fibrées de dimension trois*, Astérisque No. 235.

[Sco73b] G.P. Scott (1973). Compact submanifolds of 3-manifolds, *J. Lond. Math. Soc.* **7**, 246–250.

[Ser85a] C. Series (1985). The geometry of Markoff numbers, *Math. Intelligencer* **7**, 20–29.

[Sul81a]   D. Sullivan (1981). On the ergodic theory at infinity of an arbitrary discrete group of hyperbolic motions. In *Riemann Surfaces and Related Topics: Proceedings of the 1978 Stony Brook Conference, Ann. Math. Studies* **97**, Princeton.

[Sul85]   D. Sullivan (1985). Quasiconformal homeomorphisms and dynamics II: Structural stability implies hyperbolicity for Kleinian groups, *Acta Math.* **155**, 243–260.

[Sul86]   D. Sullivan (1986). Quasiconformal homeomorphisms in dynamics, topology and geometry. In *Proceedings of the International Conference of Mathematicians, Amer. Math. Soc.*, 1216–1228.

[Thu79]   W.P. Thurston (1979). The geometry and topology of 3-manifolds, *Princeton University Lecture Notes*.
http://www.msri.org/publications/books/gt3m

[Thu82]   W.P. Thurston (1982). Three dimensional manifolds, Kleinian groups and hyperbolic geometry, *Bull. Amer. Math. Soc.* **6**, 357–381.

[Thu86a]   W.P. Thurston (1986). Hyperbolic structures on 3-manifolds I: deformation of acylindrical manifolds, *Ann. Math.* **124**, 203–246.

[Thu86b]   W.P. Thurston (1986). Hyperbolic structures on 3-manifolds II: surface groups and 3-manifolds which fiber over the circle, *arXiv:math.GT*.

Yair N. Minsky

Department of Mathematics
SUNY at Stony Brook
Stony Brook, NY 11794-3651
USA

yair@math.sunysb.edu

AMS Classification: 30F40, 57M50

Keywords: hyperbolic 3-manifold, Kleinian group, ending lamination, complex of curves

Kleinian Groups and Hyperbolic 3-Manifolds   Y. Komori, V. Markovic & C. Series (Eds.)
Lond. Math. Soc. Lec. Notes **299**, 41–73   Cambridge Univ. Press, 2003

Harmonic deformations of hyperbolic 3-manifolds

Craig D. Hodgson and Steven P. Kerckhoff

Abstract

This paper gives an exposition of the authors' harmonic deformation theory for
3-dimensional hyperbolic cone-manifolds. We discuss topological applications to
hyperbolic Dehn surgery as well as recent applications to Kleinian group theory.
A central idea is that local rigidity results (for deformations fixing cone angles)
can be turned into effective control on the deformations that do exist. This leads
to precise analytic and geometric versions of the idea that hyperbolic structures
with short geodesics are close to hyperbolic structures with cusps. The paper also
outlines a new harmonic deformation theory which applies whenever there is a
sufficiently large embedded tube around the singular locus, removing the previous
restriction to cone angles at most 2π.

1. Introduction

The local rigidity theorem of Weil [Wei60] and Garland [Gar67] for complete, finite
volume hyperbolic manifolds states that there is no non-trivial deformation of such
a structure through *complete* hyperbolic structures if the manifold has dimension at
least 3. If the manifold is closed, the condition that the structures be complete is automatically satisfied. However, if the manifold is non-compact, there may be deformations through incomplete structures. This cannot happen in dimensions greater than
3 (Garland-Raghunathan [GRa63]); but there are always non-trivial deformations in
dimension 3 (Thurston [Thu79]) in the non-compact case.

In [HK98] this rigidity theory is extended to a class of finite volume, orientable
3-dimensional hyperbolic cone-manifolds, *i.e.* hyperbolic structures on 3-manifolds
with cone-like singularities along a knot or link. The main result is that such structures
are locally rigid if the cone angles are fixed, under the extra hypothesis that all cone
angles are at most 2π. There is a smooth, incomplete structure on the complement of
the singular locus; by completing the metric the singular cone-metric is recovered. The
space of deformations of (generally incomplete) hyperbolic structures on this open
manifold has non-zero dimension, so there will be deformations if the cone angles
are allowed to vary. An application of the implicit function theorem shows that it is

possible to deform the structure so that the metric completion is still a 3-dimensional hyperbolic cone-manifold, and it is always possible to deform the cone-manifold to make arbitrary (small) changes in the cone angles. In fact, the collection of cone angles locally parametrizes the set of cone-manifold structures.

A (smooth) finite volume hyperbolic 3-manifold with cusps is the interior of a compact 3-manifold with torus boundary components. Filling these in by attaching solid tori produces a closed manifold; there is an infinite number of topologically distinct ways to do this, parametrized by the isotopy classes of the curves on the boundary tori that bound disks in the solid tori. These curves are called the "surgery curves". The manifold with cusps can be viewed as a cone-manifold structure with cone angles 0 on any of these closed manifolds. If it is possible to increase the cone angle from 0 to 2π, this constructs a smooth hyperbolic structure on this closed manifold. This process is called *hyperbolic Dehn surgery*. Thurston ([Thu79]) proved that hyperbolic Dehn surgery fails for at most a finite number of choices of surgery curves on each boundary component.

The proof of local rigidity puts strong constraints on those deformations of hyperbolic cone-manifolds that *do* exist. It is possible to control the change in the geometric structure when the cone angles are deformed a fixed amount. Importantly, this control depends only on the geometry in a tubular neighborhood around the singular locus, not on the rest of the 3-manifold. In particular, it provides geometric and analytic control on the hyperbolic Dehn surgery process. This idea is developed in [HK02].

That paper provides a quantitative version of Thurston's hyperbolic Dehn surgery theorem. Applications include the first universal bounds on the number of non-hyperbolic Dehn fillings on a cusped hyperbolic 3-manifold, and estimates on the changes in volume and core geodesic length during hyperbolic Dehn filling.

The local rigidity theory of [HK98] was generalized by Bromberg ([Brm00]) to include geometrically finite hyperbolic cone-manifolds. Recently, there have been some very imaginative and interesting applications of the deformation theory of geometrically finite hyperbolic cone-manifolds to well-known problems in Kleinian groups. In particular, the reader is referred to [BB02b] in this volume for a description of some of these results and references to others.

Our purpose here is to provide a brief outline of the main ideas and results from [HK98] and [HK02] and how they are related to the Kleinian group applications. As noted above, the central idea is that rigidity results can be turned into effective control on the deformations that do exist. However, we wish to emphasize a particular consequence that provides the common theme between [HK02] and the Kleinian group applications in [Brm02a], [Brm02b], [BB02a], [BB02b] and [BBES]. As a corollary of the control provided by effective rigidity, it is possible to give precise analytic and geometric meaning to the familiar idea that hyperbolic structures with short geodesics are "close" to ones with cusps. Specifically, it can be shown that a structure with a sufficiently short geodesic can be deformed through hyperbolic cone-manifolds to a complete structure, viewed as having cone angle 0. Furthermore, the total change in the structure can be proved to be arbitrarily small for structures with arbitrarily short geodesics. Most importantly this control is independent of the manifolds involved, depending only on the lengths and cone angles.

There are varied reasons for wanting to find such a family of cone-manifolds. It is conjectured that any closed hyperbolic 3-manifold can be obtained by hyperbolic Dehn surgery on some singly cusped, finite volume hyperbolic 3-manifold. If this were true, it could have useful implications in 3-dimensional topology. In [HK02] it is proved that it is true for any such closed 3-manifold whose shortest geodesic has length at most 0.162. (See Theorem 5.6 in Section 5.)

In [Brm02a] Bromberg describes a construction that, remarkably, allows one to replace an incompressible, geometrically infinite end with a short geodesic by a geometrically finite one by gluing in a wedge that creates a cone angle of 4π along the short geodesic. Pushing the cone angle back to 2π provides an approximation of the structure with a geometrically infinite end by one with a geometrically finite end. In [Brm02a] and [BB02a], Bromberg and then Brock and Bromberg give proofs of important cases of the Density Conjecture in this way. This work is described in [BB02b].

In general, a sequence of Kleinian groups with geodesics that are becoming arbitrarily short (or a single Kleinian group with a sequence of arbitrarily short geodesics) is very difficult to analyze. Things are often simpler when the lengths are actually *equal* to 0; i.e., when they are cusps. Thus, if structures with short geodesics can be uniformly compared with structures where they have become cusps, this can be quite useful. One example where this idea has been successfully employed is [BBES]. It seems likely that there will be others in the near future.

Note that the application to the Density Conjecture above involves cone angles between 4π and 2π whereas the theory in [HK98] and [Brm00] requires cone angles to be at most 2π. Thus, this application actually depends on a new version of the deformation theory ([HK]) which applies to all cone angles, as long as there is a tube of a certain radius around the singular locus.

Because of its connection with these Kleinian group results, we use the current paper as an opportunity to outline the main points in this new theory. It is based on a boundary value problem which is used to construct infinitesimal deformations with the same essential properties as those utilized in [HK98] and [HK02]. Explaining those properties and how they are used occupies Sections 2 to 5. The discussion of Kleinian groups and the new deformation theory are both contained in Section 6, the final section of the paper.

The relaxation of the cone angle restriction has implications for hyperbolic Dehn surgery. Some of these are also described in the final section. (See Theorems 6.2 and 6.3.)

2. Deformations of hyperbolic structures

A standard method for analyzing families of structures or maps is to look at the infinitesimal theory where the determining equations simplify considerably. To this end, we first describe precisely what we mean by a 1-parameter family of hyperbolic structures on a manifold. Associated to the derivative of such a family are various analytic, algebraic, and geometric objects which play a central role in this theory. It is useful to be able to move freely among the interpretations provided by these objects and we attempt to explain the relationships between them.

The initial portion of this analysis is quite general, applying to hyperbolic structures in any dimension or, even more broadly, to structures modeled on a Lie group acting transitively and analytically on a manifold.

A hyperbolic structure on an n-manifold X is determined by local charts modeled on $\mathbb{H}^n$ whose overlap maps are restrictions of global isometries of $\mathbb{H}^n$. These determine, via analytic continuation, a map $\Phi : \tilde{X} \to \mathbb{H}^n$ from the universal cover $\tilde{X}$ of X to $\mathbb{H}^n$, called the *developing map*, which is determined uniquely up to post-multiplication by an element of $G = \mathrm{isom}(\mathbb{H}^n)$. The developing map satisfies the equivariance property $\Phi(\gamma m) = \rho(\gamma)\Phi(m)$, for all $m \in \tilde{X}$, $\gamma \in \pi_1(X)$, where $\pi_1(X)$ acts on $\tilde{X}$ by covering transformations, and $\rho : \pi_1(X) \to G$ is the *holonomy representation* of the structure. The developing map also determines the hyperbolic metric on $\tilde{X}$ by pulling back the hyperbolic metric on $\mathbb{H}^n$. (See [Thu97] and [Rat94] for a complete discussion of these ideas.)

We say that two hyperbolic structures are *equivalent* if there is a diffeomorphism f from X to itself taking one structure to the other. We will use the term "hyperbolic structure" to mean such an equivalence class. A *1-parameter family*, X_t, of hyperbolic structures defines a 1-parameter family of developing maps $\Phi_t : \tilde{X} \to \mathbb{H}^n$, where two families are equivalent under the relation $\Phi_t \equiv k_t \Phi_t \tilde{f}_t$ where k_t are isometries of $\mathbb{H}^n$ and $\tilde{f}_t$ are lifts of diffeomorphisms f_t from X to itself. We assume that k_0 and $\tilde{f}_0$ are the identity, and denote Φ_0 as Φ. All of the maps here are assumed to be smooth and to vary smoothly with respect to t.

The tangent vector to a smooth family of hyperbolic structures will be called an *infinitesimal deformation*. The derivative at $t = 0$ of a 1-parameter family of developing maps $\Phi_t : \tilde{X} \to \mathbb{H}^n$ defines a map $\dot{\Phi} : \tilde{X} \to T\mathbb{H}^n$. For any point $m \in \tilde{X}$, $\Phi_t(m)$ is a curve in $\mathbb{H}^n$ describing how the image of m is moving under the developing maps; $\dot{\Phi}(m)$ is the initial tangent vector to the curve.

We will identify $\tilde{X}$ locally with $\mathbb{H}^n$ and $T\tilde{X}$ locally with $T\mathbb{H}^n$ via the initial developing map Φ. Note that this identification is generally not a global diffeomorphism unless the hyperbolic structure is complete. However, it is a *local* diffeomorphism, providing identification of small open sets in $\tilde{X}$ with ones in $\mathbb{H}^n$.

In particular, each point $m \in \tilde{X}$ has a neighborhood U where $\Psi_t = \Phi^{-1} \circ \Phi_t : U \to \tilde{X}$ is defined, and the derivative at $t = 0$ defines a vector field on $\tilde{X}$, $v = \dot{\Psi} : \tilde{X} \to T\tilde{X}$. This vector field determines the infinitesimal variation in developing maps since $\dot{\Phi} = d\Phi \circ v$, and also determines the infinitesimal variation in metric as follows. Let g_t be the hyperbolic metric on $\tilde{X}$ obtained by pulling back the hyperbolic metric on $\mathbb{H}^n$ via Φ_t and put $g_0 = g$. Then $g_t = \Psi_t^* g$ and the infinitesimal variation in metrics $\dot{g} = \frac{dg_t}{dt}\big|_{t=0}$ is the Lie derivative, $\mathscr{L}_v g$, of the initial metric g along v.

Riemannian covariant differentiation of the vector field v gives a $T\tilde{X}$ valued 1-form on $\tilde{X}$, $\nabla v : T\tilde{X} \to T\tilde{X}$, defined by $\nabla v(x) = \nabla_x v$ for $x \in T\tilde{X}$. We can decompose ∇v at each point into a symmetric part and skew-symmetric part. The *symmetric part*, $\tilde{\eta} = (\nabla v)_{sym}$, represents the infinitesimal change in metric, since

$$\dot{g}(x,y) = \mathscr{L}_v g(x,y) = g(\nabla_x v, y) + g(x, \nabla_y v) = 2g(\tilde{\eta}(x), y)$$

for $x, y \in T\tilde{X}$. In particular, $\tilde{\eta}$ descends to a well-defined TX-valued 1-form η on X. The *skew-symmetric* part $(\nabla v)_{skew}$ is the *curl* of the vector field v; its value at $m \in \tilde{X}$ describes the infinitesimal rotation about m induced by v.

To connect this discussion of infinitesimal deformations with cohomology theory, we consider the Lie algebra $\mathfrak{g}$ of $G = \text{isom}(\mathbb{H}^n)$ as vector fields on $\mathbb{H}^n$ representing infinitesimal isometries of $\mathbb{H}^n$. Pulling back these vector fields via the initial developing map Φ gives locally defined infinitesimal isometries on $\tilde{X}$ and on X.

Let $\tilde{E}, E$ denote the vector bundles over $\tilde{X}, X$ respectively of (germs of) infinitesimal isometries. Then we can regard $\tilde{E}$ as the product bundle with total space $\tilde{X} \times \mathfrak{g}$, and E is isomorphic to $(\tilde{X} \times \mathfrak{g})/\sim$ where $(m, \zeta) \sim (\gamma m, Ad\rho(\gamma) \cdot \zeta)$ with $\gamma \in \pi_1(X)$ acting on $\tilde{X}$ by covering transformations and on $\mathfrak{g}$ by the adjoint action of the holonomy $\rho(\gamma)$. At each point p of $\tilde{X}$, the fiber of $\tilde{E}$ splits as a direct sum of infinitesimal pure translations and infinitesimal pure rotations about p; these can be identified with $T_p\tilde{X}$ and $so(n)$ respectively. The hyperbolic metric on $\tilde{X}$ induces a metric on $T_p\tilde{X}$ and on $so(n)$. A metric can then be defined on the fibers of $\tilde{E}$ in which the two factors are orthogonal; this descends to a metric on the fibers of E.

Given a vector field $v : \tilde{X} \to T\tilde{X}$, we can lift it to a section $s : \tilde{X} \to \tilde{E}$ by choosing an "osculating" infinitesimal isometry $s(m)$ which best approximates the vector field v at each point $m \in \tilde{X}$. Thus $s(m)$ is the unique infinitesimal isometry whose translational part and rotational part at m agree with the values of v and $curl\ v$ at m. (This is the "canonical lift" as defined in [HK98].) In particular, if v is itself an infinitesimal isometry of $\tilde{X}$ then s will be a constant section.

Using the equivariance property of the developing maps it follows that s satisfies an "automorphic" property: for any fixed $\gamma \in \pi_1(X)$, the difference $s(\gamma m) - Ad\rho(\gamma)s(m)$ is a *constant* infinitesimal isometry, given by the variation $\dot{\rho}(\gamma)$ of holonomy isometries $\rho_t(\gamma) \in G$ (see Prop 2.3(a) of [HK98]). Here $\dot{\rho} : \pi_1(X) \to \mathfrak{g}$ satisfies the cocyle condition $\dot{\rho}(\gamma_1 \gamma_2) = \dot{\rho}(\gamma_1) + Ad\rho(\gamma_1)\dot{\rho}(\gamma_2)$, so it represents a class in group cohomology $[\dot{\rho}] \in H^1(\pi_1(X); Ad\rho)$, describing the variation of holonomy representations ρ_t.

Regarding s as a vector-valued function with values in the vector space $\mathfrak{g}$, its differential $\tilde{\omega} = ds$ satisfies $\tilde{\omega}(\gamma m) = Ad\rho(\gamma)\tilde{\omega}(m)$ so it descends to a closed 1-form ω on X with values in the bundle E. Hence it determines a de Rham cohomology class $[\omega] \in H^1(X; E)$. This agrees with the group cohomology class $[\dot{\rho}]$ under the de Rham isomorphism $H^1(X; E) \cong H^1(\pi_1(X); Ad\rho)$. Also, we note that the translational part of ω can be regarded as a TX-valued 1-form on X. Its symmetric part is exactly the form η defined above (see Prop 2.3(b) of [HK98]), describing the infinitesimal change in metric on X.

On the other hand, a family of hyperbolic structures determines only an equivalence class of families of developing maps and we need to see how replacing one family by an equivalent family changes both the group cocycle and the de Rham cocycle. Recall that a family equivalent to Φ_t is of the form $k_t \Phi_t \tilde{f}_t$ where k_t are isometries of $\mathbb{H}^n$ and $\tilde{f}_t$ are lifts of diffeomorphisms f_t from X to itself. We assume that k_0 and $\tilde{f}_0$ are the identity.

The k_t term changes the path ρ_t of holonomy representations by conjugating by k_t. Infinitesimally, this changes the cocycle $\dot{\rho}$ by a coboundary in the sense of group cohomology. Thus it leaves the class in $H^1(\pi_1(X); Ad\rho)$ unchanged. The diffeomorphisms f_t amount to choosing a different map from X_0 to X_t. But f_t is isotopic to $f_0 = $ identity, so the lifts $\tilde{f}_t$ don't change the group cocycle at all. It follows that equivalent families of hyperbolic structures determine the same group cohomology class.

If, instead, we view the infinitesimal deformation as represented by the E-valued 1-form ω, we note that the infinitesimal effect of the isometries k_t is to add a constant to $s : \tilde{X} \to \tilde{E}$. Thus, ds, its projection ω, and the infinitesimal variation of metric are all unchanged. However, the infinitesimal effect of the $\tilde{f}_t$ is to change the vector field on $\tilde{X}$ by the lift of a globally defined vector field on X. This changes ω by the derivative of a *globally defined* section of E. Hence, it doesn't affect the de Rham cohomology class in $H^1(X;E)$. The corresponding infinitesimal change of metric is altered by the Lie derivative of a globally defined vector field on X.

3. Infinitesimal harmonic deformations

In the previous section, we saw how a family of hyperbolic structures leads, at the infinitesimal level, to both a group cohomology class and a de Rham cohomology class. Each of these objects has certain advantages and disadvantages. The group cohomology class is determined by its values on a finite number of group generators and the equivalence relation, dividing out by coboundaries which represent infinitesimal conjugation by a Lie group element, is easy to understand. Local changes in the geometry of the hyperbolic manifolds are not encoded, but important global information like the infinitesimal change in the lengths of geodesics is easily derivable from the group cohomology class. However, the chosen generators of the fundamental group may not be related in any simple manner to the hyperbolic structure, making it unclear how the infinitesimal change in the holonomy representation affects the geometry of the hyperbolic structure. Furthermore, it is usually hard to compute even the dimension of $H^1(\pi_1(X);Ad\rho)$ by purely algebraic means and much more difficult to find explicit classes in this cohomology group.

The de Rham cohomology cocycle does contain information about the local changes in metric. The value of the corresponding group cocycle applied to an element $\gamma \in \pi_1(X)$ can be computed simply by integrating an E-valued 1-form representing the de Rham class around any loop in the homotopy class of γ that element; this is the definition of the de Rham isomorphism map. However, it is generally quite difficult to find such a 1-form that is sufficiently explicit to carry out this computation. Furthermore, the fact that any de Rham representative can be altered, within the same cohomology class, by adding an exact E-valued 1-form, (which can be induced by any smooth vector field on X), means that the behavior on small open sets is virtually arbitrary, making it hard to extrapolate to information on the global change in the hyperbolic metric.

In differential topology, one method for dealing with the large indeterminacy within a *real-valued* cohomology class is to use Hodge theory. The existence and uniqueness of a closed and co-closed (harmonic) 1-form within a cohomology class for a closed Riemannian manifold is now a standard fact. Similar results are known for complete manifolds and for manifolds with boundary, where uniqueness requires certain asymptotic or boundary conditions on the forms. By putting a natural metric on the fibers of the bundle E, the same theory extends to the de Rham cohomology groups, $H^1(X;E)$, that arise in the deformation theory of hyperbolic structures. The fact that these forms are harmonic implies that they satisfy certain nice elliptic linear partial differential equations. In particular, for a harmonic representative $\omega \in H^1(X;E)$, the infinitesimal change in metric η, which appears as the symmetric portion of the translational part of ω, satisfies equations of this type. As we will see in the next section, these are the key to the infinitesimal rigidity of hyperbolic structures.

For manifolds with hyperbolic metrics, the theory of harmonic maps provides a non-linear generalization of this Hodge theory, at least for closed manifolds. For non-compact manifolds or manifolds with boundary, the asymptotic or boundary conditions needed for this theory are more complicated than those needed for the Hodge theory. However, at least the relationship described below between the defining equations of the two theories continues to be valid in this general context.

It is known that, given a map $f : X \to X'$ between closed hyperbolic manifolds, there is a unique harmonic map homotopic to f. (In fact, this holds for *negatively curved* manifolds. See [ES64].) Specifically, if $X = X'$ and f is homotopic to the identity, the identity map is this unique harmonic map. Associated to a 1-parameter family X_t of hyperbolic structures on X is a 1-parameter family of developing maps from the universal cover $\tilde{X}$ of X to $\mathbb{H}^n$. Using these maps to pull back the metric on $\mathbb{H}^n$ defines a 1-parameter family of metrics on $\tilde{X}$, and dividing out by the group of covering transformations determines a family of hyperbolic metrics g_t on X. However, a hyperbolic structure only determines an *equivalence class* of developing maps. Because of this equivalence relation, the metrics, g_t are only determined, for each fixed t, up to pull-back by a diffeomorphism of X. For the smooth family of hyperbolic metrics g_t on X, we consider the identity map as a map from X, equipped with the metric g_0, to X, equipped with the metric g_t. For $t = 0$, the identity map is harmonic, but in general it won't be harmonic. Choosing the unique harmonic map homotopic to the identity for each t and using it to pull back the metric g_t defines a new family of metrics beginning with g_0. (For small values of t the harmonic map will still be a diffeomorphism.) By uniqueness and the behavior of harmonic maps under composition with an isometry, the new family of metrics depends only on the family of hyperbolic structures. In this way, we can pick out a canonical family of metrics from the family of equivalence classes of metrics.

If we differentiate this "harmonic" family of metrics associated to a family of hyperbolic structures at $t = 0$, we obtain a symmetric 2-tensor which describes the infinitesimal change of metric at each point of X. Using the underlying hyperbolic metric, a symmetric 2-tensor on X can be viewed as a symmetric TX-valued 1-form.

This is precisely the form η described above which is the symmetric portion of the translational part of the Hodge representative $\omega \in H^1(X;E)$, corresponding to this infinitesimal deformation of the hyperbolic structure.

Thus, the Hodge representative in the de Rham cohomology group corresponds to an infinitesimal harmonic map. The corresponding infinitesimal change of metric has the property that it is L^2-orthogonal to the trivial variations of the initial metric given by the Lie derivative of compactly supported vector fields on X.

We now specialize to the case of interest in this paper, 3-dimensional hyperbolic cone-manifolds. We recall some of the results and computations derived in [HK98]. Let M_t be a smooth family of hyperbolic cone-manifold structures on a 3-dimensional manifold M with cone angles α_t along a link Σ, where $0 \leq \alpha_t \leq 2\pi$. By the Hodge theorem proved in [HK98], the corresponding infinitesimal deformation at time $t = 0$ has a unique Hodge representative whose translational part is a TX-valued 1-form η on $X = M - \Sigma$ satisfying

$$D^*\eta = 0, \tag{3.1}$$

$$D^*D\eta = -\eta. \tag{3.2}$$

Here $D : \Omega^1(X;TX) \rightarrow \Omega^2(X;TX)$ is the exterior covariant derivative, defined, in terms of the Riemannian connection from the hyperbolic metric on X, by

$$D\eta(v,w) = \nabla_v\eta(w) - \nabla_w\eta(v) - \eta([v,w])$$

for all vectors fields v, w on X, and $D^* : \Omega^2(X;TX) \rightarrow \Omega^1(X;TX)$ is its formal adjoint. Further, η and $*D\eta$ determine symmetric and traceless linear maps $T_xX \rightarrow T_xX$ at each point $x \in X$.

Inside an embedded tube $U = U_R$ of radius R around the singular locus Σ, η has a decomposition:

$$\eta = \eta_m + \eta_l + \eta_c \tag{3.3}$$

where η_m, η_l are "standard" forms changing the holonomy of peripheral group elements, and η_c is a correction term with $\eta_c, D\eta_c$ in L^2.

We think of η_m as an ideal model for the infinitesimal deformation in a tube around the singular locus; it is completely determined by the rate of change of cone angle. Its effect on the complex length $\mathscr{L}$ of any peripheral element satisfies

$$\frac{d\mathscr{L}}{d\alpha} = \frac{\mathscr{L}}{\alpha}.$$

In particular, the (real) length ℓ of the core geodesic satisfies

$$\frac{d\ell}{d\alpha} = \frac{\ell}{\alpha} \tag{3.4}$$

for this model deformation.

This model is then "corrected" by adding η_l to get the actual change in complex length of the core geodesic and then by adding a further term η_c that doesn't change the holonomy of the peripheral elements at all, but is needed to extend the deformation in the tube U over the rest of the manifold X.

One special feature of the 3-dimensional case is the *complex structure* on the Lie algebra $\mathfrak{g} \cong sl_2\mathbb{C}$ of infinitesimal isometries of $\mathbb{H}^3$. The infinitesimal rotations fixing a point $p \in \mathbb{H}^3$ can be identified with $su(2) \cong so(3)$, and the infinitesimal pure translations at p correspond to $isu(2) \cong T_p\mathbb{H}^3$. Geometrically, if $t \in T_p\mathbb{H}^3$ represents an infinitesimal translation, then it represents an infinitesimal rotation with axis in the direction of t. Thus, on a hyperbolic 3-manifold X we can identify the bundle E of (germs of) infinitesimal isometries with the *complexified* tangent bundle $TX \otimes \mathbb{C}$.

In [HK98] it was shown that the corresponding harmonic 1-form ω with values in the infinitesimal isometries of $\mathbb{H}^3$ can be written in this complex notation as:

$$\omega = \eta + i*D\eta. \tag{3.5}$$

There is decomposition of ω in the neighborhood U analogous to that (3.3) of η as

$$\omega = \omega_m + \omega_l + \omega_c, \tag{3.6}$$

where only ω_m and ω_l change the peripheral holonomy and ω_c is in L^2.

The fact that the hyperbolic structure on $X = M - \Sigma$ is incomplete makes the existence and uniqueness of a Hodge representative substantially more subtle than the standard theory for complete hyperbolic structures (including structures on closed manifolds). Certain conditions on the behavior of the forms as they approach the singular locus are required. This makes the theory sensitive to the value of the cone angle at the singularity; in particular, this is where the condition that the cone angle be at most 2π arises. The fact that ω_c is in L^2 is a reflection of these asymptotic conditions. In the final section of this paper, we discuss a new version of this Hodge theory, involving boundary conditions on the boundary of a tube around the singular locus, that removes the cone angle condition, replacing it with a lower bound on the radius of the tube.

4. Effective Rigidity

In this section we explain how the equations satisfied by the harmonic representative of an infinitesimal deformation lead to local rigidity results. We then come to one of our primary themes, that the arguments leading to local rigidity can be made computationally effective. By this we mean that even when there *does* exist a non-trivial deformation of a hyperbolic structure, the same equations can be used to bound both the geometric and analytic effect of such a deformation. This philosophy carries over into many different contexts, but here we will continue to focus on finite volume 3-dimensional hyperbolic cone-manifolds.

The first step is to represent an infinitesimal deformation by a Hodge (harmonic) representative ω in the cohomology group $H^1(X;E)$, as discussed in the previous section. If X is any hyperbolic 3-manifold, the symmetric real part of this representative is a 1-form η with values in the tangent bundle of X, satisfying the Weitzenböck-type formula:

$$D^*D\eta + \eta = 0$$

where D is the exterior covariant derivative on such forms and D^* is its adjoint. First, suppose X is closed. Taking the L^2 inner product of this formula with η and integrating by parts gives the formula

$$\|D\eta\|_X^2 + \|\eta\|_X^2 = 0.$$

(Here $\|\eta\|_X^2$ denotes the square of the L^2 norm of η on X.) Thus $\eta = 0$ and the deformation is trivial. This is the proof of local rigidity for closed hyperbolic 3-manifolds, using the methods of Calabi [Cal61], Weil [Wei60] and Matsushima–Murakami [MMu63].

When X has boundary or is non-compact, there will be a Weitzenböck boundary term b:

$$\|D\eta\|_X^2 + \|\eta\|_X^2 = b. \tag{4.1}$$

If the boundary term is non-positive, the same conclusion holds: the deformation is trivial. When $X = M - \Sigma$, where M is a hyperbolic cone-manifold with cone angles at most 2π along its singular set Σ, it was shown in [HK98] that, for a deformation which leaves the cone angle fixed, it is possible to find a representative as above for which the boundary term goes to zero on the boundary of tubes around the singular locus whose radii go to zero. Again, such an infinitesimal deformation must be trivial, proving *local rigidity rel cone angles*.

On the other hand, Thurston has shown ([Thu79, Chap. 5]) that there *are* non-trivial deformations of the (incomplete) hyperbolic structures on $X = M - \Sigma$. By local rigidity rel cone angles such a deformation must change the cone angles, implying that it is always possible to alter the cone angles by a small amount. Using the implicit function theorem it is further possible to show that the variety of representations $\pi_1(X) \to PSL_2(\mathbb{C})$ is smooth near the holonomy representation of such a hyperbolic cone-manifold. This leads to a local parametrization of hyperbolic cone-manifolds by cone angles.

Theorem 4.1 ([HK98]). *For a 3-dimensional hyperbolic cone-manifold with singularities along a link with cone angles $\leq 2\pi$, there are no deformations of the hyperbolic structure keeping the cone angles fixed. Furthermore, the nearby hyperbolic cone-manifold structures are parametrized by their cone angles.*

The *argument* for local rigidity rel cone angles actually provides further information about the boundary term. To explain this, we need to give a more detailed description of some of the work in [HK98]. This will provide not only a fuller explanation of the proof that there are no deformations fixing the cone angles, but also additional information about the deformations that *do* occur.

Assume that η represents a non-trivial infinitesimal deformation. Recall that, inside a tube around the singular locus, η can be decomposed as $\eta = \eta_m + \eta_l + \eta_c$, where only η_m changes the cone angle. Leaving the cone angle unchanged is equivalent to the vanishing of η_m. As we shall see below, the boundary term for η_m by itself is positive. Roughly speaking, η_m contributes positive quantities to the boundary term, while everything else gives negative contributions. (There are also cross-terms which are easily handled.) The condition that the entire boundary term be positive not only implies that the η_m term *must* be non-zero (which is equivalent to local rigidity rel cone angles), but also puts strong restrictions on the η_l and η_c terms. This is the underlying philosophy for the estimates in this section.

In order to implement this idea, we need to derive a formula for the boundary term in (4.1) as an integral over the boundary of X. For details we refer to [HK98].

Let U_r denote a tubular neighborhood of radius r about the singular locus of M and let $X = M - U_r$; it will always be assumed that r is small enough so that U_r will be embedded. Let T_r denote the boundary torus of U_r, oriented by the normal $\frac{\partial}{\partial r}$, (which is the *inward* normal for X). For any TX-valued 1-forms α, β we define

$$b_r(\alpha, \beta) = \int_{T_r} *D\alpha \wedge \beta. \tag{4.2}$$

Note that in this integral, $*D\alpha \wedge \beta$ denotes the real valued 2-form obtained using the wedge product of the form parts, and the geometrically defined inner product on the vector-valued parts of the TX-valued 1-forms $*D\alpha$ and β.

As above, we express the Hodge E-valued 1-form as $\omega = \eta + i * D\eta$ where $D^* D\eta + \eta = 0$. Fix a radius R, remove the tubular neighborhood U_R, and denote $M - U_R$ by X. Then one computes that the Weitzenböck boundary term b in (4.1) equals $b_R(\eta, \eta)$ (see Proposition 1.3 and p. 36 of [HK98]). This implies:

Lemma 4.2.

$$b_R(\eta, \eta) = ||\eta||_X^2 + ||D\eta||_X^2 = ||\omega||_X^2. \qquad (4.3)$$

In particular, we see that $b_R(\eta, \eta)$ is *non-negative*. Writing $\eta = \eta_0 + \eta_c$ where $\eta_0 = \eta_m + \eta_l$, we analyze the contribution from each part. First, using the Fourier decomposition for η_c obtained in [HK98], it turns out that the cross-terms vanish so that the boundary term is simply the sum of two boundary terms:

$$b_R(\eta, \eta) = b_R(\eta_0, \eta_0) + b_R(\eta_c, \eta_c). \qquad (4.4)$$

Next, we see that the contribution, $b_R(\eta_c, \eta_c)$, from the part of the "correction term" that doesn't affect the holonomy of the peripheral elements, is *non-positive*. In fact,

Proposition 4.3.

$$b_R(\eta_c, \eta_c) = -(||\eta_c||_{U_R}^2 + ||D\eta_c||_{U_R}^2) = -||\omega_c||_{U_R}^2. \qquad (4.5)$$

We have assumed that ω_c is harmonic in a neighborhood of U_R so the same argument applied above to η can be applied to η_c on this neighborhood. Consider a region N between tori at distances r, R from Σ with $r < R$. As before, integration by parts over this region implies that the difference $b_r(\eta_c, \eta_c) - b_R(\eta_c, \eta_c)$ equals $||\omega_c||_N^2$. Then the main step is to show that $\lim_{r \to 0} b_r(\eta_c, \eta_c) = 0$. This follows from the proof of rigidity rel cone angles (in section 3 of [HK98]), since η_c represents an infinitesimal deformation which doesn't change the cone angle.

Combining (4.4) with (4.3) and (4.5), we obtain:

$$b_R(\eta_0, \eta_0) = ||\omega||_{M-U_R}^2 + ||\omega_c||_{U_R}^2. \qquad (4.6)$$

In particular, this shows that

$$0 \le b_R(\eta_0, \eta_0), \qquad (4.7)$$

and that

$$||\omega||_{M-U_R}^2 \le b_R(\eta_0, \eta_0). \qquad (4.8)$$

Remark 4.4. We emphasize that the only place in the derivation of (4.7) and (4.8) that we have used the analysis near the singular locus from [HK98] is in the proof of Proposition 4.3. Furthermore, all that is required from this Proposition is the fact that $b_R(\eta_c, \eta_c)$ is *non-positive*. This, together with (4.3) and (4.4), implies both (4.7) and (4.8). In the final section of this paper, we describe another method for finding a Hodge representative for which $b_R(\eta_c, \eta_c)$ is non-positive. This method requires a tube radius of at least a universal size, but no bound on the cone angle. Once this is established, all the results described here carry over immediately to the case where the tube radius condition is satisfied.

We will focus here on applications of the inequality (4.7) which is the primary use of this analysis in [HK02]. The work of Brock and Bromberg discussed in these proceedings ([BB02b]) also requires the second inequality (4.8). This is discussed further in the final section.

As we show below, (4.7) implies that, in the decomposition $\eta_0 = \eta_m + \eta_l$, the η_m term must be non-trivial for a non-trivial deformation. This is equivalent to local rigidity rel cone angles. The positivity result (4.7) can also be used to find *upper bounds* on $b_R(\eta_0, \eta_0)$. On the face of it, this may seem somewhat surprising, but, as we explain below, the algebraic structure of the quadratic form $b_R(\eta_0, \eta_0)$ makes it quite straightforward to derive such bounds.

The possible harmonic forms $\eta_0 = \eta_m + \eta_l$ give a 3-dimensional real vector space W representing models for deformations of hyperbolic cone-manifold structures in a neighborhood of the boundary torus. Here η_m lies in a 1-dimensional subspace W_m containing deformations changing the cone angle, while η_l lies in the 2-dimensional subspace W_l consisting of deformations leaving the cone angle unchanged. In [HK98], we describe explicit TX-valued 1-forms giving bases for these subspaces.

Now consider the quadratic form $Q(\eta_0) = b_R(\eta_0, \eta_0)$ on the vector space W. One easily computes that in all cases, Q is *positive definite* on W_m and *negative definite* on W_l, so Q has signature $+ - -$. This gives the situation shown in figure 1.

The positivity condition (4.7) says that η_0 lies in the cone where $Q \geq 0$ for any deformation which extends over the manifold X. Further, η_m must be non-zero if the deformation is non-trivial; so the cone angle must be changed. Thus (4.7) implies local rigidity rel cone angles.

As noted above, the local parametrization by cone angles (Theorem 4.1) follows from this, and a smooth family of cone-manifold structures M_t is completely determined by a choice of parametrization of the cone angles α_t. We are free to choose this parametrization as we wish. Then the term η_m is completely determined by the derivative of the cone angle.

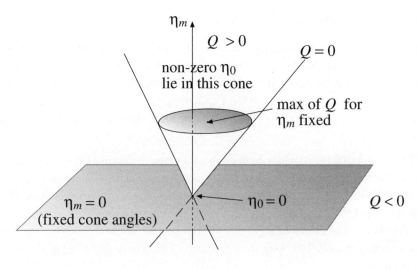

Figure 1

Once η_m is fixed, the inequality $Q(\eta_0) \geq 0$ restricts $\eta_0 = \eta_m + \eta_l$ to lie in an ellipse (as illustrated above). Since Q has a positive maximum on this compact set, this gives an explicit upper bound for $Q(\eta_0)$ for any deformation.

It turns out to be useful to parametrize the cone-manifolds by the *square* of the cone angle α; i.e., we will let $t = \alpha^2$. With this choice of parametrization we obtain:

$$b_R(\eta_m, \eta_m) = \frac{\text{area}(T_R)}{16m^4}\,(\tanh R + \tanh^3 R), \tag{4.9}$$

where $m = \alpha \sinh(R)$ is the length of the meridian on the tube boundary T_R.

Essentially, any contribution to $b_R(\eta_0, \eta_0)$ from η_l will be negative; the cross-terms only complicate matters slightly. Computation leads easily to the following upper bound:

$$b_R(\eta_0, \eta_0) \leq \frac{\text{area}(T_R)}{8m^4}. \tag{4.10}$$

Combining (4.6) and (4.10), we see that the boundary formula (4.1) leads to the estimate:

Theorem 4.5.

$$\|\omega\|^2_{M-U_R} + \|\omega_c\|^2_{U_R} \leq \frac{\text{area}(T_R)}{8m^4}, \tag{4.11}$$

where $m = \alpha \sinh(R)$ is the length of the meridian on the tube boundary T_R.

Remark 4.6. A crucial property of the inequalities (4.10) and (4.11) is their dependence only on the geometry of the boundary torus, not on the rest of the hyperbolic manifold. This is the reason that *a priori* bounds, independent of the underlying manifold can be derived by these methods. Furthermore, it is possible to find geometric conditions, like a very short core geodesic, that force the upper bounds to be very small.

In particular, (4.11) provides an upper bound on the L^2 norm of ω on the complement of the tubular neighborhood of the singular locus. Such a bound can be used to bound the infinitesimal change in geometric quantities, like lengths of geodesics, away from the singular locus. Arbitrarily short core geodesics typically lead to the conclusion that these infinitesimal changes are arbitrarily small, a useful fact when studying approximation by manifolds with short geodesics. This will be discussed further in the final section.

The principle that the contribution of η_l to $b_R(\eta_0, \eta_0)$ is essentially negative (ignoring the cross-terms) while $b_R(\eta_0, \eta_0)$ is positive also provides a bound on the size of η_l. Although η_m does change the length of the singular locus, its effect is fixed since η_m is fixed by the parametrization. The only other term affecting the length of the singular locus is η_l. Explicit computation ultimately leads to the following estimate:

Theorem 4.7 ([HK02]). *Consider any smooth family of hyperbolic cone structures on M, all of whose cone angles are at most 2π. For any component of the singular set, let ℓ denote its length and α its cone angle. Suppose there is an embedded tube of radius R around that component. Then*

$$\frac{d\ell}{d\alpha} = \frac{\ell}{\alpha}(1+E),$$

where

$$\frac{-1}{\sinh^2(R)}\left(\frac{2\sinh^2(R)+1}{2\sinh^2(R)+3}\right) \leq E \leq \frac{1}{\sinh^2(R)}. \tag{4.12}$$

Note that the 'error' term E represents the deviation from the standard model; compare (3.4).

This is the key estimate in [HK02]. The next section discusses some of the applications of this inequality to the theory of hyperbolic Dehn surgery.

5. A quantitative hyperbolic Dehn surgery theorem

We begin with a non-compact, finite volume hyperbolic 3-manifold X, which, for simplicity, we assume has a single cusp. In the general case the cusps are handled independently. The manifold X is the interior of a compact manifold which has a single torus boundary, which is necessarily incompressible. By attaching a solid torus by a diffeomorphism along this boundary torus, one obtains a closed manifold, determined up to diffeomorphism by the isotopy class of the non-trivial simple closed curve γ on the torus which bounds a disk in the solid torus. The resulting manifold is denoted by $X(\gamma)$ and γ is referred to either as the *meridian* or the *surgery curve*. The process is called *Dehn filling* (or Dehn surgery).

The set of Dehn fillings of X is thus parametrized by the set of simple closed curves on the boundary torus; after choosing a basis for $H_1(T^2, \mathbb{Z})$, these are parametrized by pairs (p, q) of relatively prime integers. Thurston ([Thu79]) proved that, for all but a finite number of choices of γ, $X(\gamma)$ has a complete, smooth hyperbolic structure. However, the proof is computationally ineffective. It gives no indication of how many non-hyperbolic fillings there are or which curves γ they might correspond to. In particular, it left open the possibility that there is no upper bound to the number of non-hyperbolic fillings as one varies over all possible X.

One approach to putting a hyperbolic metric on $X(\gamma)$ is through families of hyperbolic cone-manifolds. The complete metric on the open manifold X is deformed through incomplete metrics whose metric completions are hyperbolic cone-manifold structures on $X(\gamma)$, with the singular set equal to the core of the added solid torus. The complete structure can be considered as a cone-manifold with angle 0. The cone angle is increased monotonically, and, if the angle of 2π is reached, it defines a smooth hyperbolic metric on $X(\gamma)$.

Thurston's proof of his finiteness theorem actually shows that, for *any* non-trivial simple closed curve γ on the boundary torus, it is possible to find such a cone-manifold structure on $X(\gamma)$ for sufficiently small (possibly depending on X) values of the cone angle. This also follows from Theorem 4.1, which further implies that, for any angle at most 2π, it is always possible to change the cone angle a small amount, either to increase it or to decrease it. Locally, this can be done in a unique way since the cone angles locally parametrize the set of cone-manifold structures on $X(\gamma)$. Although there are always small variations of the cone-manifold structure, the structures may degenerate in various ways as a family of angles reaches a limit. In order to find a smooth hyperbolic metric on $X(\gamma)$ it is necessary and sufficient to show that no degeneration occurs before the angle 2π is attained.

In [HK02] the concept of convergence of metric spaces in the Gromov-Hausdorff topology (see, e.g., [Gro81], [GB99], or [CHK00]) is utilized to rule out degeneration of the metric as long as there is a lower bound on the tube radius and an upper bound on the volume. The main issue is to show that the injectivity radius in the complement of the tube around the singular locus stays bounded below. One shows that, if the injectivity radius goes to zero, a new cusp develops. Analysis of Dehn filling on this new cusp leads to a contradiction of Thurston's finiteness theorem. The argument is similar to a central step in the proof of the orbifold theorem (see, e.g., [CHK00], [BP01], or [BLP]). One result proved in [HK02] is:

Theorem 5.1. *Let M_t, $t \in [0, t_\infty)$, be a smooth path of closed hyperbolic cone-manifold structures on (M, Σ) with cone angle α_t along the singular locus Σ. Suppose $\alpha_t \to \alpha$ as $t \to t_\infty$ where $\alpha \geq 0$, that the volumes of the M_t are bounded above, and that there is a positive constant R_0 such that there is an embedded tube of radius at least R_0 around Σ for all t. Then the path extends continuously to $t = t_\infty$ so that as $t \to t_\infty$, M_t converges in the bilipschitz topology to a cone-manifold structure M_∞ on M with cone angles α along Σ.*

As a result of this theorem, we can focus on controlling the tube radius. A general principle for smooth hyperbolic manifolds is that short geodesics have large embedded tubes around them. This follows from the Margulis lemma or, alternatively in dimension 3, from Jørgensen's inequality. However, both of these results require that the holonomy representation be discrete which is almost never true for cone-manifolds. In fact, the statement that there is a universal lower bound to the tube radius around a short geodesic is easily seen to be false for cone-manifolds. To see this, consider a sequence of hyperbolic cone-manifolds whose diameters go to zero. Then, both the length of the singular locus and the tube radius go to zero. For example, S^3 with singular locus the figure-8 knot and cone angles approaching $\frac{2\pi}{3}$ from below behaves in this manner.

However, a more subtle statement is true. If, at the beginning of a family of hyperbolic cone-manifold structures, the tube radius is sufficiently large and the length of the core curve is sufficiently small, then as long as the core curve remains sufficiently small, the tube radius will be bounded from below. The precise statement is given below. It should be noted that it is actually the *product* of the cone angle and the core length that must be bounded from above.

Theorem 5.2 ([HK02]). *Let M_s be a smooth family of finite volume 3-dimensional hyperbolic cone-manifolds, with cone angles α_s, $0 \leq s < 1$, where $\lim_{s \to 1} \alpha_s = \alpha_1$. Suppose the tube radius R satisfies $R \geq 0.531$ for $s = 0$ and $\alpha_s \ell_s \leq 1.019$ holds for all s, where ℓ_s denotes the length of the singular geodesic. Then the tube radius satisfies $R \geq 0.531$ for all s.*

The proof of this theorem involves estimates on "tube packing" in cone-manifolds. We look in a certain cover of the complement of the singular locus where the tube around the singular geodesic lifts to many copies of a tube of the same radius around infinite geodesics which are lifts of the core geodesic. One of the lifts is chosen and then packing arguments are developed to show that other lifts project onto the chosen one in a manner that fills up at least a certain amount of area in the tube in the original cone-manifold. In this way, we derive a lower bound for the area of the boundary torus which depends only on the product $\alpha\ell$, assuming $R \geq 0.531$. This, in turn, further bounds the tube radius *strictly* away from 0.531 as long as $\alpha\ell$ is sufficiently small. Thus, if this product stays small the tube radius stays away from the value 0.531 so that the estimates continue to hold. The estimates derived actually prove that, as long as these bounds hold, then $R \to \infty$ as $\alpha\ell \to 0$. Furthermore, the rate at which the tube radius goes to infinity is bounded below.

This result shows that the tube radius can be bounded below by controlling the behavior of the holonomy of the peripheral elements. The cone angle is determined by the parametrization so it suffices to understand the longitudinal holonomy. In the previous section, we derived estimates (4.12) on the derivative of the core length with respect to cone angle, where the bounds depend on the tube radius. On the other hand, from Theorem 5.2, the change in the tube radius can be controlled by the product of the cone angle and the core length. Putting these results together, we arrive at differential inequalities that provide strong control on the change in the geometry of the maximal tube around the singular geodesic, including the tube radius.

A horospherical torus which is a cross-section of the cusp for the complete structure on X has an intrinsic flat structure (i.e., zero curvature metric). Any two such cross-sections differ only by scaling. Given a choice γ of surgery curve, an important quantity associated to this flat structure is the *normalized flat length* of γ, which, by definition equals the geodesic length of γ in the flat metric, scaled to have area 1. Clearly this is independent of the choice of cross-section. Its significance comes from the fact that its square (usually called the *extremal length* of γ) is the limiting value of the ratio $\frac{\alpha}{\ell}$ of the cone angle to the core length as $\alpha \to 0$ for hyperbolic cone-manifold structures on $X(\gamma)$. In particular, near the complete structure, even though ℓ and α individually approach zero, their *ratio* approaches a finite, non-zero value.

The estimate (4.12) implies that, as long as the tube radius isn't too small, $\frac{d\ell}{d\alpha}$ is approximately equal to the ratio $\frac{\ell}{\alpha}$. In the case of equality (when the error term E in (4.12) equals zero), the ratio $\frac{\ell}{\alpha}$ will remain constant. A small error term implies that the ratio doesn't change too much. If the initial value of the *reciprocal*, $\frac{\alpha}{\ell}$ is large, then $\frac{\ell}{\alpha}$ will be small and stay small as long as the tube radius doesn't get too small. But this implies that the product of the cone angle and the core length will remain small.

In turn, the packing argument then provides a lower bound to the tube radius.

Formally, this can be expressed as a differential inequality, bounding the change of core length with respect to the change in cone angle as a function of the core length and the cone angle. Solving this inequality, with initial conditions coming from the normalized length, gives a proof of the following theorem:

Theorem 5.3 ([HK02]). *Let X be a complete, finite volume orientable hyperbolic 3-manifold with one cusp and let T be a horospherical torus embedded as a cusp cross section. Fix γ, a simple closed curve on T. Let $X_\alpha(\gamma)$ denote the cone-manifold structure on $X(\gamma)$ with cone angle α along the core, Σ, of the added solid torus, obtained by increasing the angle from the complete structure. If the normalized length of γ on T is at least 7.515, then there is a positive lower bound to the tube radius around Σ in $X_\alpha(\gamma)$ for all α satisfying $2\pi \geq \alpha \geq 0$.*

Given X and T as in Theorem 5.3, choose *any* non-trivial simple closed curve γ on T. There is a maximal sub-interval $J \subseteq [0, 2\pi]$ containing 0 such that there is a smooth family M_α, where $\alpha \in J$, of hyperbolic cone-manifold structures on $X(\gamma)$ with cone angle α. Thurston's Dehn surgery theorem ([Thu79]) implies that J is non-empty and Theorem 4.1 implies that it is open. Theorem 5.1 implies that, with a lower bound on the tube radii and an upper bound on the volume, the path of M_α's can be extended continuously to the endpoint of J. Again, Theorem 4.1 implies that this extension can be made to be smooth. Hence, under these conditions J will be closed. By Schläfli's formula (see (5.1) below), the volumes decrease as the cone angles increase so they will clearly be bounded above. Theorem 5.3 provides initial conditions on γ which guarantee that there will be a lower bound on the tube radii for all $\alpha \in J$. Thus, assuming Theorems 5.3 and 5.1, we have proved:

Theorem 5.4. *Let X be a complete, finite volume orientable hyperbolic 3-manifold with one cusp, and let T be a horospherical torus which is embedded as a cross-section to the cusp of X. Let γ be a simple closed curve on T whose normalized length is at least 7.515. Then the closed manifold $X(\gamma)$ obtained by Dehn filling along γ is hyperbolic.*

This result also gives a universal bound on the number of non-hyperbolic Dehn fillings on a cusped hyperbolic 3-manifold X, independent of X.

Corollary 5.5. *Let X be a complete, orientable hyperbolic 3-manifold with one cusp. Then at most 60 Dehn fillings on X yield manifolds which admit no complete hyperbolic metric.*

When there are multiple cusps the results are only slightly weaker. Theorem 5.1 holds without change. If there are k cusps, the cone angles α_t and α are simply interpreted as k-tuples of angles. Having tube radius at least R is interpreted as meaning that there are disjoint, embedded tubes of radius R around all of components of the singular locus. The conclusion of Theorem 5.3 and hence of Theorem 5.4 holds when there are multiple cusps as long as the normalized lengths of all of the meridian curves are at least $7.515\sqrt{2} \approx 10.627$. At most 114 curves from each cusp need to be excluded. In fact, this can be refined to say that at most 60 curves need to be excluded from one cusp and at most 114 excluded from the remaining cusps. The rest of the Dehn filled manifolds are hyperbolic.

By Mostow rigidity the volume of a closed or cusped hyperbolic 3-manifold is a topological invariant. The set of volumes of hyperbolic 3-manifolds is well-ordered ([Thu79]); the hyperbolic volume gives an important measure of the complexity of the manifold. It is therefore of interest to find the smallest volume hyperbolic manifold. This is conjectured to be the Weeks manifold which can be described as a surgery on the Whitehead link and has volume ≈ 0.9427. On the other hand, the smallest volume of a hyperbolic 3-manifold with a single cusp is known ([CM01]) to equal ≈ 2.0299, the volume of the figure eight knot complement.

Thus, one can attempt to estimate the volume of a closed hyperbolic 3-manifold by comparing it to the complete structure on the non-compact manifold obtained by removing a simple closed geodesic from the closed manifold. It is conjectured that these two hyperbolic structures are always connected by a smooth family of cone-manifolds, with singular locus equal to the simple closed geodesic, with cone angle α varying from 2π in the smooth structure on the closed manifold to 0 for the complete structure on the complement of the geodesic. In such a family, Schläfli's formula implies that the derivative of the volume V satisfies the equation (see, e.g. [Hod86] or [CHK00, Theorem 3.20]):

$$\frac{dV}{d\alpha} = -\frac{1}{2}\ell, \tag{5.1}$$

where ℓ denotes the length of the singular locus. Thus, controlling of the length of the singular locus throughout the family of cone-manifolds would control the change in the volume. In [HK02] it is shown that the derivative of the length of the singular locus with respect to the cone angle is positive as long as the tube radius is at least $\text{arctanh}(1/\sqrt{3})$ and much sharper statements are proved, using the packing arguments, when the length of the singular locus is small.

It is also shown that, if the original simple closed geodesic is sufficiently short, then such a family of cone-manifolds connecting the smooth structure to the complete structure on the complement of the geodesic will always exist. To see this, note that, for sufficiently short geodesics in a smooth structure, there will always be a tube of radius greater than $\mathrm{arctanh}(1/\sqrt{3})$. Thus the core length will decrease as the cone angle decreases and, in particular, the product of the cone angle and the core length will decrease. By Theorems 5.2 and 5.1 there can be no degeneration and the complete structure will be reached. (It is not hard to show that the volume is bounded above during the deformation.)

Combining these two ideas, we can bound the volume of a closed hyperbolic 3-manifold with a sufficiently short geodesic in terms of the associated cusped 3-manifold. An example of an explicit estimate derived in this manner in [HK02] is:

Theorem 5.6. *Let M be a* closed *hyperbolic manifold whose shortest closed geodesic τ has length at most 0.162. Then the hyperbolic structure on M can be deformed to a complete hyperbolic structure on $M - \tau$ by decreasing the cone angle along τ from 2π to 0. Furthermore, the volumes of these manifolds satisfy the inequality* $\mathrm{Volume}(M) \geq \mathrm{Volume}(M - \tau) - 0.329$. *In particular,* $\mathrm{Volume}(M) \geq 1.701$ *so that it has larger volume than the closed hyperbolic manifold with the smallest known volume.*

6. Kleinian groups and boundary value theory

In this section we give a brief description of generalizations and applications of the deformation theory described in the previous sections.

The first way in which one could attempt to generalize the harmonic deformation theory is to allow finitely many infinite volume, but geometrically finite, ends in our hyperbolic cone-manifolds X. Without giving a formal definition, this structure provides ends that, asymptotically, are like the ends of smooth geometrically finite hyperbolic 3-manifolds. We further assume that there are no rank-1 cusps (so the ends are like those Kleinian groups with *compact* convex cores). The term "geometrically finite" will include this extra assumption throughout this section. Each infinite volume end of such a geometrically finite cone-manifold determines a conformal structure (in fact a complex projective structure) on a surface at infinity, which can be used to (topologically) compactify the cone-manifold. We refer to these as *boundary conformal structures*.

When there is no singular set, then the quasi-conformal deformation theory developed by Ahlfors, Bers and others (see, e.g., [AB60], [Ber60]) implies that such structures are parametrized by these conformal structures at infinity. In particular, they satisfy local rigidity relative to the boundary conformal structures; it is not possible locally to vary the hyperbolic structure without varying at least one of the conformal structures at infinity.

In [Brm00] Bromberg extends the harmonic deformation theory outlined in the previous sections to such geometrically finite cone-manifolds, assuming that the cone angles are at most 2π. In particular, it is proved there that there are no infinitesimal deformations of such structures that fix both the cone angles and the conformal structures at infinity, i.e., rigidity rel cone angles and boundary conformal structures. This generalizes the local theory both for smooth geometrically finite Kleinian groups and for finite volume cone-manifolds. However, it should be pointed out that, since the holonomy groups for these geometrically finite cone-manifold are not usually discrete, the global quasi-conformal theory on the sphere at infinity which is the basis for the smooth (Ahlfors-Bers) theory can't be used at all. So, new techniques are required. Similarly, since the deformation is only assumed to be "conformal at infinity", not trivial at infinity, one cannot assume that there is a representative of this infinitesimal deformation for which the infinitesimal change of metric is asymptotically trivial. As a result, the asymptotic behavior of the E-valued forms representing the deformation must be analyzed at these infinite volume ends as well as near the singular locus. This makes the proof of the necessary Hodge theorem much more difficult. After the Hodge theorem is proved, the final step is to show that there is a harmonic (Hodge) representative for which the contribution to the Weitzenböck boundary term goes to zero near infinity. This requires some subtle analysis.

Once this is established, the analytic results from the finite volume cone-manifold theory go through without change. In particular, the inequalities (4.7) and (4.8) still apply. Once packing arguments, which imply non-degeneration results, are proved in this context, then the Dehn filling results, such as Theorem 5.4, can be generalized (with different numerical values). Similarly, there will be an analog, for smooth, geometrically finite hyperbolic manifolds, of Theorem 5.6. Recall that this theorem says that a closed, smooth hyperbolic manifold with a sufficiently short simple closed geodesic can be connected by a family of cone-manifolds to the complete structure on the manifold obtained by removing the geodesic.

Such packing and non-degeneration results are proved by Bromberg in [Brm02b]. However, his goal and the goal of subsequent papers by (various subsets of) Brock, Bromberg, Evans and Souto (see for example, [Brm02a], [BB02a], [BBES]) is an analytically sharper version of this type of result. A goal in each of these papers is to approximate structures with very short (possibly singular) geodesics by ones with smaller cone angles, including ones with cusps. Not only is a path of cone-manifolds connecting the structures needed but this must be constructed so as to bound the distortion of the structure along the way.

As was remarked in Section 4, the results in [HK02] which we have discussed in the previous sections require only the inequality (4.7). However, the analysis in that section also led to the inequality (4.8) which bounds the L^2 norm of the harmonic representative for the infinitesimal deformation. Further calculations there also provided an upper bound (4.11) for this L^2 norm in terms of the geometry of the boundary torus. It is easy to check that, for any given non-zero cone angle, this upper bound will become arbitrarily small as the length of the singular locus goes to zero, assuming there is a lower bound to the tube radius.

This allows one to bound the L^2 norm of the harmonic representative ω of an infinitesimal deformation at such a structure, where we are assuming that the deformation is infinitesimally conformal at infinity on the infinite volume ends. Such deformations are locally parametrized by cone angles. To be consistent with the previous sections, we use the parametrization $t = \alpha^2$, where α is the cone angle. Then the following theorem holds:

Theorem 6.1 ([Brm02b], [HK]). *Let X be a geometrically finite hyperbolic cone-manifold with no rank-1 cusps and with an embedded tube of radius at least* $\mathrm{arctanh}(1/\sqrt{3})$ *around the singular locus τ. Suppose that ω is the harmonic representative of an infinitesimal deformation of X as above. Then, for any fixed cone angle $\alpha > 0$ and any $\varepsilon > 0$, there is a length $\delta > 0$, depending only on α and ε, such that, if the length of τ is less than δ, then $\|\omega\|^2_{X-U} < \varepsilon$, for some embedded tube U around τ.*

Recall that the real part of ω corresponds to the infinitesimal change in metric induced by the infinitesimal deformation. So, in particular, the above theorem gives an L^2 bound on the size of the infinitesimal change of metric. However, it is still necessary to bound the change in the hyperbolic structure in a more usable way. In [Brm02a] and [Brm02b] this was turned into a bound in the change of the projective structures at infinity. This is sufficient for the applications in those papers. For the applications in [BB02a] and [BBES], an upper bound is needed on the bilipschitz constants of maps between structures along the path of cone-manifolds. For such a bound, it is necessary to turn the L^2 bounds on the infinitesimal change of metric into pointwise bounds. It is then also necessary to extend the bilipschitz maps into the tubes in a way that still has small bilipschitz constant. This is carried out in [BB02a].

The work in [Brm02b] and [BB02a] globalizes Theorem 6.1. It implies that, under the same assumptions, it is possible to find a path of cone-manifolds from the geometrically finite cone-manifold X to the complete structure on X with the geodesic removed, and that this can be done so that all the geometric structures along the path can be made arbitrarily close to each other. In [Brm02b] the distance between the structures is measured in terms of projective structures at infinity, whereas in [BB02a] it is measured by the bilipschitz constant of maps. Since the core geodesic is removed in the cusped structure, the authors call these results "Drilling Theorems" (see [BB02b]).

These theorems provide very strong quantitative statements of the qualitative idea that hyperbolic structures with short geodesics are "close" to ones with cusps.

The careful reader will have noticed that Theorem 6.1 has no conditions on the size of the cone angles whereas the theory in [HK98] requires that the cone angles be at most 2π. As stated, this theorem and hence the full Drilling Theorem depends on a harmonic deformation theory which has no conditions on the cone angle, but only requires the above lower bound on the tube radius. Such a theory is developed in [HK].

Some uses of the Drilling Theorem (e.g., [BBES]) only involve going from a smooth structure (cone angle 2π) to a cusp, so only the analysis in [HK98] is needed. However, the proofs of the Density Conjecture in [Brm02a], [Brm02b], and [BB02a], as described in these proceedings ([BB02b]), require a path of cone-manifolds beginning with cone angle 4π and ending at cone angle 2π, so they depend on the new work in [HK].

Below we give a brief description of the boundary value problem involved in this new version of the harmonic deformation theory, as well as some applications to finite volume hyperbolic cone-manifolds.

We will assume, for simplicity, that X is a compact hyperbolic 3-manifold with a single torus as its boundary. Hyperbolic manifolds with multiple torus boundary components can be handled by using the same type of boundary conditions on each one. The theory extends to hyperbolic manifolds which also have infinite volume geometrically finite ends, whose conformal structure at infinity is assumed to be fixed, using the same techniques as in [Brm00].

We further assume that the geometry near each boundary torus is modelled on the *complement* of an open tubular neighborhood of radius R around the singular set of a hyperbolic cone-manifold. (A horospherical neighborhood of a cusp is included by allowing $R = \infty$.) In particular, we assume that each torus has an intrinsic flat metric with constant principal normal curvatures κ and $\frac{1}{\kappa}$, where $\kappa \geq 1$. The normal curvatures and the tube radius, R, are related by $\coth R = \kappa$ so they determine each other. In fact, given such a boundary structure, it can be canonically filled in. In general, the filled-in structure has "Dehn surgery type singularities" (see [Thu79, Chap. 4]), which includes cone singularities with arbitrary cone angle. We say that X has *tubular boundary*. This structure is described in more detail in [HK02].

In [HK98] specific closed E-valued 1-forms, defined in a neighborhood of the singular locus, were exhibited which had the property that some complex linear combination of them induced every possible infinitesimal change in the holonomy representation of a boundary torus. As a result, by standard cohomology theory, for any infinitesimal deformation of the hyperbolic structure, it is possible to find a closed

E-valued 1-form $\hat{\omega}$ on X which equals such a complex linear combination of these standard forms in a neighborhood of the torus boundary. This combination of standard forms corresponds to the terms $\omega_m + \omega_l$ in equation (3.6) in Section 3.

The standard forms are harmonic so the E-valued 1-form $\hat{\omega}$ will be harmonic in a neighborhood of the boundary but not generally harmonic on all of X. Since it represents a cohomology class in $H^1(X;E)$, it will be closed as an E-valued 1-form, but it won't generally be co-closed. If we denote by d_E and δ_E the exterior derivative and its adjoint on E-valued forms, then this means that $d_E\hat{\omega} = 0$, but $\delta_E\hat{\omega} \neq 0$ in general. (Note that E is a flat bundle so that d_E is the coboundary operator for this cohomology theory.)

A representative for a cohomology class can be altered by a coboundary without changing its cohomology class. An E-valued 1-form is a coboundary precisely when it can be expressed as $d_E s$, where s is an E-valued 0-form, i.e. a *global* section of E. Thus, finding a harmonic (closed and co-closed) representative cohomologous to $\hat{\omega}$ is equivalent to finding a section s such that

$$\delta_E d_E s = -\delta_E\hat{\omega}. \tag{6.1}$$

Then, $\omega = \hat{\omega} + d_E s$ satisfies $\delta_E\omega = 0$, $d_E\omega = 0$; it is a closed and co-closed (hence harmonic) representative in the same cohomology class as $\hat{\omega}$.

In [HK98] it was shown that in order to solve equation (6.1) for E-valued sections, it suffices to solve it for the "real part" of E, where we are interpreting E as the complexified tangent bundle of X as discussed at the end of Section 3. The real part of a section s of E is just a (real) section of the tangent bundle of X; i.e., a vector field, which we denote by v. The real part of $\delta_E d_E s$ equals $(\nabla^*\nabla + 2)v$, where ∇ denotes the (Riemannian) covariant derivative and ∇^* is its adjoint. The composition $\nabla^*\nabla$ is sometimes called the "rough Laplacian" or the "connection Laplacian". We will denote it by $\triangle$.

To solve the equation $\delta_E d_E s = -\delta_E\hat{\omega}$, we take the real part of $-\delta_E\hat{\omega}$, considered as a vector field, and denote it by ζ. We then solve the equation

$$(\triangle + 2)v = \zeta, \tag{6.2}$$

for a vector field v on X. As discussed in [HK98], this gives rise to a section s of E whose real part equals v which is a solution to (6.1). We denote the correction term $d_E s$ by ω_c. In a neighborhood of the boundary $\hat{\omega}$ equals a combination of standard forms, $\omega_m + \omega_l$. Thus, in a neighborhood of the boundary, we can decompose the harmonic representative ω as $\omega = \omega_m + \omega_l + \omega_c$, just as we did in (3.6) in Section 3. Note that, since $\hat{\omega}$ was already harmonic in a neighborhood of the boundary, the correction term ω_c will also be harmonic on that neighborhood.

Because ω is harmonic, it will satisfy Weitzenböck formulae as described in Section 4. As will be outlined below we will choose boundary conditions that will further guarantee that $\omega = \eta + i*D\eta$ where η is a 1-form with values in the tangent bundle of X. It decomposes as $\eta = \eta_m + \eta_l + \eta_c$ in a neighborhood of the boundary and satisfies the equation $D^*D\eta + \eta = 0$ on all of X. As before, taking an L^2 inner product and integrating by parts leads to equation (4.1). The key to generalizing the harmonic deformation theory from the previous sections is finding boundary conditions that will guarantee that the contribution to the boundary term of (4.1) from the correction term η_c will be *non-positive*. Once this is established, everything else goes through without change.

In order to control the behavior of $d_E s = \omega_c$ (hence of η_c), it is necessary to put restrictions on the domain of the operator $(\triangle + 2)$ in equation (6.2). On smooth vector fields with compact support the operator $(\triangle + 2)$ is self-adjoint. It is natural to look for boundary conditions for which self-adjointness still holds. It is possible to find boundary conditions which make this operator elliptic and self-adjoint with trivial kernel. Standard theory then implies that equation (6.2) is always uniquely solvable.

There are many choices for such boundary data. Standard examples include the conditions that either v or its normal derivative be zero, analogous to Dirichlet and Neumann conditions for the Laplacian on real-valued functions. However, none of these standard choices of boundary data have the key property that the Weitzenböck boundary term for η_c will necessarily be non-positive.

The main analytic result in [HK] is the construction of a boundary value problem that has this key additional property. We give a very brief description of the boundary conditions involved. (In particular, we will avoid discussion of the precise function spaces involved.)

The first boundary condition on the vector fields allowed in the domain of the operator $(\triangle + 2)$ in equation (6.2) is that its (3-dimensional) divergence vanish on the boundary. This means that the corresponding infinitesimal change of metric is volume preserving at points on the boundary. The combination of standard forms, $\omega_m + \omega_l$, also induces infinitesimally volume preserving deformations. Once we prove the existence of a harmonic E-valued 1-form ω whose correction term comes from a vector field satisfying this boundary condition, we can conclude that ω is infinitesimally volume preserving at the boundary. If we denote by div the function measuring the infinitesimal change of volume at a point, then this means that div vanishes at the boundary. However, for any harmonic E-valued 1-form div satisfies the equation:

$$\triangle \operatorname{div} = -4 \operatorname{div}, \tag{6.3}$$

where $\triangle$ here denotes the laplacian on functions given locally by the sum of the *negatives* of the second derivatives.

A standard integration by parts argument shows that any function satisfying such an equation and vanishing at the boundary must be identically zero. Thus we can conclude that the deformation induced by ω is infinitesimally volume preserving at *every* point in X. This allows us to conclude that ω can be written as $\omega = \eta + i*D\eta$ where η satisfies $D^*D\eta + \eta = 0$. (See Proposition 2.6 in [HK98].) The computation of the Weitzenböck boundary term now proceeds as before.

The second boundary condition is more complicated to describe. Recall that the boundary of X has the same structure as the boundary of a tubular neighborhood of a (possibly singular) geodesic. In particular, it has a neighborhood which is foliated by tori which are equidistant from the boundary. In a sufficiently small neighborhood, these surfaces are all embedded and, on each of them, the nearest point projection to the boundary is a diffeomorphism. If we denote by u the (2-dimensional) tangential component of the vector field v at the boundary, we can use these projection maps to pull back u to these equidistant surfaces. We denote the resulting extension of u to the neighborhood of the boundary by $\hat{u}$.

In dimension 3, the *curl* of a vector field is again a 3-dimensional vector field. The second boundary condition is that the (2-dimensional) tangential component of $\operatorname{curl} v$ agree with that of $\operatorname{curl} \hat{u}$ on the boundary. Note that, on the boundary, the normal component of $\operatorname{curl} v$ equals the curl of the 2-dimensional vector field u (this curl is a function) so it automatically agrees with that of $\hat{u}$.

As a partial motivation for this condition, consider a vector field which generates an infinitesimal isometry in a neighborhood of the boundary which preserves the boundary as a set. Geometrically, it just translates the boundary and all of the nearby equidistant surfaces along themselves. Thus, this vector field is tangent to the boundary and to all the equidistant surfaces. It has the property that it equals the vector field $\hat{u}$ defined above as the extension of its tangential boundary values. Thus, the above condition can be viewed as an attempt to mirror properties of infinitesimal isometries preserving the boundary.

To see why it might be natural to put conditions on the *curl* of v, rather than on v itself, consider the real-valued 1-form τ dual (using the hyperbolic metric) to v. Then, $\delta\tau$ and $*d\tau$ correspond, respectively, to the divergence and curl of v, where d denotes exterior derivative, δ its adjoint, and $*$ is the Hodge star-operator. Our boundary conditions can be viewed as conditions on the exterior derivative and its adjoint applied to this dual 1-form.

However, the ultimate justification for these boundary conditions is that they lead to a Weitzenböck boundary term with the correct properties, as long as the tube radius is sufficiently large. A direct geometric proof of this fact is still lacking, as is an understanding of the geometric significance of the value of the required lower bound on the tube radius. Nonetheless, the fact that the contribution to the Weitzenböck boundary from the correction term ω_c is always non-positive when it arises from a vector field satisfying these boundary conditions can be derived by straightforward (though somewhat intricate) calculation. All the results from the previous harmonic theory follow immediately. For example, we can conclude:

Theorem 6.2 ([HK]). *For a finite volume hyperbolic cone-manifold with singularities along a link with tube radius at least* $\operatorname{arctanh}(1/\sqrt{3}) \approx 0.65848$, *there are no deformations of the hyperbolic structure keeping the cone angles fixed. Furthermore, the nearby hyperbolic cone-manifold structures are parametrized by their cone angles.*

In the statement of this theorem, when the singular link has more than one component, having tube radius at least R means that there are disjoint embedded tubes of radius R around all the components.

Besides being able to extend local rigidity rel cone angles and parametrization by cone angles, the boundary value theory permits all of the estimates involved in the effective rigidity arguments to go through. In particular, inequalities (4.7), (4.8), and (4.12) continue to hold. The packing arguments require no restriction on cone angles so that the proofs of the results on hyperbolic Dehn surgery (e.g., Theorems 5.4 and 5.5) go through unchanged.

In order to give an efficient description of the conclusions of these arguments in this more general context we first extend some previous definitions. Recall that, if X has a complete finite volume hyperbolic structure with one cusp and T is an embedded horospherical torus, the normalized length of a simple closed curve γ on T is defined as the length of the geodesic isotopic to γ in the flat metric on T, scaled to have area 1. A *weighted* simple closed curve is just a pair, (λ, γ), where λ is a positive real number. Its normalized length is then defined to be λ times the normalized length of γ. If a basis is chosen for $H_1(T, \mathbb{Z})$, the set of isotopy classes of non-contractible simple closed curves on T corresponds to pairs (p, q) of relatively prime integers. Then a weighted simple closed curve (λ, γ) can be identified with the point $(\lambda p, \lambda q) \in \mathbb{R}^2 \cong H_1(T, \mathbb{Z}) \otimes \mathbb{R} \cong H_1(T, \mathbb{R})$. It is easy to check that the notion of normalized length extends by continuity to any $(x, y) \in H_1(T, \mathbb{R})$.

The *hyperbolic Dehn surgery space* for X (denoted $\mathcal{HDS}(X)$) is a subset of $H_1(T,\mathbb{R}) \cup \infty$ which serves as a parameter space for (generally incomplete) hyperbolic structures on X (with certain restrictions on the structure near its end). In particular, if we view a weighted simple closed curve (λ, γ) as an element of $H_1(T,\mathbb{R})$, then saying that it is in $\mathcal{HDS}(X)$ means that there is a hyperbolic cone-manifold structure on the manifold obtained from X by doing Dehn filling with γ as meridian which has the core curve as the singular locus with cone angle $\frac{2\pi}{\lambda}$. The point at infinity corresponds to the complete hyperbolic structure on X.

Thurston's hyperbolic Dehn surgery theorem (see [Thu79]) states that $\mathcal{HDS}(X)$ always contains an open neighborhood of ∞. This, in particular, implies that it contains all but a finite number of pairs (p,q) of relatively prime integers, which implies that all but a finite number of the manifolds obtained by (topological) Dehn surgery are hyperbolic. However, since most of these pairs are clustered "near" infinity, the statement that it contains an open neighborhood of infinity is considerably stronger. Again, Thurston's proof is not effective; it provides no information about the size of any region contained in hyperbolic Dehn surgery space. The following theorem, whose proof is analogous to that of Theorem 5.4 provides such information:

Theorem 6.3 ([HK]). *Let X be a complete, finite volume orientable hyperbolic 3-manifold with one cusp, and let T be a horospherical torus which is embedded as a cross-section to the cusp of X. Let $(x,y) \in H_1(T,\mathbb{R})$ have normalized length at least 7.583. Then there is a hyperbolic structure on X with Dehn surgery coefficient (x,y). In particular, the hyperbolic Dehn surgery space for X contains the complement of the ellipse around the origin determined by the condition that the normalized length of (x,y) is less than 7.583. Furthermore, the volumes of hyperbolic structures in this region differ from that of X by at most 0.306.*

Remark 6.4. The homology group $H_1(T,\mathbb{R})$ can be naturally identified with the universal cover of T so the flat metric on T, normalized to have area 1, induces a flat metric on $H_1(T,\mathbb{R})$. Then the ellipse in the above theorem becomes a metric disk of radius 7.583. From this point of view, the theorem provides a universal size region in $\mathcal{HDS}(X)$ (the complement of a "round disk" of radius 7.583), which is *independent* of X. However, it is perhaps more interesting to note that, if $H_1(T,\mathbb{R})$ is more naturally identified with $H_1(T,\mathbb{Z}) \otimes \mathbb{R}$, then this region actually reflects the *shape* of T.

Finally we note that these techniques also provide good estimates on the change in geometry during hyperbolic Dehn filling as in Theorem 6.3. For example, Schläfli's formula (5.1) together with control on the length of the singular locus (as in Theorem 4.7) leads to explicit upper and lower bounds for the decrease in volume, ΔV. These bounds are independent of the cusped manifold X, and can be viewed as refinements of the asymptotic formula of Neumann and Zagier [NZ85]:

$$\Delta V \sim \frac{\pi^2}{L(x,y)^2} \quad \text{as} \ L(x,y) \to \infty,$$

where $L(x,y)$ denotes the normalized length of the Dehn surgery coefficient $(x,y) \in H_1(T;\mathbb{R})$. The details will appear in [HK].

References

[AB60]   L. Ahlfors and L. Bers (1960). Riemann's mapping theorem for variable metrics, *Ann. Math.* **72**, 385–404.

[Ber60]   L. Bers (1960). Simultaneous uniformization, *Bull. Amer. Math. Soc.* **66**, 94–97.

[BP01]   M. Boileau and J. Porti (2001). Geometrization of 3-orbifolds of cyclic type, (Appendix A by M. Heusener and Porti), *Astérisque* **272**, Soc. Math. de France.

[BLP]   M. Boileau, B. Leeb and J. Porti. Geometrization of 3-dimensional orbifolds II, *arXiv:math.GT/0010185*.

[BB02a]   J. Brock and K. Bromberg (2002). On the density of geometrically finite Kleinian groups, *arXiv:math.GT/0212189*.

[BB02b]   J. Brock and K. Bromberg (2002). Cone-manifolds and the Density Conjecture, *this volume*.

[BBES]   B. Brock, K. Bromberg, R. Evans and J. Souto. Tameness on the boundary and Alhfors' measure conjecture, *arXiv:math.GT/0211022*.

[Brm00]   K. Bromberg (2000). Rigidity of geometrically finite hyperbolic cone-manifolds, *arXiv:math.GT/0009149*.

[Brm02a]   K. Bromberg (2002). Projective structures with degenerate holonomy and the Bers density conjecture, *arXiv:math.GT/0211402*.

[Brm02b]   K. Bromberg (2002). Hyperbolic cone-manifolds, short geodesics and Schwarzian derivatives, *arXiv:math.GT/0211401*.

[Cal61]   E. Calabi (1961). On compact riemannian manifolds with constant curvature I, *Amer. Math. Soc. Proceedings of Symposia in Pure Mathematics* **3**, 155–180.

[CM01]   C. Cao and R. Meyerhoff (2001). The Orientable Cusped Hyperbolic 3-manifolds of Minimum Volume, *Invent. Math.* **146**, no. 3, 451–478.

[CHK00] D. Cooper, C.D. Hodgson and S. Kerckhoff (2000). Three-dimensional Orbifolds and Cone-manifolds, *Memoirs of Math. Soc. of Japan*, **5**.

[ES64]   J. Eells and J. Sampson (1964). Harmonic mappings of Riemannian manifolds, *Amer. J. Math.* **86**, 109–160.

[Gar67] H. Garland (1967). A rigidity theorem for discrete subgroups, *Trans. Amer. Math. Soc.* **129**, 1–25.

[GB99]   M. Gromov and S.M. Bates (1999). *Metric Structures for Riemannian and Non-Riemannian Spaces*, Birkhäuser.

[GRa63] H. Garland and M.S. Raghunathan (1963). Fundamental domains for lattices in (R)-rank 1 semi-simple Lie groups, *Ann. Math.* **78**, 279–326.

[Gro81] M. Gromov (1981). *Structures métriques pour les variétés riemanniennes, edited by J. Lafontaine and P. Pansu.*, Textes Mathématiques 1. CEDIC.

[GT87]   M. Gromov and W. Thurston (1987). Pinching constants for hyperbolic manifolds, *Invent. Math.* **89**, 1–12.

[HK98]   C.D. Hodgson and S.P. Kerckhoff (1998). Rigidity of hyperbolic cone-manifolds and hyperbolic Dehn surgery, *J. Diff. Geom.* **48**, 1–59.

[HK02]   C.D. Hodgson and S.P. Kerckhoff (2002). Universal bounds for hyperbolic Dehn surgery, *arXiv:math.GT/0204345*.

[HK]   C.D. Hodgson and S.P. Kerckhoff. The shape of hyperbolic Dehn surgery space, *in preparation*.

[Hod86] C.D. Hodgson (1986). *Degeneration and regeneration of geometric structures on 3–manifolds*, Ph.D. thesis, Princeton Univ..

[MMu63] Y. Matsushima and S. Murakami (1963). Vector bundle valued harmonic forms and automorphic forms on a symmetric riemannian manifold, *Ann. Math.* **78**, 365–416.

[NZ85]   W.D. Neumann and D. Zagier (1985). Volumes of hyperbolic 3-manifolds, *Topology* **24**, 307–332.

[Rat94] J.G. Ratcliffe (1994). *Foundations of Hyperbolic Manifolds*, Springer-Verlag.

[Thu79] W.P. Thurston (1979). The geometry and topology of three-manifolds, *Princeton University Lecture Notes.*
http://www.msri.org/publications/books/gt3m/

[Thu97] W.P. Thurston (1997). *Three-dimensional Geometry and Topology, Volume 1*, Princeton Univ. Press.

[Wei60] A. Weil (1960). On discrete subgroups of Lie groups, *Ann. Math.* **72**, 369–384.

Craig D. Hodgson

Department of Mathematics and
Statistics
University of Melbourne
Victoria 3010
Australia

cdh@ms.unimelb.edu.au

Steven P. Kerckhoff

Department of Mathematics
Stanford University
Stanford, CA 94305
USA

spk@math.stanford.edu

AMS Classification: 57M50, 57N10, 30F40

Keywords: hyperbolic 3-manifolds, harmonic deformations, Dehn surgery

Kleinian Groups and Hyperbolic 3-Manifolds   Y. Komori, V. Markovic & C. Series (Eds.)
Lond. Math. Soc. Lec. Notes **299**, 75–93   Cambridge Univ. Press, 2003

Cone-manifolds and the density conjecture

Jeffrey F. Brock and Kenneth W. Bromberg

Abstract

We give an expository account of our proof that each cusp-free hyperbolic 3-manifold M with finitely generated fundamental group and incompressible ends is an algebraic limit of geometrically finite hyperbolic 3-manifolds.

1. Introduction

The aim of this paper is to outline and describe new constructions and techniques we hope will provide a useful tool to study deformations of hyperbolic 3-manifolds. An initial application addresses the following conjecture.

Conjecture 1.1 (Bers–Sullivan–Thurston. The Density Conjecture). Each complete hyperbolic 3-manifold M with finitely generated fundamental group is an algebraic limit of geometrically finite hyperbolic 3-manifolds.

Algebraic convergence of M_n to M refers to convergence in the *algebraic deformation space* or in the topology of convergence on generators of the holonomy representations

$$\rho_n \colon \pi_1(M) \to \mathrm{PSL}_2(\mathbb{C}) = \mathrm{Isom}^+(\mathbb{H}^3).$$

The approximating manifolds $M_n = \mathbb{H}^3/\rho_n(\pi_1(S))$ are *geometrically finite* if the *convex core* of M_n, the minimal convex subset homotopy equivalent to M_n, has finite volume. We give an expository account of our progress toward Conjecture 1.1 [BB02].

Theorem 1.2. *Let M be a complete hyperbolic 3-manifold with no cusps, finitely generated fundamental group, and incompressible ends. Then M is an algebraic limit of geometrically finite hyperbolic 3-manifolds.*

Our result represents an initial step in what we hope will be a general geometrically finite approximation theorem for *topologically tame* complete hyperbolic 3-manifolds, namely, for each such manifold M that is homeomorphic to the interior of a compact 3-manifold.

Indeed, the clearly essential assumption in our argument is that M is tame; we make direct use of the following theorem due to Bonahon and Thurston (see [Bon86, Thu79]).

Theorem 1.3 (Bonahon–Thurston). *Each cusp-free complete hyperbolic 3-manifold M with finitely generated fundamental group and incompressible ends is geometrically and topologically tame.*

The tameness of a complete hyperbolic 3-manifold with finitely generated fundamental group reduces to a consideration of its *ends* since every such 3-manifold M contains a *compact core*, namely, a compact submanifold $\mathcal{M}$ whose inclusion is a homotopy equivalence. Each end e of M is associated to a component E of $M \setminus \mathrm{int}(\mathcal{M})$, which we typically refer to as an "end" of M, assuming an implicit choice of compact core. An end E is *incompressible* if the inclusion of E induces an injection $\pi_1(E) \hookrightarrow \pi_1(M)$. The end E is *geometrically finite* if it has compact intersection with the convex core. Otherwise, it is *degenerate*.

For a degenerate end E, *geometric* tameness refers to the existence of a family of simple closed curves on the closed surface $S = \partial \mathcal{M} \cap E$ whose geodesic representatives leave every compact subset of E. Using interpolations of pleated surfaces, Thurston showed that a geometrically tame end is homeomorphic to $S \times \mathbb{R}^+$, so M is topologically tame if all its ends are geometrically finite or geometrically tame (R. Canary later proved the equivalence of these notions [Can93]).

1.1. Approximating the ends

Our approach to Theorem 1.2 will be to approximate the manifold M *end by end*. Such an approach is justified by an *asymptotic isolation* theorem (Theorem 1.7) that isolates the geometry of the ends of M from one another when M is obtained as a limit of geometrically finite manifolds. Each degenerate end E of M has one of two types: E has either

   I. *bounded geometry:* there is a uniform lower bound to the length of the shortest geodesic in E, or

   II. *arbitrarily short geodesics:* there is some sequence γ_n of geodesics in E whose length is tending to zero.

Historically, it is the latter category of ends that have been persistently inscrutable (they are known to be generic [McM91, CCHS01]). Our investigation of such ends begins with another key consequence of tameness, due to J. P. Otal (see [Ota95], or his article [Ota02] in this volume). Before discussing this result, we introduce some terminology.

If E is an incompressible end of M, the cover $\tilde{M}$ corresponding to $\pi_1(E)$ is homotopy equivalent to the surface $S = \partial \mathcal{M} \cap E$. Thus, $\tilde{M}$ sits in the *algebraic deformation space* $AH(S)$, namely, hyperbolic 3-manifolds M equipped with homotopy equivalences, or *markings*, $f: S \to M$ up to isometries that preserve marking and orientation (see [Thu86b], [McM96]). The space $AH(S)$ is equipped with the algebraic topology, or the topology of convergence of holonomy representations, as described above. Theorem 1.3 guarantees each $M \in AH(S)$ is homeomorphic to $S \times \mathbb{R}$; Otal's theorem provides deeper information about how short geodesics in M sit in this product structure.

Theorem 1.4 (Otal [Ota95]). *Let M lie in $AH(S)$. There is an $\varepsilon_{\text{knot}} > 0$ so that if $\mathscr{A}$ is any collection of closed geodesics so that for each $\gamma \in \mathscr{A}$ we have*

$$\ell_M(\gamma) < \varepsilon_{\text{knot}}$$

then there exists a collection of distinct real numbers $\{t_\gamma \mid \gamma \in \mathscr{A}\}$ and an ambient isotopy of $M \cong S \times \mathbb{R}$ taking each γ to a simple curve in $S \times \{t_\gamma\}$.

Said another way, sufficiently short curves in M are simple, unknotted and pairwise unlinked with respect to the product structure $S \times \mathbb{R}$ on M.

Otal's theorem directly facilitates the *grafting* of tame ends that carry sufficiently short geodesics. This procedure, introduced in [Brm02b], uses embedded end-homotopic annuli in a degenerate end to perform 3-dimensional version of grafting from the theory of projective structures (see e.g. [McM98, GKM00]). In section 3 we will describe how successive graftings about short curves in an end E of M can be used to produce a sequence of projective structures with holonomy $\pi_1(M)$ whose underlying conformal structures X_n reproduce the asymptotic geometry of the end E in a limit.

Our discussion of ends E with bounded geometry relies directly on a large body of work of Y. Minsky [Min93, Min94, Min00, Min01] which has recently resulted in the following *bounded geometry theorem*.

Theorem 1.5 (Minsky. Bounded Geometry Theorem). *Let M lie in $AH(S)$, and assume M has a global lower bound to its injectivity radius* inj: $M \to \mathbb{R}^+$. *If $N \in AH(S)$ has the same end-invariant as that of M then $M = N$ in $AH(S)$.*

In other words, there is an orientation preserving isometry $\varphi: M \to N$ that respects the homotopy classes of the markings on each. The "end invariant" $\nu(M)$ refers to a union of invariants, each associated to an end E of M. Each invariant is either a Riemann surface in the *conformal boundary* ∂M that compactifies the end, or an *ending lamination*, namely, the support $|\mu|$ of a limit $[\mu]$ of simple closed curves γ_n whose geodesic representatives in M that exit the end E (here $[\mu]$ is the limit of $[\gamma_n]$ in Thurston's projective measured lamination space $\mathscr{PL}(S)$ [Thu79, Thu86b]).

Minsky's theorem proves Theorem 1.2 for each M with a lower bound to its injectivity radius, since given any end invariant $v(M)$ there is some limit M_∞ of geometrically finite manifolds with end invariant $v(M_\infty) = v(M)$ (see [Ohs90, Bro00]).

1.2. Realizing ends on a Bers boundary

Grafting ends with short geodesics and applying Minsky's results to ends with bounded geometry, we arrive at a *realization* theorem for ends of manifolds $M \in AH(S)$ in some *Bers compactification*.

Theorem 1.6 (Ends are Realizable). *Let $M \in AH(S)$ have no cusps. Then each end of M is realized in a Bers compactification.*

We briefly explain the idea and import of the theorem. The subset of $AH(S)$ consisting of geometrically finite cusp-free manifolds is the *quasi-Fuchsian locus* $QF(S)$. In [Ber60] Bers exhibited the parameterization

$$Q: \text{Teich}(S) \times \text{Teich}(S) \to QF(S)$$

so that $Q(X,Y)$ contains X and Y in its conformal boundary; $Q(X,Y)$ *simultaneously uniformizes* the pair (X,Y). Fixing one factor, we obtain the *Bers slice* $B_Y = \{Q(X,Y) \mid Y \in \text{Teich}(S)\}$, which Bers proved to be precompact. The resulting compactification $\overline{B_Y} \subset AH(S)$ for Teichmüller space has frontier ∂B_Y, a *Bers boundary* (see [Ber70a]).

We say an end E of $M \in AH(S)$ is *realized by Q in the Bers compactification* $\overline{B_Y}$ if there is a manifold $Q \in \overline{B_Y}$ and a marking preserving bi-Lipschitz embedding $\phi: E \to Q$ (see Definition 4.2).

The cusp-free manifold $M \in AH(S)$ is *singly-degenerate* if exactly one end of M is compactified by a conformal boundary component Y. In this case, the main theorem of [Brm02b] establishes that M itself lies in the Bers boundary ∂B_Y, which was originally conjectured by Bers [Ber70a]. Theorem 1.6 generalizes this result to the relative setting of a given incompressible end of M, allowing us to pick candidate approximates for a given M working end-by-end.

1.3. Candidate approximates

To see explicitly how candidate approximates are chosen, let M have finitely generated fundamental group and incompressible ends. For each end E of M, Theorem 1.6 allows us to choose $X_n(E)$ so that the limit of $Q(X_n(E),Y)$ in $\overline{B_Y}$ realizes the end E. Then we simply let M_n be the geometrically finite manifold homeomorphic to M determined by specifying the data

$$(X_n(E_1), \ldots, X_n(E_m)) \in \text{Teich}(\partial \mathscr{M})$$

where $\mathscr{M}$ is a compact core for M; $\text{Teich}(\partial \mathscr{M})$ naturally parameterizes such manifolds (see section 5). The union $X_n(E_1) \cup \ldots \cup X_n(E_m)$ constitutes the conformal boundary ∂M_n.

To conclude that the limit of M_n is the original manifold M, we must show that limiting geometry of each end of M_n does not depend on limiting phenomena in the other ends. We show ends of M_n are *asymptotically isolated*.

Theorem 1.7 (Asymptotic Isolation of Ends). *Let N be a complete cusp-free hyperbolic 3-manifold with finitely generated fundamental group and incompressible ends. Let M_n converge algebraically to N. Then up to bi-Lipschitz diffeomorphism, the end E of M depends only on the corresponding sequence $X_n(E) \subset \partial M_n$.*

(See Theorems 4.1 and 5.1 for a more precise formulation).

When $N \in AH(S)$ is singly-degenerate, the theorem is well known (for example see [McM96, Prop. 3.1]). For N not homotopy equivalent to a surface, the cover corresponding to each end of N is singly-degenerate, so the theorem follows in this case as well.

The ideas in the proof of Theorem 1.7 when N is doubly-degenerate represent a central focus of this paper. In this case, the cover of N associated to each end is again the manifold N and thus not singly-degenerate, so the asymptotic isolation is no longer immediate. The situation is remedied by a new technique in the cone-deformation theory called the *drilling theorem* (Theorem 2.3).

This drilling theorem allows us to "drill out" a sufficiently short curves in a geometrically finite cusp-free manifold with bounded change to the metric outside of a tubular neighborhood of the drilling curve. When quasi-Fuchsian manifolds $Q(X_n, Y_n)$ converge to the cusp-free limit N, any short geodesic γ in N may be drilled out of each $Q(X_n, Y_n)$.

The resulting drilled manifolds $Q_n(\gamma)$ converge to a limit $N(\gamma)$ whose higher genus ends are bi-Lipschitz diffeomorphic to those of N. In the manifold $N(\gamma)$, the rank-2 cusp along γ serves to insulate the geometry of the ends from one another, giving the necessary control. (When there are no short curves, Minsky's theorem again applies).

The drilling theorem manifests the idea that the thick part of a hyperbolic 3-manifold with a short geodesic looks very similar to the thick part of the hyperbolic 3-manifold obtained by removing that curve. We employ the cone-deformation theory of C. Hodgson and S. Kerckhoff to give analytic control to this qualitative picture.

1.4. Plan of the paper

In what follows we will give descriptions of each facet of the argument. Our descriptions are expository in nature, in the interest of conveying the main ideas rather than detailed specific arguments (which appear in [BB02]). We will focus on the case when M is homotopy equivalent to a surface, which presents the primary difficulties, treating the general case briefly at the conclusion.

In section 2 we provide an overview of techniques in the deformation theory of hyperbolic cone-manifolds we will apply, providing bounds on the metric change outside a tubular neighborhood of the cone-singularity under a change in the cone-angle. In section 3 we describe the grafting construction and how it produces candidate approximates for the ends of M with arbitrarily short geodesics. Section 4 describes the asymptotic isolation theorem (Theorem 1.7), the realization theorem for ends (Theorem 1.6), and finally how these results combine to give a proof of Theorem 1.2 when M lies in $AH(S)$. The general case is discussed in section 5.

1.5. Acknowledgments

The authors would like to thank Craig Hodgson and Steve Kerckhoff for their support and for providing much of the analytic basis for our results, Dick Canary and Yair Minsky for their input and inspiration, and Caroline Series for her role in organizing the 2001 Warwick conference and for her solicitation of this article.

2. Cone-deformations

Over the last decade, Hodgson and Kerckhoff have developed a powerful rigidity and deformation theory for 3-dimensional hyperbolic cone-manifolds [HK98]. While their theory was developed initially for application to closed hyperbolic cone-manifolds, work of the second author (see [Brm00]) has generalized this rigidity and deformation theory to *infinite volume* geometrically finite manifolds.

The cone-deformation theory represents a key technical tool in Theorem 1.2. Let N be a compact, hyperbolizable 3-manifold with boundary; assume that ∂N does not contain tori for simplicity. Let c be a simple closed curve in the interior of N. A *hyperbolic cone-metric* is a hyperbolic metric on the interior of $N \setminus c$ that completes to a singular metric on all of the interior of N. Near c the metric has the form

$$dr^2 + \sinh^2 r d\theta^2 + \cosh^2 r dz^2$$

where θ is measured modulo the *cone-angle*, α.

Just as $\mathbb{H}^3$ is compactified by the Riemann sphere, complete infinite volume hyperbolic 3-manifolds are often compactified by projective structures. If a hyperbolic cone-metric is so compactified it is *geometrically finite without rank-one cusps*. As we have excised the presence of rank-one cusps in our hypotheses, we simply refer to such metrics as geometrically finite.

A projective structure on ∂N has an underlying conformal structure; we often refer to ∂N together with its conformal structure as the *conformal boundary of N*.

Theorem 2.1. *Let M_α denote N with a 3-dimensional geometrically finite hyperbolic cone-metric with cone-angle α at c. If the cone-singularity has tube-radius at least $\sinh^{-1}\left(\sqrt{2}\right)$, then nearby cone-metrics are locally parameterized by the cone-angle and the conformal boundary.*

Here, the *tube-radius* about c is the radius of the maximally embedded metric tube about c in M_α.

This local parameterization theorem was first proven by Hodgson and Kerckhoff for closed manifolds with cone-angle less than 2π and no assumption on the size of the tube radius [HK98]. In the thesis of the second author [Brm00], Hodgson and Kerckhoff's result was generalized to the setting of general geometrically finite cone-manifolds, where the conformal boundary may be non-empty. The replacement of the cone-angle condition with the tube-radius condition is recent work of Hodgson and Kerckhoff (see [HK02a] in this volume).

Theorem 2.1 allows us to decrease the cone-angle while keeping the conformal boundary fixed at least for cone-angle near α. We need more information if we wish to decrease the cone-angle all the way to zero.

Theorem 2.2 ([Brm02a]). *Let M_α be a 3-dimensional geometrically finite hyperbolic cone-metric with cone-angle α. Suppose that the cone-singularity c has tube-radius at least $\sinh^{-1}\left(\sqrt{2}\right)$. Then there exists an $\varepsilon > 0$ depending only on α such that if the length of c is less than ε there exists a one-parameter family M_t of geometrically finite cone-metrics with cone-angle t and conformal boundary fixed for all $t \in [0,\alpha]$.*

2.1. The drilling theorem

When the cone-angle α is 2π the hyperbolic cone-metric M_α is actually a smooth hyperbolic metric. When the cone-angle is zero the hyperbolic cone-metric is also a smooth complete metric; the curve c, however, has receded to infinity leaving a rank-two cusp, and the complete hyperbolic metric lives on the interior of $N \setminus c$. We call $N \setminus c$ with its complete hyperbolic metric M_0 the *drilling along c* of M_α.

Applying the analytic tools and estimates developed by Hodgson and Kerckhoff [HK02b], we obtain infinitesimal control on the metric change outside a tubular neighborhood of the cone-singularity under a change in the cone-angle. Letting $U_t \subset M_t$ denote a standard tubular neighborhood of the cone-singularity we obtain the following *drilling theorem*, which summarizes the key geometric information emerging from these estimates.

Theorem 2.3 (The Drilling Theorem). *Suppose M_α is a geometrically finite hyperbolic cone-metric satisfying the conditions of Theorem 2.2, and let M_t be the resulting family of cone-metrics. Then for each $K > 1$ there exists an $\varepsilon' > 0$ depending only on α and K such that if the length of c is less than ε', there are diffeomorphisms of pairs*

$$\phi_t : (M_\alpha \setminus U_\alpha, \partial U_\alpha) \longrightarrow (M_t \setminus U_t, \partial U_t)$$

so that ϕ_t is K-bi-Lipschitz for each $t \in [0, \alpha]$, and ϕ_t extends over U_α to a homeomorphism for each $t \in (0, \alpha]$.

3. Grafting short geodesics

A simple closed curve γ in $M \in AH(S)$ is *unknotted* if it is isotopic in M to a simple curve γ_0 in the "level surface" $S \times \{0\}$ in the product structure $S \times \mathbb{R}$ on M. For such a γ, there is a bi-infinite annulus A containing γ representing its free homotopy class so that A is isotopic to $\gamma_0 \times \mathbb{R}$. Let A^+ denote the sub-annulus of A exiting the positive end of M, let A^- denote the sub-annulus of A exiting the negative end.

The *positive grafting* $\mathrm{Gr}^+(\gamma, M)$ *of M along γ* is the following surgery of M along the *positive grafting annulus* A^+.

1. Let $M_{\mathbb{Z}}$ denote the cyclic cover of M associated to the curve γ. Let

$$F : S^1 \times [0, \infty) \to A^+$$

be a parameterization of the grafting annulus and let $F_{\mathbb{Z}}$ be its lift to $M_{\mathbb{Z}}$.

2. Cutting M along A^+ and $M_{\mathbb{Z}}$ along $A_{\mathbb{Z}}^+ = F_{\mathbb{Z}}(S^1 \times [0, 1))$, the complements $M \setminus A^+$ and $M_{\mathbb{Z}} \setminus A_{\mathbb{Z}}^+$ each have two isometric copies of the annulus in their metric completions $\overline{M \setminus A^+}$ and $\overline{M_{\mathbb{Z}} \setminus A_{\mathbb{Z}}^+}$: the *inward annulus* inherits an orientation from F that agrees with the orientation induced by the positive orientation on $M \setminus A^+$ and the *outward annulus* inherits the opposite orientations from F and $M \setminus A^+$. The complement $M_{\mathbb{Z}} \setminus A_{\mathbb{Z}}^+$ also contains an *inward* and *outward* copy of $A_{\mathbb{Z}}^+$ in its metric completion.

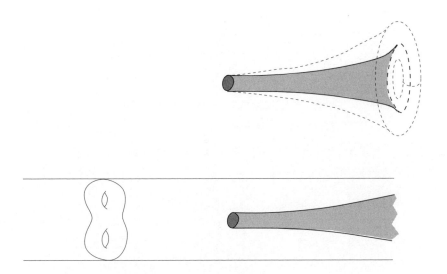

Figure 1: The grafting annulus and its lift.

3. Let F^{in} and F^{out} denote the natural parameterizations of the inward and outward annulus for the metric completion of $M \setminus A^+$ induced by F and let $F_{\mathbb{Z}}^{\mathrm{in}}$ and $F_{\mathbb{Z}}^{\mathrm{out}}$ be similarly induced by $F_{\mathbb{Z}}$.

4. Let ϕ be the isometric gluing of the inward annulus for $\overline{M_{\mathbb{Z}} \setminus A_{\mathbb{Z}}^+}$ to the outward annulus for $\overline{M \setminus A^+}$ and the outward annulus of $\overline{M_{\mathbb{Z}} \setminus A_{\mathbb{Z}}^+}$ to the inward annulus of $\overline{M \setminus A^+}$ so that

$$\phi(F^{\mathrm{in}}(x,t)) = F_{\mathbb{Z}}^{\mathrm{out}}(x,t) \quad \text{and} \quad \phi(F^{\mathrm{out}}(x,t)) = F_{\mathbb{Z}}^{\mathrm{in}}(x,t)$$

(the map ϕ on the geodesic $\widetilde{\gamma} \subset M_{\mathbb{Z}}$ should just be the restriction covering map $M_{\mathbb{Z}} \to M$).

The result $\mathrm{Gr}^+(M, \gamma)$ of positive grafting along γ is no longer a smooth manifold since its metric is not smooth at γ, but $\mathrm{Gr}^+(M, \gamma)$ inherits a smooth hyperbolic metric from M and $M_{\mathbb{Z}}$ away from γ.

3.1. Graftings as cone-manifolds.

Otal's theorem (Theorem 1.4) guarantees that a sufficiently short closed geodesic γ^* is unknotted. In this case, the positive grafting $\mathrm{Gr}^+(M, \gamma^*)$ along the closed geodesic γ^* is well defined, and the singularity has a particularly nice structure: since the singularity is a geodesic, the smooth hyperbolic structure on $\mathrm{Gr}^+(M, \gamma^*) \setminus \gamma^*$ extends to

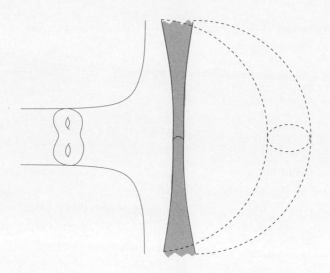

Figure 2: Grafting: glue the wedge $M_{\mathbb{Z}} \setminus A_{\mathbb{Z}}^{+}$ along the completion of $M \setminus A^{+}$.

a hyperbolic cone-metric on $\mathrm{Gr}^{+}(M, \gamma^{*})$ with cone-singularity γ^{*} and cone-angle 4π at γ^{*} (cf. [Brm02b]).

3.2. Simultaneous grafting

We would like to apply the cone-deformation theory of section 2 to the grafting $\mathrm{Gr}^{+}(M, \gamma^{*})$. The deformation theory applies, however, only to geometrically finite hyperbolic cone-manifolds. The grafting $\mathrm{Gr}^{+}(M, \gamma^{*})$ alone may not be geometrically finite if the manifold M is doubly-degenerate. Indeed, in the doubly-degenerate case positive grafting produces a geometrically finite *positive* end, but to force geometric finiteness of *both* ends, we must perform negative grafting as well.

Let γ and β be two simple unknotted curves in M that are also *unlinked*: γ is isotopic to a level surface in the complement of β. Then γ is homotopic either to $+\infty$ or to $-\infty$ in the complement of β. Assume the former. Then we may choose a positive grafting annulus A_{γ}^{+} for γ and a negative grafting annulus A_{β}^{-} for β and perform *simultaneous grafting* on M: we simply perform the grafting surgery on A_{γ}^{+} and A_{β}^{-} at the same time.

By Otal's theorem, when γ^{*} and β^{*} are sufficiently short geodesics in the hyperbolic 3-manifold M, they are simple, unknotted and unlinked. If γ^{*} is homotopic to $+\infty$ in $M \setminus \beta^{*}$, the simultaneous grafting

$$\mathrm{Gr}^{\pm}(\beta^{*}, \gamma^{*}, M)$$

produces a hyperbolic cone-manifold with two cone-singularities, one at γ^* and one at β^*, each with cone-angle 4π.

We then prove the following theorem.

Theorem 3.1 (Simultaneous Graftings). *Let γ^* and β^* be two simple closed geodesics in M as above. Then the simultaneous grafting $\mathrm{Gr}^{\pm}(\beta^*, \gamma^*, M)$ is a geometrically finite hyperbolic cone-manifold.*

The proof applies the theory of geometric finiteness for variable negative curvature developed by Brian Bowditch [Bow94] [Bow95], to a variable negative curvature smoothing $\mathcal{M}$ of $\mathrm{Gr}^{\pm}(\beta^*, \gamma^*, M)$ at its cone-singularities. Using these results, we obtain the following version of Canary's geometric tameness theorem [Can93] for Riemannian 3-manifolds with curvature pinched between two negative constants, or *pinched negative curvature* (we omit the cusped case as usual).

Theorem 3.2 (Geometric Tameness for Negative Curvature). *Each end E of the topologically tame 3-manifold $\mathcal{M}$ with pinched negative curvature and no cusps satisfies the following dichotomy: either*

1. *E is geometrically finite: E has finite volume intersection with the convex core of $\mathcal{M}$, or*

2. *E is simply degenerate: there are essential, non-peripheral simple closed curves γ_n on the surface S cutting off E whose geodesic representatives exit every compact subset of E.*

In our setting, any simple closed curve η on S whose geodesic representative η^* avoids the cone-singularities of $\mathrm{Gr}^{\pm}(\beta^*, \gamma^*, M)$ projects to a closed geodesic $\pi(\eta^*)$ in M under the natural local isometric covering

$$\pi \colon \mathrm{Gr}^{\pm}(\beta^*, \gamma^*, M) \setminus \beta^* \sqcup \gamma^* \to M.$$

The projection π extends to a homotopy equivalence across $\beta^* \sqcup \gamma^*$, so the image $\pi(\eta^*)$ is the geodesic representative of η in M. Though π is not proper, we show that any sequence η_n of simple closed curves on S whose geodesic representatives in $\mathrm{Gr}^{\pm}(\beta^*, \gamma^*, M)$ leave every compact subset must have the property that $\pi(\eta_n^*)$ leaves every compact subset of M. This contradicts bounded diameter results from Thurston's theory of pleated surfaces [Thu79], which guarantee that realizations of $\pi(\eta_n^*)$ by pleated surfaces remain in a compact subset of M. The contradiction implies that grafted ends are geometrically finite, proving Theorem 3.1.

The simultaneous grafting $\mathrm{Gr}^+(\beta^*,\gamma^*,M)$ has two components in its *projective boundary at infinity* to which the hyperbolic cone-metric extends. Already, we can give an outline of the proof of Theorem 1.2 in the case that each end of the doubly-degenerate manifold $M \in AH(S)$ has arbitrarily short geodesics.

Here are the steps:

I. Let $\{\gamma_n^*\}$ be arbitrarily short geodesics exiting the positive end of M and let $\{\beta_n^*\}$ be arbitrarily short geodesics exiting the negative end of M. Assume γ_n^* is homotopic to $+\infty$ in $M \setminus \beta_n^*$.

II. The simultaneous graftings

$$\mathrm{Gr}^\pm(\beta_n^*,\gamma_n^*,M) = M_n^c$$

have projective boundary with underlying conformal structures X_n on the negative end of M_n^c and Y_n on the positive end of M_n^c.

III. By Theorem 3.1 the manifolds M_n^c are geometrically finite hyperbolic cone-manifolds (with no cusps, since M has no cusps).

IV. Applying Theorem 2.3, we may deform the cone-singularities at γ_n^* and β_n^* back to 2π fixing the conformal boundary of M_n^c to obtain quasi-Fuchsian hyperbolic 3-manifolds $Q(X_n,Y_n)$.

V. Since the lengths of γ_n^* and β_n^* are tending to zero, the metric distortion of the cone-deformation outside of tubular neighborhoods of the cone-singularities is tending to zero. Since the geodesics γ_n^* and β_n^* are exiting the ends of M, larger and larger compact subsets of M are more and more nearly isometric to large compact subsets of $Q(X_n,Y_n)$ for n sufficiently large. Convergence of $Q(X_n,Y_n)$ to M follows.

Next, we detail our approach to the general doubly-degenerate case, which handles ends with bounded geometry and ends with arbitrarily short geodesics transparently.

4. Drilling and asymptotic isolation of ends

It is peculiar that manifolds $M \in AH(S)$ of *mixed type*, namely, doubly-degenerate manifolds with one bounded geometry end and one end with arbitrarily short geodesics, present some recalcitrant difficulties that require new techniques. Here is an example of the type of phenomenon that is worrisome:

4.1. Example

Consider a sequence $Q(X_n, Y)$ tending to a limit Q_∞ in the Bers slice B_Y for which Q_∞ is partially degenerate, and for which Q_∞ has arbitrarily short geodesics. Allowing Y to vary in Teichmüller space, we obtain a *limit Bers slice* B_∞ *associated to the sequence* $\{X_n\}$ (this terminology was introduced by McMullen [McM98]). The limit Bers slice B_∞ is an embedded copy of $\mathrm{Teich}(S)$ in $AH(S)$ consisting of manifolds

$$M(Y') = \lim_{n \to \infty} Q(X_n, Y') \quad \text{where } Y' \text{ lies in } \mathrm{Teich}(S).$$

Each $M(Y')$ has a degenerate end that is bi-Lipschitz diffeomorphic to Q_∞ (see, e.g., [McM96, Prop. 3.1]), but the bi-Lipschitz constant depends on Y'.

If, for example, δ is a simple closed curve on S and $\tau^n(Y) = Y_n$ is a divergent sequence in $\mathrm{Teich}(S)$ obtained via an iterated Dehn twist τ about δ, a subsequence of $\{M(Y_n)\}_{n=1}^\infty$ converges to a limit M_∞, but there is no *a priori* reason for the degenerate end of M_∞ to be bi-Lipschitz diffeomorphic to that of $M(Y)$. The limiting geometry of the ends compactified by Y_n could, in principle, bleed over into the degenerate end, causing its asymptotic structure to change in the limit. (We note that such phenomena would violate Thurston's *ending lamination conjecture* since M_∞ has the same ending lamination associated to its degenerate end as does $M(Y)$).

4.2. Isolation of ends

For a convergent sequence of quasi-Fuchsian manifolds $Q(X_n, Y_n) \to N$, we seek some way to isolate the limiting geometry of the ends of $Q(X_n, Y_n)$ as n tends to infinity. Our strategy is to employ the drilling theorem in a suitably chosen family of convergent approximates $Q(X_n, Y_n) \to N$ for which a curve γ is short in $Q_n = Q(X_n, Y_n)$ for all n. We prove that drilling γ out of each Q_n to obtain a drilled manifold $Q_n(\gamma)$ produces a sequence converging to a drilled limit $N(\gamma)$ whose higher genus ends are bi-Lipschitz diffeomorphic to those of N.

An application of the covering theorem of Thurston and Canary [Thu79, Can96] then demonstrates that the limiting geometry of the negative end of N depends only on the sequence $\{X_n\}$ and the limiting geometry of the positive end of N depends only on the sequence $\{Y_n\}$.

When N has no such short geodesic γ, the ends depend only on the end invariant $\nu(N)$, since in this case N has bounded geometry and Theorem 1.5 applies. These arguments are summarized in the following isolation theorem for the asymptotic geometry of N (cf. Theorem 1.7).

Theorem 4.1 (Asymptotic Isolation of Ends). *Let $Q(X_n, Y_n) \in AH(S)$ be a sequence of quasi-Fuchsian manifolds converging algebraically to the cusp-free limit manifold N. Then, up to marking and orientation preserving bi-Lipschitz diffeomorphism, the positive end of N depends only on the sequence $\{Y_n\}$ and the negative end of N depends only on the sequence $\{X_n\}$.*

We now argue that as a consequence of Theorem 4.1 we need only show that each end of a doubly-degenerate manifold M arises as the end of a singly-degenerate manifold lying in a Bers boundary.

Definition 4.2. Let E be an end of a complete hyperbolic 3-manifold M. If E admits a marking and orientation preserving bi-Lipschitz diffeomorphism to an end E' of a manifold Q lying in a Bers compactification, we say E is *realized in a Bers compactification by Q*.

If, for example, the positive end E^+ of M is realized by Q_∞^+ on the Bers boundary ∂B_X then there are by definition surfaces $\{Y_n\}$ so that $Q(X, Y_n)$ converges to Q_∞^+, so E^+ depends only on $\{Y_n\}$ up to bi-Lipschitz diffeomorphism. Arguing similarly, if E^- is realized by Q_∞^- on the Bers boundary ∂B_Y, the approximating surfaces $\{X_n\}$ for which $Q(X_n, Y) \to Q_\infty^-$ determine E^- up to bi-Lipschitz diffeomorphism.

By an application of Theorem 4.1, if the manifolds $Q(X_n, Y_n)$ converge to a cusp-free limit N, then the negative end E_N^- is bi-Lipschitz diffeomorphic to E^- and the positive end E_N^+ is bi-Lipschitz diffeomorphic to E^+. We may glue bi-Lipschitz diffeomorphisms

$$\psi^- : E_N^- \to E^- \quad \text{and} \quad \psi^+ : E_N^+ \to E^+$$

along the remaining compact part to obtain a global bi-Lipschitz diffeomorphism

$$\psi : N \to M$$

that is marking and orientation preserving. By applying Sullivan's rigidity theorem [Sul81a], ψ is homotopic to an isometry, so $Q(X_n, Y_n)$ converges to M.

4.3. Realizing ends in Bers compactifications

To complete the proof of Theorem 1.2, then, we seek to realize each end of the doubly-degenerate manifold M on a Bers boundary; we restate Theorem 1.6 here.

Theorem 4.3 (Ends are Realizable). *Let $M \in AH(S)$ have no cusps. Then each end of M is realized in a Bers compactification.*

In the case that M has a conformal boundary component Y, the theorem asserts that M lies within the Bers compactification $\overline{B_Y}$. This is the main result of [Brm02b], which demonstrates all such manifolds are limits of quasi-Fuchsian manifolds.

We are left to attend to the case when M is doubly-degenerate. As one might expect, the discussion breaks into cases depending on whether an end E has bounded geometry or arbitrarily short geodesics. We discuss the positive end of M; one argues symmetrically for the negative end.

1. If a bounded geometry end E has ending lamination ν, choose a measured lamination μ with support ν and a sequence of weighted simple closed curves $t_n \gamma_n \to \mu$. Choose Y_n so that $\ell_{Y_n}(\gamma_n) < 1$.

2. If γ_n^* are arbitrarily short geodesics exiting the end E, we apply the drilling theorem to $\mathrm{Gr}^{\pm}(\gamma_0, \gamma_n, M)$ to send the cone-angles at γ_0^* and γ_n^* to 2π. The result is a sequence $Q(X, Y_n)$ of quasi-Fuchsian manifolds.

We wish to show that after passing to a subsequence $Q(X, Y_n)$ converges to a limit Q_∞ that realizes E on the Bers boundary ∂B_X.

Bounded geometry. When E has bounded geometry, we employ [Min00] to argue that its end invariant ν has *bounded type*. This condition ensures that any end with ν as its end invariant has bounded geometry. The condition $\ell_{Y_n}(\gamma_n) < 1$ guarantees that $\ell_{Y_n}(t_n \gamma_n) \to 0$ so that any limit Q_∞ of $Q(X, Y_n)$ has ν as its end-invariant (by [Bro00], applying [Ber70a, Thm. 3]). We may therefore apply a relative version of Minsky's ending lamination theorem for bounded geometry (see [Min94], and an extension due to Mosher [Mos01] that treats the case when the manifold may not possess a *global* lower bound to its injectivity radius) to conclude that Q_∞ realizes E.

Arbitrarily short geodesics. If E has an exiting sequence $\{\gamma_n\}$ of arbitrarily short geodesics, we argue using Theorem 2.3 that $Q(X, Y_n)$ converges in the Bers boundary ∂B_X to a limit Q_∞ that realizes E.

4.4. Binding realizations

As a final detail we mention that to apply Theorem 4.1, we require a *convergent* sequence $Q(X_n, Y_n) \to N$ so that the limit $Q^- = \lim Q(X_n, Y_0)$ realizes the negative end E^- of M and the limit $Q^+ = \lim Q(X_0, Y_n)$ realizes the positive end E^+.

By an application of [Bro01], the realizations described in our discussion of Theorem 1.6 produce surfaces $\{X_n\}$ and $\{Y_n\}$ that converge up to subsequence to laminations in Thurston's compactification of Teichmüller space that *bind the surface S*.

Thus, an application of Thurston's *double limit theorem* (see [Thu86b, Thm. 4.1], [Ota96]) implies that $Q(X_n, Y_n)$ converges to a cusp-free limit N after passing to a subsequence.

5. Incompressible ends

We conclude the paper with a brief discussion of the proof of Theorem 1.2 when M is not homotopy equivalent to a closed surface.

Since M has incompressible ends, Theorem 1.3 implies that M is homeomorphic to the interior of a compact 3-manifold N. Equipped with a homotopy equivalence or *marking* $f : N \to M$, the manifold M determines an element of the *algebraic deformation space $AH(N)$* consisting of all such marked hyperbolic 3-manifolds up to isometries preserving orientation and marking, equipped with the topology of algebraic convergence.

By analogy with the quasi-Fuchsian locus, the subset $AH(N)$ consisting of M' that are geometrically finite, cusp-free and homeomorphic to M is parameterized by the product of Teichmüller spaces

$$\text{Teich}(\partial N) = \prod_{X \subset \partial N} \text{Teich}(X).$$

In this situation, the cover $\tilde{M}$ corresponding to an end E of M lies in $AH(S)$. Theorem 1.6 guarantees that if E is degenerate it is realized on a Bers boundary; indeed, since M is cusp-free and M is not homotopy equivalent to a surface, it follows that $\tilde{M}$ is itself singly-degenerate, so Theorem 1.6 guarantees that $\tilde{M}$ lies in a Bers compactification.

The remaining part of Theorem 1.2, then, follows from the following version of Theorem 1.7.

Theorem 5.1 (Asymptotic Isolation of Ends II). *Let M be a cusp-free complete hyperbolic 3-manifold with incompressible ends homeomorphic to* $\text{int}(N)$. *Let $M_n \to M$ in $AH(N)$ be a sequence of cusp-free geometrically finite hyperbolic manifolds so that each M_n is homeomorphic to M. Let $(E^1, \ldots, E^m)$ denote the ends of M, and let $\partial M_n = X_n^1 \sqcup \ldots \sqcup X_n^m$ be the corresponding points in* $\text{Teich}(\partial N)$. *Then, up to marking preserving bi-Lipschitz diffeomorphism, E^j depends only on the sequence $\{X_n^j\}$.*

In the case not already covered by Theorem 4.1, the covers of M_n corresponding to a fixed boundary component are quasi-Fuchsian manifolds $Q(Y_n, X_n^j)$. Their limit is the singly-degenerate cover of M corresponding to E^j, so the surfaces Y_n range in a compact subset of Teichmüller space.

Again, it follows that the marked bi-Lipschitz diffeomorphism type of the end E does not depend on the surfaces Y_n. Theorem 1.2 then follows in this case from an application of Theorem 1.6 to each end degenerate end E of M, after an application of Sullivan's rigidity theorem [Sul81a].

References

[Abi77]   W. Abikoff (1977). Degenerating families of Riemann surfaces, *Ann. Math.* **105**, 29–44.

[BB02]   J. Brock and K. Bromberg (2002). On the density of geometrically finite Kleinian groups, *arXiv:math.GT/0212189*.

[Ber60]   L. Bers (1960). Simultaneous uniformization, *Bull. Amer. Math. Soc.* **66**, 94–97.

[Ber70a]   L. Bers (1970). On boundaries of Teichmüller spaces and on Kleinian groups I, *Ann. Math.* **91**, 570–600.

[Bon86]   F. Bonahon (1986). Bouts des variétés hyperboliques de dimension 3, *Ann. Math.* **124**, 71–158.

[Bow94]   B. Bowditch (1994). Some results on the geometry of convex hulls in manifolds of pinched negative curvature, *Comment. Math. Helv.* **69**, 49–81.

[Bow95]   B. Bowditch (1995). Geometrical finiteness with variable negative curvature, *Duke Math. J.* **77**, 229–274.

[Brm00]   K. Bromberg (2000). Rigidity of geometrically finite hyperbolic cone-manifolds, *arXiv:math.GT/0009149*.

[Brm02a]   K. Bromberg (2002). Hyperbolic cone manifolds, short geodesics and Schwarzian derivatives, *arXiv:math.GT/0211401*.

[Brm02b]   K. Bromberg (2002). Projective structures with degenerate holonomy and the Bers density conjecture, *arXiv:math.GT/0211402*.

[Bro00]   J. Brock (2000). Continuity of Thurston's length function, *Geom. and Funct. Anal.* **10**, 741–797.

[Bro01]   J. Brock (2001). Boundaries of Teichmüller spaces and geodesic laminations, *Duke Math. J.* **106**, 527–552.

[Can93]   R.D. Canary (1993). Ends of hyperbolic 3-manifolds, *J. Amer. Math. Soc.* **6**, 1–35.

[Can96]   R.D. Canary (1996). A covering theorem for hyperbolic 3-manifolds and its applications, *Topology* **35**, 751–778.

[CCHS01] R. Canary, M. Culler, S. Hersonsky and P. Shalen (2001). Approximations by maximal cusps in boundaries of deformation spaces, *preprint, submitted for publication*.

[GKM00]   D. Gallo, M. Kapovich and A. Marden (2000). The monodromy groups of Schwarzian equations on closed Riemann surfaces, *Ann. Math.* **151**, 625–704.

[HK98]   C. Hodgson and S. Kerckhoff (1998). Rigidity of hyperbolic cone-manifolds and hyperbolic Dehn surgery, *J. Diff. Geom.* **48**, 1–59.

[HK02a]   C. Hodgson and S. Kerckhoff (2002). Harmonic deformations of hyperbolic 3-manifolds, *arXiv:math.GT0301226 and this volume*.

[HK02b]   C. Hodgson and S. Kerckhoff (2002). Universal bounds for hyperbolic Dehn surgery, *arXiv:math.GT/0204345*.

[McM91]   C. McMullen (1991). Cusps are dense, *Ann. Math.* **133**, 217–247.

[McM96]   C. McMullen (1996). Renormalization and 3-manifolds which fiber over the circle, *Ann. Math. Studies* **142**, Princeton University Press.

[McM98]   C. McMullen (1998). Complex earthquakes and Teichmüller theory, *J. Amer. Math. Soc.* **11**, 283–320.

[Min93]   Y.N. Minsky (1993). Teichmüller geodesics and ends of hyperbolic 3-manifolds, *Topology* **32**, 625–647.

[Min94]   Y.N. Minsky (1994). On rigidity, limit sets and end invariants of hyperbolic 3-manifolds, *J. Amer. Math. Soc.* **2**, 539–588.

[Min00]   Y.N. Minsky (2000). Kleinian groups and the complex of curves, *Geometry and Topology* **4**, 117–148.

[Min01]   Y.N. Minsky (2001). Bounded geometry for Kleinian groups, *Invent. Math.* **146**, 143–192.

[Mos01] L. Mosher (2001). Stable Teichmüller quasigeodesics and ending laminations, *arXiv:math.GT/0107035*.

[Ohs90] K. Ohshika (1990). Ending laminations and boundaries for deformation spaces of Kleinian groups, *J. Lond. Math. Soc.* **42**, 111–121.

[Ota95] J.-P. Otal (1995). Sur le nouage des géodésiques dans les variétés hyperboliques, *C. R. Acad. Sci. Paris. Sèr. I Math.* **320**, 847–852.

[Ota96] J.-P. Otal (1996). *Le théorème d'hyperbolisation pour les variétés fibrées de dimension trois*, Astérisque No. 235.

[Ota02] J.-P. Otal (2002). Les géodésiques fermées d'une variété hyperbolique en tant que nœds, *this volume*.

[Sul81a] D. Sullivan (1981). On the ergodic theory at infinity of an arbitrary discrete group of hyperbolic motions. In *Riemann Surfaces and Related Topics: Proceedings of the 1978 Stony Brook Conference, Ann. Math. Studies* **97**, Princeton.

[Thu79] W.P. Thurston (1979). The geometry and topology of three-manifolds, *Princeton University Lecture Notes*.
http://www.msri.org/publications/books/gt3m/

[Thu86b] W.P. Thurston (1986). Hyperbolic structures on 3-manifolds II: Surface groups and 3-manifolds which fiber over the circle, *arXiv:math.GT/9801045*.

Jeffrey F. Brock

Department of Mathematics
University of Chicago
5734 S. University Ave
Chicago, IL 60637-1546
USA

brock@math.uchicago.edu

Kenneth W. Bromberg

Mathematics 253-37
Caltech
Pasadena CA 91125
USA

bromberg@its.caltech.edu

AMS Classification: 30F40, 37F30, 37F15

Keywords: Kleinian group, hyperbolic 3-manifold, projective structure, cone-manifold, hyperbolic dynamics

Kleinian Groups and Hyperbolic 3-Manifolds   Y. Komori, V. Markovic & C. Series (Eds.)
Lond. Math. Soc. Lec. Notes **299**, 95–104   Cambridge Univ. Press, 2003

Les géodésiques fermées d'une variété hyperbolique en tant que nœuds

Jean-Pierre Otal

Résumé

Le but de cette note est de compléter certains arguments contenus dans [Ota95], en particulier le théorème A de cette note qui établissait que les géodésiques fermées de longueur suffisamment courte dans une variété hyperbolique ayant le type d'homotopie d'une surface compacte sont "non nouées". Nous considèrerons aussi des variétés hyperboliques plus générales, et donnerons une condition portant sur le cœur de Nielsen d'une telle variété pour qu'une géodésique fermée y soit non nouée.

Closed geodesics in a hyperbolic manifold, viewed as knots

Abstract

The goal of this note is to complete some arguments given in [Ota95], in particular in Theorem A of that paper which stated that the closed geodesics which are sufficiently short in a hyperbolic 3-manifold homotopic equivalent to a closed surface are "unknotted". We will consider also more general hyperbolic 3-manifolds, and give a condition on the Nielsen core of such a manifold insuring that a closed geodesic be unknotted.

1. Introduction

Définition 1.1. Soit S une surface (pas nécessairement compacte) et $f : S \to M$ un plongement dans une variété M de dimension 3. On dit qu'une courbe fermée sans points doubles $\gamma \subset M$ est *non nouée par rapport à* $f : S \to M$ si le plongement f est proprement isotope à un plongement f' telle que γ soit contenue dans $f'(S)$.

Si γ est une courbe non nouée par rapport à f, alors l'élément du groupe fondamental $\pi_1(M)$ qu'elle représente est évidemment contenu dans un conjugué de l'image

$f^*(\pi_1(S))$; donc la question de savoir si une géodésique est nouée ou pas par rapport à un plongement, ne se pose que pour les géodésiques avec cette propriété.

Soit M une variété orientable de dimension 3 qui est *hyperbolique*, c'est-à-dire qui porte une métrique riemannienne complète de courbure -1. On définit *la partie ε-mince de M* comme l'ensemble $M^{]0,\varepsilon]}$ des points de M par lesquels il passe une courbe fermée non homotope à 0 et de longueur inférieure à ε. Rappelons qu'il existe une constante $\varepsilon(3) > 0$ [Thu97, Théorème 4.5.6] telle que, si $\varepsilon \leq \varepsilon(3)$, chaque composante connexe de $M^{]0,\varepsilon]}$ est ou bien *un cusp*, isométrique au quotient d'une horoboule par un groupe parabolique, ou bien *un tube de Margulis*, isométrique au quotient d'un voisinage de rayon constant d'une géodésique de l'espace hyperbolique $\mathbb{H}^3$ par un groupe cyclique de transformations loxodromiques le long de cette géodésique. En particulier, toute géodésique fermée *primitive* (i.e. qui n'est pas homotope à une puissance non nulle d'un lacet) et de longueur inférieure à $\varepsilon(3)$ est une courbe plongée.

Dans cette note, nous allons montrer que certaines géodésiques fermées dans une variété hyperbolique de dimension 3 sont non nouées par rapport à des plongements naturels. Nous étudierons d'abord le cas où M est homéomorphe à un produit $S \times \mathbb{R}$, où S est une surface de genre $g \geq 2$; d'après un théorème de F. Bonahon, ceci correspond exactement au cas où M à le type d'homotopie de S. Les problèmes de nouage seront alors considérés pour le plongement, canonique à isotopie près, de S dans $M \simeq S \times \mathbb{R}$. Nous considèrerons dans une deuxième partie le cas où la topologie de la variété M est quelconque mais nous supposerons alors que M est différente de son cœur de Nielsen N. Les problèmes de nouage que nous étudierons seront alors par rapport au plongement du bord ∂N dans M.

2. Les géodésiques courtes dans une variété hyperbolique homéomorphe à $S \times \mathbb{R}$

La note [Ota95] contenait une démonstration du théorème suivant.

Théorème A. *Soit M une variété hyperbolique de dimension 3 ayant le type d'homo-topie d'une surface compacte orientable S de genre g. Il existe une constante $c(g) > 0$ telle que si $\gamma^* \subset M$ est une géodésique fermée de longueur inférieure à $c(g)$, alors γ^* est non nouée.*

Démonstration. Nous allons redonner brièvement le début de l'argument, puis détaillerons un point de topologie de dimension 3. Par un résultat de Thurston et Bonahon, on sait que M est homéomorphe à $S \times \mathbb{R}$. D'après une inégalité de Thurston [Thu86b], la géodésique γ^* est homotope à une courbe *simple* γ tracée sur $S \times \{0\}$.

Considérons une *surface plissée* $f : S \to M$ (où nous identifions S avec $S \times \{0\}$) dont le lieu de plissage contient γ. Ceci signifie que f est homotope à l'inclusion $S \to S\times] - \infty, \infty[$, que pour une certaine métrique hyperbolique sur S, f est une isométrie et que pour cette métrique, la courbe γ est envoyée isométriquement sur γ^*.

Supposons $l(\gamma^*) \le \varepsilon(3)$ et soit $\mathcal{T}$ le tube de Margulis, composante de la partie $\varepsilon(3)$-mince de M qui contient γ^*. Si la longueur de γ^* est très courte, alors $\mathcal{T}$ est un voisinage régulier de γ de rayon très grand ($\simeq K \log \frac{\varepsilon(3)}{l(\gamma)}$ pour une constante K indépendante de M). Il existe donc $c(g)$ telle que si $l(\gamma^*) \le c(g)$, alors le méridien de $\mathcal{T}$ a une aire supérieure à $2\pi|\chi(S)|$, l'aire pour la métrique hyperbolique de S. Ceci implique que les intersections de l'image $f(S)$ avec $\mathcal{T}$ ne sont pas "essentielles". On peut alors homotoper f en une nouvelle application f' telle que l'intersection de $f'(S)$ avec $\mathcal{T}$ est un anneau plongé, image par f d'un voisinage régulier A de γ sur S ; de plus, par sa construction, la nouvelle surface $f'(S)$ est contenue dans un voisinage régulier de la réunion de $f(S)$ et de $\partial\mathcal{T}$ (cf. [Ota95]).

Le théorème A ci-dessus découlera du lemme suivant. Ce résultat est implicite dans la démonstration du théorème 2.1 de [FHS83], mais, comme il n'est pas cité sous la forme que nous allons donner, nous en détaillerons la démonstration, renvoyant à [FHS83] lorsque les arguments sont les mêmes.

Lemme 2.1. *Soient S une surface fermée orientable de genre g, $\Sigma \subset S$ une sous-surface incompressible, M une variété de dimension 3, orientable et irréductible. Soit $f : S \to M$ une application en position normale telle que :*

* *f est une équivalence d'homotopie ;*

* *la restriction $f|_\Sigma$ est un plongement ;*

* *$f^{-1}(f(\Sigma)) = \Sigma$.*

Alors f est homotope à un plongement f' tel que $f'|_\Sigma = f|_\Sigma$. De plus, pour tout voisinage régulier de $f(S)$, l'application f' peut être choisie de sorte que $f'(S)$ soit contenue dans ce voisinage.

Démonstration. On fixe un voisinage régulier N de $f(S)$ dans M qui contient un voisinage régulier de $f(\Sigma)$ difféomorphe à $f(\Sigma) \times [-1,1]$. Puisque $f : S \to M$ est une équivalence d'homotopie, elle induit une injection $H_i(S, \mathbb{Z}_2) \to H_i(N, \mathbb{Z}_2)$ pour $i = 1, 2$. Lorsque l'application induite par f de $H_1(S, \mathbb{Z}_2) \to H_1(N, \mathbb{Z}_2)$ est surjective, on pose $N = N_0$. Sinon, on définit par récurrence une suite de variétés N_i. Puisque l'application induite par f n'est pas surjective, il existe un revêtement connexe, de degré 2, $p_1 : N_1' \to N_0 = N$ tel que $p_1^*(\pi_1(N_1'))$ contient $f^*(\pi_1(S))$. L'application

$f_0 = f$ se relève alors en une application en position normale $f_1 : S \to N_1'$; de plus, puisque la restriction $f_0|_\Sigma$ est un plongement, f_1 se prolonge en un relèvement du voisinage $f_0(\Sigma) \times [-1, 1]$. On définit alors N_1 comme un voisinage régulier de $f_1(S)$ dans N_1'; on peut supposer que, près de $f_1(\Sigma)$, N_1 coïncide avec le relevé du voisinage $f_0(\Sigma) \times [-1, 1]$. Puisque f_1 induit une injection de $H_1(S, \mathbb{Z}_2) \to H_1(N_1', \mathbb{Z}_2)$, elle induit aussi une injection $H_1(S, \mathbb{Z}_2) \to H_1(N_1, \mathbb{Z}_2)$.

Lorsque f_1 induit une *surjection* de $H_1(S, \mathbb{Z}_2) \to H_1(N_1, \mathbb{Z}_2)$, on arrête la construction. Sinon, on recommence le raisonnement. On suppose constuit pour $i \leq k$ une suite de revêtements connexes, de degré 2, $p_i : N_i' \to N_{i-1}$, une suite de sous-variétés $N_i \subset N_i'$ et des applications $f_i : S \to N_i$ telles que $f_{i-1} = p_i \circ f_i$. La variété N_k est définie comme un voisinage régulier de $f_k(S)$ dans N_k'; par récurrence, ce voisinage régulier N_k peut être choisi de sorte que près de $f_k(\Sigma)$, il coïncide avec le relevé de $f_k(\Sigma) \times [-1, 1]$ (alors son image sous la composition $p_1 \circ p_2 \circ \cdots \circ p_k$ sera égale, près de $f(\Sigma)$, à $f(\Sigma) \times [-1, 1]$). Lorsque f_k induit un isomorphisme de $H_1(S, \mathbb{Z}_2) \to H_1(N_k, \mathbb{Z}_2)$, on arrête. Sinon, il existe un revêtement connexe de degré 2 N_{k+1}' de N_k auquel on peut relever f_k. L'argument classique de Papakyriakopoulos – le nombre de courbes de points doubles sur S que f_i identifie deux-à-deux décroît strictement comme fonction de i (cf. [Hem76]) – entraîne alors que la tour de revêtements ainsi construite est finie, c'est-à-dire qu'il existe k tel que f_k induise un isomorphisme de $H_1(S, \mathbb{Z}_2) \to H_1(N_k, \mathbb{Z}_2)$. On montre comme dans [FHS83, Lemme 2.3] que $N_k - f_k(S)$ peut être coloriée avec 2 couleurs. Le coloriage induit sur le bord définit une partition en deux sous-ensembles A_k et B_k, dont chacun est la réunion de sphères et d'une surface de genre g [FHS83, Lemma 2.4] ; on note F_k la surface de genre g contenue dans A_k. Elle a les propriétés suivantes :

- $p_k|_{F_k}$ est une injection de $H_*(F_k, \mathbb{Z}_2) \to H_*(N_{k-1}, \mathbb{Z}_2)$ (car la composée $p_1 \circ p_2 \circ \cdots \circ p_k : F_k \to M$ est une équivalence d'homologie $\mathbb{Z}_2$);

- l'image de F_k par la projection p_k contient $f_{k-1}(\Sigma) \times \{\pm 1\}$, disons $f_{k-1}(\Sigma) \times \{1\}$;

- si τ_k est l'involution du revêtement $N_k' \to N_{k-1}$, les deux surfaces F_k et $\tau_k(F_k)$ s'intersectent transversalement.

La projection $p_k : F_k \to N_{k-1}$ ne fait qu'identifier deux par deux des paires de courbes plongées disjointes. Soient c_1 et c_2 deux telles courbes; puisque $p_k|_{F_k}$ induit une injection en homologie, c_1 et c_2 sont homologues sur F_k, où elles bordent donc une sous-surface dont le bord est nécessairement disjoint de $f_k(\Sigma) \times \{1\}$. Après découpage de F_k le long de $c_1 \cup c_2$ et identification des courbes obtenues par le difféomorphisme $(p_k|_{c_1})^{-1} \circ p_k|_{c_2}$ de c_2 vers c_1, on obtient une nouvelle surface F_k', difféomorphe à F_k sur laquelle p_k induit une nouvelle application de $\bar{p}_k : F_k' \to N_{k-1}$ dont l'image est égale à $p_k(F_k)$ (notons qu'il existe deux possibilités pour réaliser cette opération de chirurgie, mais qu'une seule conduira à une surface F'_k difféomorphe à F_k). Ensuite, par une perturbation de l'application $\bar{p}_k$, supportée dans un voisinage de $c_1 \cup c_2$, on élimine ces deux courbes de points doubles. On obtient alors une nouvelle application $p'_k : F'_k \to N_{k-1}$ qui identifie "moins" de paires de courbes simples. Notons que par construction, $p'_k(F'_k)$ coïncide avec $p_k(F_k)$ dans un voisinage de la sous-surface $f_{k-1}(\Sigma) \times \{1\}$.

Montrons que p'_k induit une injection de $H_*(F'_k, \mathbb{Z}_2)$ dans $H_*(N_{k-1}, \mathbb{Z}_2)$. En effet, $[p'_k(F'_k)] = [p_k(F_k)]$ dans $H_2(N_{k-1}, \mathbb{Z}_2)$ et la classe $[p_k(F_k)]$ n'est pas nulle puisque son image par $p_1 \circ \cdots \circ p_{k-1}$ est $[f(S)]$, le générateur de $H_2(M, \mathbb{Z}_2)$. Donc $p_1 \circ \cdots \circ p_{k-1} \circ p'_k : F'_k \to M \simeq S$ est une application de degré non nul (modulo 2) entre surfaces de même genre $g \geq 1$: une telle application induit nécessairement une surjection au niveau du groupe fondamental. Par conséquent, c'est un isomorphisme au niveau du H_1. Donc p'_k induit bien un injection en homologie $\mathbb{Z}_2$.

On peut alors recommencer la construction précédente et éliminer les autres paires de courbes de points doubles; on aboutit alors à une surface F_{k-1} de genre g *plongée* dans N_{k-1}. La restriction de la composée $p_1 \circ \cdots \circ p_{k-1}$ à F_{k-1} induit un isomorphisme entre $H_1(F_{k-1}, \mathbb{Z}_2)$ et $H_1(M, \mathbb{Z}_2)$. Par construction, elle coïncide avec $p_k(F_k)$ hors d'un voisinage de l'image des lignes de points doubles ; en particulier elle contient la surface $f_{k-1}(\Sigma) \times \{1\}$ et donc son image par la composée $p_1 \circ \cdots \circ p_{k-1}$ contient $f(\Sigma) \times \{1\}$. On se ramène au cas où F_{k-1} est de plus transverse à la surface $\tau_{k-1}(F_{k-1})$ quitte à faire une perturbation de F_{k-1} disjointe de $f_{k-1}(\Sigma) \times \{1\}$. On peut alors recommencer la construction et descendre la tour. Au bas de la tour, on trouve une surface F_0 de genre g, plongée dans $N \subset M$; cette surface contient $f(\Sigma) \times \{1\}$ et l'inclusion dans M est une équivalence d'homologie à coefficients $\mathbb{Z}_2$. D'après [FHS83, Lemma 2.5], c'est une équivalence d'homotopie. Un théorème classique de Stallings assure que F_0 est isotope à S. Comme $f(\Sigma) \times \{1\} \subset F_0$, le lemme est démontré. $\square$

Ceci termine la démonstration du Théorème A. $\square$

Nous allons montrer maintenant le Théorème B de [Ota95]. Si M est une variété hyperbolique ayant le type d'homotopie d'une surface compacte de genre g, alors M est homéomorphe au produit $S \times \mathbb{R}$. Soit $L \subset M$ un *entrelacs*, c'est-à-dire une réunion localement finie de courbes plongées, deux-à-deux disjointes (on autorise que L ait une infinité de composantes). On dira qu'un entrelacs L est *un entrelacs non noué* s'il existe un homéomorphisme entre M et le produit $S \times] - \infty, \infty[$ tel que chaque composante de L soit contenue dans l'une des surfaces $S \times \{i\}$, pour $i \in \mathbb{Z}$.

Théorème B. *Soit M est une variété hyperbolique de dimension trois ayant le type d'homotopie d'une surface compacte de genre g. Il existe une constante $c(g) > 0$ telle que l'ensemble L, réunion des géodésiques fermées primitives de M de longueur inférieure à $c(g)$ est non noué.*

Démonstration. Soit $c(g)$ la constante du théorème A. Puisque $c(g)$ est inférieure à la constante $\varepsilon(3)$, on sait que le lemme de Margulis entraîne que deux géodésiques primitives de longueur inférieure à $c(g)$ sont disjointes ou confondues. L'ensemble $L \subset M$, réunion des géodésiques fermées de longueur inférieure à $c(g)$ est donc bien un entrelacs : le nombre de ses composantes connexes est fini quand le rayon d'injectivité de M est supérieur à $c(g)$ dans les deux bouts de M, infini (resp. bi-infini) quand il est inférieur à $c(g)$ dans un bout (resp. dans les deux bouts) de M.

Notons E^+ et E^- les deux bouts de M ; notons $\mathscr{T}_i$ le tube de Margulis autour de γ_i^*, c'est-à-dire la composante de la partie $\varepsilon(3)$-mince qui contient γ_i^*.

Lorsque L a un nombre fini de composantes connexes, on indexe celles-ci par un ensemble fini. Si L a un nombre infini de connexes, on les indexera par $\mathbb{N}$ ou par $\mathbb{Z}$ selon que les composantes de L sont toutes contenues dans le même bout ou pas; on convient alors que si i tend vers $\pm\infty$, la suite des géodésiques γ_i^* tend vers le bout $E^\pm$. Soit γ_i une courbe fermée simple sur S qui représente la classe d'homotopie de γ_i^* et soit $f_i : S \to M$ une surface plissée qui réalise γ_i. Par le Théorème de compacité des surfaces plissées [CEG87], la suite d'applications f_i quitte tout compact quand $|i| \to \infty$. L'homotopie de f_i réalisée dans la première partie de la démonstration du théorème A, produit une application $f_i' : S \to M$ en position normale telle que $f_i'(S)$ intersecte $\mathscr{T}_i$ le long d'un anneau plongé d'âme $\gamma_i^* = f_i(\gamma_i)$; par une homotopie supportée dans le complémentaire de la préimage de cet anneau, on peut en outre supposer que $f_i'(S)$ n'intersecte pas d'autres tubes de Margulis que $\mathscr{T}_i$. Par construction $f_i'(S)$ est contenue dans un voisinage régulier de la réunion de $f_i(S)$ et des tubes $\mathscr{T}_j$ que $f_i(S)$ intersecte. En particulier, f_i' tend vers $E^\pm$ dans M quand $i \to \pm\infty$. Le Lemme 1 permet de remplacer chaque surface $f_i'(S)$ par une surface plongée incompressible S_i qui contient γ_i^* ; on peut supposer que les surfaces S_i sont transverses l'une par rapport à l'autre. Toujours d'après le Lemme 1, S_i est contenue dans un voisinage régulier de $f_i'(S)$; en particulier, l'intersection de deux surfaces distinctes $S_i \cap S_j$ est disjointe de L. Nous allons maintenant isotoper chaque surface S_i en une nouvelle surface plongée Σ_i, de sorte que les surfaces Σ_i soient deux à deux disjointes et que Σ_i contienne γ_i^*. Le lemme de Stallings permettra alors de conclure : ce lemme produit en effet un homéomorphisme entre $S \times \mathbb{R}$ et M qui envoie la réunion des Σ_i dans $S \times \mathbb{Z}$. Pour construire les surfaces Σ_i, nous utiliserons le résultat suivant (cf. [FHS83, Lemme 4.1]):

Lemme 2.2. *Soit M une variété irréductible de dimension 3. Soient S_1 et S_2 deux surfaces compactes incompressibles qui sont homotopes dans M et qui s'intersectent* *transversalement. Alors si $S_1 \cap S_2 \neq \emptyset$, il existe des surfaces $F_1 \subset S_1$ et $F_2 \subset S_2$, avec $\partial F_1 = \partial F_2$, telles que $F_1 \cup F_2$ borde un produit $F_1 \times [0,1]$, d'intérieur disjoint de $S_1 \cup S_2$.*

Voyons comment ce lemme permet d'éliminer les intersections de deux surfaces distinctes S_k et S_l tout en préservant la surface S_k. Pour cela, on considère les surfaces $F \subset S_k$ et $F' \subset S_l$ fournies par le Lemme 2; selon que γ_l^* est contenue ou pas dans F', on remplace S_l par $F' \cup (S_k - F)$ ou bien par $F \cup (S_l - F')$. La surface obtenue est isotope à S_l, contient γ_l^* par construction, et après une perturbation, donne une surface dont l'intersection avec S_k a un nombre de composantes connexes strictement inférieur à celui de $S_k \cap S_l$. Après un nombre fini de telles modifications, on obtient une surface Σ_l qui contient γ_l^* et qui est disjointe de S_k.

Lorsque L a un nombre fini de composantes connexes $\{\gamma_1^*, \cdots, \gamma_k^*\}$, on transforme par récurrence la suite (S_i) en une suite (Σ_i). Pour cela, on pose $\Sigma_0 = S_0$, et on transforme chaque S_i, $i \neq 0$ par la construction précédente; on définit Σ_1 comme la surface obtenue à partir de S_1. Alors les nouvelles surfaces obtenues à partir de S_i, pour $i \neq 0$, $i \neq 1$ sont disjointes de $\Sigma_0 \cup \Sigma_1$ et on recommence.

Lorsque L a un nombre infini de composantes connexes, on extrait d'abord de la suite (S_i) une sous-suite (S_{n_k}) telle que les surfaces S_i qui rencontrent S_{n_k} ne rencontrent pas les autres surfaces S_{n_l}, pour $k \neq l$ et aussi telle que (n_k) tend vers $\pm\infty$ avec k. On modifie alors comme précédemment les surfaces S_l qui intersectent S_{n_k}, pour les rendre disjointes de S_{n_k}. On obtient alors une suite de surfaces plongées (S_i'), telle que $\gamma_i^* \subset S_i'$, $S_{n_k}' = S_{n_k}$ et telle que chaque surface S_{n_k}' est disjointe des autres surfaces S_j', $j \neq n_k$. Deux surfaces S_{n_k}', S_{n_l}' sont dites *consécutives*, lorsque la composante connexe bornée de $M - (S_{n_k}' \cup S_{n_l}')$ ne contient pas d'autres surfaces S_{n_p}' pour $p \neq k$, $p \neq l$. La composante connexe bordée par deux surfaces consécutives ne contient qu'un nombre fini de surfaces de la suite S_i', puisqu'il n'y a qu'un nombre fini de composantes de L dans un compact donné. Une nouvelle application du lemme 2 aux surfaces S_i' contenues dans cette composante permet de les remplacer par des surfaces plongées Σ_i deux-à-deux disjointes, disjointes aussi de $S_{n_k}' \cup S_{n_l}'$. En effectuant cette construction pour toutes les paires de surfaces consécutives, et en posant $S_{n_k}' = \Sigma_{n_k}$, on obtient la suite de surfaces (Σ_i) cherchée. $\qquad\square$

3. Le cas des variétés à bord compressible

Maintenant, nous ne supposons plus que M a le type d'homotopie d'une surface compacte. Rappelons que si M est une variété hyperbolique de dimension 3, *le cœur de Nielsen de M* est le plus petit convexe fermé $N \subset M$ tel que l'inclusion $N \to M$ soit une équivalence d'homotopie. Nous allons montrer le résultat suivant.

Théorème 3. *Soit M une variété hyperbolique; soit N ⊂ M le cœur de Nielsen de M. Soit γ* géodésique fermée primitive de M telle que γ* est librement homotope à une courbe fermée simple tracée sur ∂N de longueur inférieure à ε(3). Alors, γ* est non nouée par rapport au plongement ∂N → M*

Remarque 3.1. Si M est l'intérieur d'une variété compacte $\overline{M}$, les composantes connexes de ∂N peuvent être déformées, par une isotopie propre, de sorte à être contenues dans $\partial \overline{M}$; alors sous les hypothèses du Théorème 3, $\gamma^* \subset \overline{M}$ est non nouée par rapport au plongement $\partial \overline{M} \to \overline{M}$. Notons que cette situation a lieu en particulier lorsque M est géométriquement finie.

Remarque 3.2. La métrique induite sur ∂N par la métrique de M est hyperbolique. Pour tout constante η inférieure à une constante $\varepsilon(2) > 0$, on a une décomposition de ∂N en partie η-mince et partie η-épaisse et les composantes de la partie η-mince sont des anneaux. Toutefois, sans hypothèses d'incompressibilité sur ∂N, un tel anneau peut être homotope à 0 dans M. Le Théorème 3 signifie que si $\gamma \subset \partial N$ est une courbe simple, âme d'un anneau composante de la partie $\varepsilon(3)$-mince de ∂N, et si γ est librement homotope à une géodésique primitive γ^* de $\pi_1(M)$, alors γ^* est non nouée par rapport au plongement $\partial N \to M$.

Signalons qu'il n'est pas toujours vrai que l'âme d'une composante de la partie mince représente un élément primitif de $\pi_1(M)$. Du point de vue topologique, un exemple typique de variété M telle qu'il existe une courbe simple essentielle $\gamma \subset \partial M$ divisible dans ∂M (i.e. $[\gamma] = [\gamma^*]^k$ avec $k \geq 2$), est lorsqu'une composante connexe de la sous-variété caractéristique de M est difféomorphe à un tore solide T : l'âme de ce tore est homotope à γ^* et $T \cap \partial M$ est un anneau d'âme γ.

Un autre exemple est celui où M est un I-fibré tordu au-dessus d'une surface non orientable F ; notons $p : M \to F$ cette fibration. Si $\gamma^* \subset F$ est un lacet qui renverse l'orientation, la préimage $p^{-1}(\gamma^*)$ est un ruban de Möbius dont le bord est une courbe simple $\gamma \subset \partial M$: dans $\pi_1(M)$, on a $[\gamma] = [\gamma^*]^2$. Cette variété peut être uniformisée par un groupe géométriquement fini, extension de degré 2 d'un groupe quasi-fuchsien de telle sorte que γ soit l'âme d'un tube Margulis de ∂N, de longueur aussi petite qu'on le veut.

Démonstration. Il n'y a rien à démontrer lorsque γ^* intersecte ∂N, car alors on voit facilement que γ^* est entièrement contenue dans ∂N. Sinon, γ^* est contenue dans l'intérieur de N et nous allons montrer qu'il existe un anneau plongé dans N, transversalement à ∂N et dont le bord est la réunion de γ et de γ^* ; le Théorème 3 en découlera. L'hypothèse sur la courbe γ signifie que γ est contenue dans le tube de Margulis $\mathcal{T}$, composante connexe de la partie $\varepsilon(3)$-mince de M contenant γ^*. Choisissons un tore solide $\mathcal{V}$, voisinage de γ^* dans M de rayon constant et suffisamment petit, de sorte que $\mathcal{V}$ soit contenu dans l'intérieur de N. Considérons le revêtement $\pi : \mathcal{X} \to M$ associé au sous-groupe cyclique $\langle[\gamma^*]\rangle$ de $\pi_1(M)$ engendré par la classe d'homotopie de γ^*. Topologiquement, $\mathcal{X}$ est un tore solide : $\mathcal{X} \simeq \mathbb{H}^3/\langle[\gamma^*]\rangle$. Il résulte des propriétés de la partie $\varepsilon(3)$-mince rappelées au début du texte que $\mathcal{T}$ se relève homéomorphiquement dans $\mathcal{X}$ en un tore solide $\mathcal{T}'$; en particulier π se restreint à $\mathcal{T}'$ en un homéomorphisme. On note γ', $(\gamma^*)'$, $\mathcal{V}'$ les relevés de γ, γ^* et $\mathcal{V}$.

Considérons la projection orthogonale $\pi : \mathcal{X} \to \mathcal{V}'$ qui associe à tout point de $\mathcal{X}$ le point de $\mathcal{V}'$ qui en est le plus proche. La réunion des segments $[z, \pi(z)]$ lorsque z décrit γ' est un anneau *plongé* (puisque γ' est une courbe plongée dans le bord d'un convexe qui contient $\mathcal{V}'$). L'image de cet anneau par la projection de revêtement p est un anneau A', plongé dans $\overline{N - \mathcal{V}}$ et qui joint γ à une courbe $p \circ \pi(\gamma')$ contenue $\partial\mathcal{V}$. L'hypothèse que la géodésique γ^* est primitive signifie que $p \circ \pi(\gamma')$ est homotope à l'âme γ^* du tore solide $\mathcal{V}$. Cette homotopie peut être réalisée par un anneau A'', plongé dans $\mathcal{V}$, joignant $p \circ \pi(\gamma')$ et γ^*. En recollant les deux anneaux obtenus A' et A'' le long de $p \circ \pi(\gamma')$, on obtient un anneau plongé qui vérifie les conditions du Théorème 3. $\qquad\square$

Références

[CEG87] R.D. Canary, D.B.A. Epstein and P. Green (1987). Notes on notes of Thurston. In *Analytical and geometric aspects of hyperbolic space (LMS lecture notes 111)*, 3–92.

[FHS83] M. Freedman, J. Hass and P. Scott (1983). Least area incompressible surfaces in 3-manifolds, *Invent. Math.* **71**, 600–642.

[Hem76] J. Hempel (1976). *3-manifolds*, Ann. Math. Studies **86**.

[Ota95] J.-P. Otal (1995). Sur le nouage des géodésiques dans les variétés hyperboliques, *C. R. Acad. Sci. Paris. Sèr. I Math.* **320**, 847–852.

[Thu97] W.P. Thurston (1997). *Three-dimensional geometry and topology*, Princeton University Press.

[Thu86b] W.P. Thurston (1986). Hyperbolic structures on 3-manifolds II: Surface groups and 3-manifolds which fiber over the circle, *arXiv:math.GT/9801045*.

Jean-Pierre Otal

Unité de Mathématiques Pures et Appliquées
École Normale Supérieure de Lyon
46 Allée d'Italie
69364 Lyon, France

jpotal@umpa.ens-lyon.fr

AMS Classification: 57M07, 57M05

Keywords: hyperbolic manifold, Kleinian group

Kleinian Groups and Hyperbolic 3-Manifolds   Y. Komori, V. Markovic & C. Series (Eds.)

Lond. Math. Soc. Lec. Notes **299**, 105–129   Cambridge Univ. Press, 2003

Ending laminations in the Masur domain

Gero Kleineidam[1] *and Juan Souto*[2]

Abstract

We study the relationship between the geometry and the topology of the ends of a hyperbolic 3-manifold M whose fundamental group is not a free group. We prove that a compressible geometrically infinite end of M is tame if there is a Masur domain lamination which is not realized by a pleated surface. It is due to Canary that, in the absence of rank-1-cusps, this condition is also necessary.

1. Introduction

Marden [Mar74] proved that every geometrically finite hyperbolic 3-manifold is *tame*, i.e. homeomorphic to the interior of a compact manifold, and he conjectured that this holds for any hyperbolic 3-manifold M with finitely generated fundamental group. By a theorem of Scott [Sco73b], M contains a *core*, a compact submanifold C such that the inclusion of C into M is a homotopy equivalence. Moreover, any two cores are homeomorphic by a homeomorphism in the correct homotopy class [MMS85].

So the discussion of the tameness of M boils down to a discussion of the ends of M. The ends are in a bijective correspondence with the boundary components of the compact core C. An end E is said to be *tame* if it has a neighborhood homeomorphic to the product of the corresponding boundary component ∂E of C with the half-line. Hence M is tame if its ends are.

Since M is aspherical, either $\pi_1(M) = 1$ or every boundary component of C is a closed surface of genus at least 1. Moreover, if a boundary component ∂E is a torus, then either $\pi_1(M)$ is abelian or the corresponding end E is a rank-2-cusp in M and is hence tame. From now on, we will assume that $\pi_1(M)$ is not abelian.

It is due to Bonahon [Bon86] that an end E is tame if $\partial E \subset \partial C$ is incompressible in C. Furthermore, in the case that E contains no cusps, he showed that E is either geometrically finite or simply degenerate. Here, a *simply degenerate* end is an end E such that there is a sequence of simple closed curves α_i on ∂E whose geodesic

[1]Supported by the Sonderforschungsbereich 611.

[2]Supported by the Sonderforschungsbereich 611.

representatives in M converge to E. The sequence (α_i) converges to a lamination on ∂E which is called the *ending lamination* of E. The ending lamination is not realized in M by any pleated surface.

For a compressible end E, i.e. an end with compressible ∂E, there are many laminations on ∂E which are not realized, e.g. every simple closed curve which is compressible in C. Otal [Ota88] studied the *Masur domain* of ∂E, an open set of projective measured laminations on ∂E, and proved that the support of every Masur domain lamination is realized in M if the end E is geometrically finite and has no cusps. It is due to Kerckhoff [Ker90] that the Masur domain has full measure.

On the other hand, Canary [Can93b] proved that if a compressible geometrically infinite end E without rank-1-cusps is tame, then it is simply degenerate and the ending lamination is the support of a Masur domain lamination. Again, the ending lamination is not realized by a pleated surface. We prove a converse to Canary's result in the case that $\pi_1(M)$ is not a free group.

Theorem 1.1. *Let E be a compressible end of a complete oriented hyperbolic 3-manifold M whose fundamental group is finitely generated but not free. If a Masur domain lamination on ∂E is not realized in M, then the end E is tame.*

We briefly describe the strategy of the proof of Theorem 1.1 in the case that the Masur domain lamination λ on ∂E which is not realized in M is minimal arational, i.e. every component of $\partial E - \lambda$ is simply connected. The end E lifts homeomorphically to a cover M_E of M which has a compact core homeomorphic to a compression body. Here, a *compression body* is a compact, irreducible 3-manifold N which has a distinguished compressible boundary component $\partial_e N$, called the *exterior boundary*, whose fundamental group surjects onto $\pi_1(N)$. Every component of the *interior boundary* $\partial_{int}N = \partial N - \partial_e N$ is incompressible; in particular, E is the only compressible end of M_E. We show that for any sequence (γ_i) of simple closed curves on $\partial_e N$ which converge to the minimal arational lamination λ there is a sequence of pleated surfaces which realize γ_i in M_E. A generalization of a Compactness Theorem of pleated surfaces by Otal then shows that these surfaces exit every compact set. This implies, by a theorem of the second author [Sou02], that M_E is tame, and hence that E is tame.

Theorem 1.1 and the discussion above imply.

Corollary 1.2. *Let M be a complete oriented hyperbolic 3-manifold which does not contain rank-1-cusps and whose fundamental group is finitely generated but not free. Then M is tame if and only if for every compressible end E of M which is not geometrically finite there is a Masur domain lamination on ∂E which is not realized in M.*

We will also apply Theorem 1.1 to establish the tameness of certain algebraic limits of geometrically finite hyperbolic structures on a compression body N. Let $\rho_0 : \pi_1(N) \to \mathrm{PSL}_2\mathbb{C}$ be a geometrically finite representation which is induced by a homeomorphism from the interior of N to $M_{\rho_0} = \mathbb{H}^3/\rho_0(\pi_1(N))$ such that M_{ρ_0} does not contain rank-1-cusps. Denote by P the collection of the toroidal components of ∂N and by $\mathscr{T}(\partial N - P)$ the product of the Teichmüller space $\mathscr{T}(\partial_e N)$ and the Teichmüller spaces of the non-toroidal interior boundary components of N. Then, by Ahlfors-Bers theory, the space $QH(\rho_0)$ of quasi-conformal deformations of ρ_0 is the quotient of $\mathscr{T}(\partial N - P)$ by a discrete group. We refer to Anderson [And98] for a beautiful survey on the deformation theory of Kleinian groups.

Thurston [FLP79] compactified the Teichmüller space $\mathscr{T}(\partial_e N)$ by the space $\mathscr{PML}$ of projective classes of measured laminations on $\partial_e N$. Recall that the Masur domain is an open full measure subset of $\mathscr{PML}$. We will say that a sequence (ρ_i) in $QH(\rho_0)$ *converges into the Masur domain* if it lifts under the Ahlfors-Bers covering $\mathscr{T}(\partial N - P) \to QH(\rho_0)$ to a sequence $(S_i^e, S_i^{int})_i$ in $\mathscr{T}(\partial N - P) = \mathscr{T}(\partial_e N) \times \mathscr{T}(\partial_{int} N - P)$ such that $(S_i^e)_i$ converges in the Thurston compactification to a measured lamination in the Masur domain of $\partial_e N$. We impose no restrictions on the conformal structures S_i^{int} on the interior boundary.

Theorem 1.3. *Let N be a compression body which is not a handlebody and let $(\rho_i) \subset QH(\rho_0)$ be a sequence which converges algebraically to $\rho : \pi_1(N) \to \mathrm{PSL}_2\mathbb{C}$. If (ρ_i) converges into the Masur domain, then $M_\rho = \mathbb{H}^3/\rho(\pi_1(N))$ is tame.*

Recently Brock, Bromberg, Evans and Souto [BBES] proved that every algebraic limit is tame if its discontinuity domain is not empty. Remark that this assumption does not necessarily hold in the setting of Theorem 1.3.

Remark 1.4. The referee pointed out that Theorem 1.3 generalizes by quite similar arguments to all boundary-compressible manifolds other than the handlebody.

The paper is organized as follows: In Section 2 we introduce some background material and some notation. Section 3 is devoted to the study of Masur domain laminations. In Section 4 we prove a compactness theorem for pleated surfaces realizing Masur domain laminations. Theorem 1.1 and Theorem 1.3 are proved in Section 5.

The results of this paper are extensions of results of Otal [Ota88] which unfortunately have never been published.

We thank Ursula Hamenstädt, Jean-Pierre Otal and Cyril Lecuire for their interest in our work.

We thank the referee for a very careful reading and for pointing out several uncertainties. Moreover, the referee remarked that Theorem 1.3 and Proposition 5.6 may be generalized by substantially the same arguments to a wider class of manifolds.

2. Preliminaries

Let $M = \mathbb{H}^3/\Gamma$ be a complete oriented hyperbolic 3-manifold such that $\Gamma \simeq \pi_1(M)$ is a finitely generated, torsion free, discrete subgroup of $PSL_2\mathbb{C} = Isom_+\mathbb{H}^3$. In the sequel we will assume, without further comment, that the fundamental group of M is not abelian; equivalently, it contains a free subgroup of rank two.

For $\varepsilon > 0$, the ε-thin part $M^{<\varepsilon}$ of M is the set of all points in M where the injectivity radius is less than ε. The complement $M^{\geq\varepsilon} = M - M^{<\varepsilon}$ is the ε-thick part. The Margulis lemma asserts that there is a constant $\mu_3 > 0$, which does not depend on M, such that for all $\varepsilon < \mu_3$ every component U of $M^{<\varepsilon}$ has abelian fundamental group $\pi_1(U)$ and U is either a Margulis tube around a short geodesic or a cusp of rank 1 or 2.

The ε-thin part can have infinitely many components for all ε, but Sullivan's finiteness theorem [Sul81] asserts that the number of cusps is finite. Equivalently, the group Γ has only finitely many conjugacy classes of maximal parabolic subgroups.

We say that Γ is *minimally parabolic* if every maximal parabolic subgroup is a free abelian group of rank 2. A representation ρ is minimally parabolic if its image is.

2.1. Compact cores

It is a fundamental result of Scott [Sco73b] that every hyperbolic 3-manifold has a compact *core*, i.e. a compact 3-dimensional submanifold C such that the inclusion $C \hookrightarrow M$ is a homotopy equivalence. Furthermore, it is due to McCullough, Miller and Swarup [MMS85] that for any two compact cores C_1, C_2 there is a homeomorphism $h: C_1 \to C_2$ such that $i_2 \circ h$ is homotopic to i_1 where i_j is the inclusion of C_j into M.

The ends of the manifold M are in bijective correspondence to the boundary components of the compact core [Bon86]. We will denote by ∂E the boundary component of C which corresponds to an end E of M.

An end E is *tame* if it has a neighborhood which is homeomorphic to $\partial E \times \mathbb{R}$. Recall that M is tame if it is homeomorphic to the interior of a compact manifold with boundary; equivalently if every end is tame. Bonahon [Bon86] proved

Theorem 2.1 (Bonahon). *Let M be a complete oriented hyperbolic 3-manifold. If the following condition (*) holds, then M is tame:*

(*) *For every splitting of $\pi_1(M)$ as a non-trivial free product $A * B$ or as an HNN-extension $A *$ there is a parabolic element in $\pi_1(M)$ which cannot be conjugated into A.*

2.2. Compression bodies

A *compression body* N is a compact, irreducible 3-manifold which has a boundary component $\partial_e N$ whose fundamental group surjects onto $\pi_1(N)$; $\partial_e N$ is said to be the *exterior boundary* of N. We are interested in those compression bodies whose fundamental group is not abelian. If the exterior boundary is incompressible, then N is homeomorphic to the trivial interval bundle over $\partial_e N$; such a compression body is said to be *trivial*. In the case that N is non-trivial, the exterior boundary is the unique compressible component of ∂N. A *meridian* is a simple closed curve m on $\partial_e N$ which is compressible in N but essential in $\partial_e N$. By Dehn's lemma [Jac80], every meridian bounds an embedded disk in N.

Every non-trivial compression body N contains a non-empty collection $\mathscr{D}$ of disjoint, non-parallel, properly embedded essential disks $(D_i, \partial D_i) \subset (N, \partial_e N)$ such that $N - \mathscr{D}$ is either homeomorphic to a closed ball or to the disjoint union of trivial compression bodies. The compression body N is a *handlebody* if $N - \mathscr{D}$ is a closed ball; equivalently $\pi_1(N)$ is a free group. If N is not a handlebody, then the *interior boundary* $\partial_{int} N = \partial N - \partial_e N$ is not empty.

The compression body N is *small* if there is a properly embedded essential disk $(D, \partial D) \subset (N, \partial_e N)$ such that $N - D$ is either homeomorphic to a trivial compression body or to the disjoint union of two trivial compression bodies. Note that the exterior boundary of a non-trivial compression body contains infinitely many meridians if the compression body is not small.

Given a compression body N, let $Mod(N)$ be the group of all isotopy classes of orientation preserving diffeomorphisms of N and let $Mod_0(N)$ be the subgroup of those elements which are homotopic to the identity; $Mod_0(N)$ is trivial if N is trivial. For a non-trivial compression body N, we can identify the group $Mod(N)$ with a subgroup of the mapping class group of the exterior boundary $\partial_e N$. A diffeomorphism $\phi : \partial_e N \to \partial_e N$ extends to a diffeomorphism $\Phi : N \to N$ if and only if it maps meridians to meridians. Remark that two compression bodies are homeomorphic if and only if they are homotopy equivalent. For more on compression bodies see [MMi86].

2.3. Function groups

For every compression body N, there is a geometrically finite and minimally parabolic representation $\rho \colon \pi_1(N) \to \mathrm{PSL}_2\mathbb{C}$ such that the interior of N is homeomorphic to $M_\rho = \mathbb{H}^3/\rho(\pi_1(N))$ by a homeomorphism which induces ρ; we say that ρ *uniformizes* N. The image of ρ is a *function group*, i.e. there is an invariant component of the discontinuity domain Ω_ρ of the action of $\rho(\pi_1(N))$ on $\hat{\mathbb{C}}$ [Mak88]. Recall that the discontinuity domain is the complement in $\hat{\mathbb{C}}$ of the limit set Λ_ρ. We can identify the Riemann surface $\Omega_\rho/\rho(\pi_1(N))$ with the union of all non-toroidal components of ∂N. Under this identification the exterior boundary $\partial_e N$ corresponds to an invariant component of the discontinuity domain. In particular, the invariant component in Ω_ρ is unique if N is non-trivial.

2.4. Boundary groups

Let M be a hyperbolic 3-manifold with finitely generated fundamental group, C be a compact core of M and E an end of M. Bonahon [Bon83, Appendix B] proved that there is a compression body $B_E \subset C$ whose exterior boundary is ∂E and such that $\pi_1(B_E)$ injects into $\pi_1(C)$; furthermore, B_E is unique up to isotopy. The compression body B_E is said to be the *relative compression body* associated to the end E. If the relative compression body B_E is a handlebody for some end E of M, then $B_E = C$ and hence $\pi_1(C) = \pi_1(M)$ is free. The end E is said to be *incompressible* if the compression body B_E is trivial; otherwise it is *compressible*. The *boundary group* of the end E is the subgroup $\pi_1(B_E)$ of $\pi_1(M)$; it is well defined up to conjugacy.

Given an end E, let M_E be the cover of M determined by the boundary group $\pi_1(B_E)$. The end E and the compression body B_E lift homeomorphically to M_E; in particular, tameness of M_E implies tameness of the end E of M. Therefore, the study of the ends of a hyperbolic manifold can be reduced to the study of its boundary groups.

It follows from Bonahon's theorem that every incompressible end is tame; we are interested in the compressible ends. It is not known if every compressible end is tame but the second author proved the following criterion for the tameness of a compressible end (see [Sou02]).

Theorem 2.2 (Tameness Criterion). *Let E be a compressible end of a complete orientable hyperbolic 3-manifold M such that $\pi_1(M)$ is finitely generated but not free. If there is a sequence $(S_i)_i$ of surfaces which are homotopic in M to ∂E and which converge to E when i tends to ∞, then the end E is tame.*

We say that a sequence (S_i) of surfaces converges to an end E of M if for every neighborhood U of E there is i_0 with $S_i \subset U$ for all $i \geq i_0$. Note that the homotopies considered in the theorem are not assumed to be proper.

2.5. Laminations on surfaces

A (geodesic) lamination on a closed hyperbolic surface S is a compact subset of S which can be decomposed as a disjoint union of simple geodesics, called leaves. A *multicurve* is a lamination which contains only compact leaves. A lamination is called *minimal* if every half-leaf is dense; every minimal lamination is connected. A minimal lamination is *minimal arational* if its complementary regions are simply-connected. Each lamination can be decomposed as a union of finitely many minimal laminations, called minimal components, and finitely many non-compact isolated leaves. The set of laminations is compact with respect to the topology induced by the Hausdorff distance. We will refer to this topology as the *Hausdorff topology*.

A *measured lamination* is a lamination with a transverse measure of full support. The support of a measured lamination is a finite union of minimal components, in particular it does not contain isolated non-compact leaves. A measured lamination is *maximal* if its support is not a proper subset of the support of another measured lamination. If the support of a measured lamination is a simple closed curve γ, then the measure is a Dirac measure of weight $a > 0$ and we denote it by $a\gamma$; such measured laminations are called *weighted curves*.

There is a topology on the set $\mathscr{ML}$ of measured laminations on S which is induced by the intersection form $i : \mathscr{ML} \times \mathscr{ML} \to \mathbb{R}_+$. The set of weighted curves is dense in $\mathscr{ML}$ with respect to this topology.

Rescaling the measure provides an action of $\mathbb{R}_+$ on $\mathscr{ML}$. The quotient space is compact and is denoted by $\mathscr{PML}$. The elements in $\mathscr{PML}$ are called *projective measured laminations* on S. If a sequence of projective measured laminations converges to λ in $\mathscr{PML}$ and to a lamination $\lambda_{\mathscr{H}}$ in the Hausdorff topology, then the support of λ is contained in $\lambda_{\mathscr{H}}$.

The *Teichmüller space* of S is denoted by $\mathscr{T}(S)$. Thurston compactified $\mathscr{T}(S)$ by the space $\mathscr{PML}$. A sequence (S_i) in $\mathscr{T}(S)$ converges to the projective class of a measured lamination $\lambda \in \mathscr{ML}$ if there is a sequence (ε_i) of positive real numbers which converge to 0 and with

$$\lim_{i \to \infty} \varepsilon_i a_i l_{S_i}(\gamma_i) = i(\mu, \lambda)$$

for every $\mu \in \mathscr{ML}$ and every sequence of weighted curves $(a_i \gamma_i)$ which converges to μ in the topology of $\mathscr{ML}$. Here, $l_{S_i}(\gamma_i)$ is the length in S_i of the geodesic homotopic to γ_i. We will often identify curves with their free homotopy classes.

It is an important fact that the compactification of $\mathcal{T}(S)$ by $\mathcal{PML}$ reflects the geometric behavior of divergent sequences in $\mathcal{T}(S)$: If a sequence (S_i) in $\mathcal{T}(S)$ converges to $\lambda \in \mathcal{PML}$, then there is a sequence (γ_i) of simple closed curves on S converging to λ in $\mathcal{PML}$ and such that

$$l_{S_i}(\gamma_i)/l_S(\gamma_i) \to 0 \text{ for } i \to \infty.$$

See Otal [Ota96, Appendix] for more about laminations and measured laminations and Fathi-Laudenbach-Poénaru [FLP79] for a detailed exposition of Thurston's compactification of Teichmüller space.

2.6. Pleated surfaces

Let M be a hyperbolic 3-manifold, S_0 a hyperbolic surface and $h : S_0 \to M$ a map. A *pleated surface* is a length preserving map $f : S \to M$ from a hyperbolic surface $S \in \mathcal{T}(S_0)$ to M homotopic to h and such that every point $p \in S$ is contained in a geodesic segment which is mapped isometrically. A lamination λ on S_0 is *realized* in M if there is a pleated surface $f : S \to M$ that maps every leaf of λ to a geodesic in M.

Proposition 2.3. *Let λ be a lamination which is realized by a pleated surface $f : S \to M$. Then λ has a neighborhood U with respect to the Hausdorff topology on the set of laminations such that:*

1. *For every simple closed curve $\gamma \in U$, the geodesic in M homotopic to $f(\gamma)$ is contained in the 1-neighborhood of $f(\lambda)$.*

2. *There is $L \in (0, 1)$ with*

$$l_S(\gamma) \geq l_M(f(\gamma)) \geq L l_S(\gamma)$$

for every simple closed geodesic $\gamma \in U$. Here $l_M(f(\gamma))$ is the length of the geodesic in M freely homotopic to $f(\gamma)$.

Since Proposition 2.3 is well-known, we only give a brief sketch of the proof. As in the easy part of [Bro00, Lemma 5.2] there is, for every $\varepsilon > 0$, a train-track τ which carries λ and admits an ε-nearly-straight realization in M (see [Bro00] for details and definitions). The train-track τ determines a neighborhood $U(\tau)$ of λ in the space of laminations with respect to the Hausdorff topolgy. For sufficiently small ε, the neighborhood $U(\tau)$ has the desired properties. See also [Bon86, Chapter 5] and [CEG87, pp.87-88].

Pleated surfaces are discussed in [Bro00, CEG87, Ota88].

3. Laminations on the exterior boundary

In this section, we study topological properties of laminations on the exterior boundary $\partial_e N$ of a compression body N. We assume without further comment that N is not trivial. Many of the results in this section are due to Otal [Ota88] but we reproduce some of them here because, unfortunately, Otal's *Thèse d'Etat* has not been published.

A meridian may be seen as an element in the space $\mathscr{P}\mathscr{M}\mathscr{L}$ of projective measured lamination on $\partial_e N$. The set of projective classes of weighted multicurves of meridians in $\mathscr{P}\mathscr{M}\mathscr{L}$ will be denoted by $\mathscr{M}$ and its closure in $\mathscr{P}\mathscr{M}\mathscr{L}$ by $\mathscr{M}'$. The following lemma shows that laminations in $\mathscr{M}'$ can differ essentially from meridians.

Lemma 3.1. *[Ota88] Let N be a compression body which is not small. A lamination μ is the support of a lamination in $\mathscr{M}'$ if one of the following conditions holds:*

1. *$\mu \subset \partial_e N$ is the union of two disjoint, non-parallel simple closed curves which are freely homotopic in N, or*

2. *μ is a simple closed curve representing a divisible element in $\pi_1(N)$, or*

3. *μ is a simple closed curve which is freely homotopic to a simple closed curve on the interior boundary of N.*

Proof. Assume that the first condition holds for μ. Then there is a properly embedded annulus $(A, \partial A) \subset (N, \partial_e N)$ with $\partial A = \mu$ [Jac80]; let $\delta_A : N \to N$ be the Dehn-twist about A. Since N is not small, there is a meridian m which intersects both components of μ. The sequence of meridians $(\delta_A^i(m))$ converges to a measured lamination whose support is μ.

The proof is analogous for the second and third conditions. $\qquad\square$

Remark that if a lamination μ on the exterior boundary of a small compression body satisfies one of the three conditions above, then there is a meridian m with $i(\mu, m) = 0$.

Definition 3.2. If N is not a small compression body, then the *Masur domain* of $\partial_e N$ is

$$\mathscr{O} := \{\lambda \in \mathscr{P}\mathscr{M}\mathscr{L} \mid i(\lambda, \mu) > 0 \text{ for all } \mu \in \mathscr{M}'\}.$$

If N is small, then the Masur domain $\mathscr{O}$ is the set of all $\lambda \in \mathscr{P}\mathscr{M}\mathscr{L}$ with $i(\lambda, \mu) > 0$ for every $\mu \in \mathscr{P}\mathscr{M}\mathscr{L}$ such that there is $\nu \in \mathscr{M}'$ with $i(\mu, \nu) = 0$.

We will say that $\lambda \in \mathscr{M}\mathscr{L}$ is in $\mathscr{O}$ (resp. $\mathscr{M}'$) if its projective class is in $\mathscr{O}$ (resp. $\mathscr{M}'$).

It is a topological condition that a measured lamination belongs to $\mathcal{O}$: Two measured laminations with same support are either both in $\mathcal{O}$ or none is.

The Masur domain $\mathcal{O}$ has been studied by Masur [Mas86] and Otal [Ota88] and they proved

Theorem 3.3. *The Masur domain $\mathcal{O}$ is open and invariant under the action of $Mod(N)$ on $\mathcal{PML}$. Moreover, the action of $Mod(N)$ on $\mathcal{O}$ is properly discontinuous.*

Otal further studied how laminations in $\mathcal{O}$ reflect topological properties of the compression body N. It follows from Lemma 3.1 that

Lemma 3.4. *[Ota88, 1.15] The complement $\partial_e N - \lambda$ of a multicurve $\lambda \in \mathcal{O}$ is incompressible and acylindrical.*

Corollary 3.5. *[Ota88] Let λ_1 and λ_2 be two multicurves in the Masur domain. If λ_1 and λ_2 are freely homotopic in N, then there exists $\psi \in Mod_0(N)$ with $\psi(\lambda_1) = \lambda_2$.*

Proof. First there is a homotopy equivalence between the pared manifolds (N, λ_1) and (N, λ_2) which is, in N, homotopic to the identity. Now by Lemma 3.4, the pared manifolds (N, λ_1) and (N, λ_2) are incompressible and acylindrical. Hence a theorem of Johannson [Joh79] implies that the homotopy equivalence is homotopic to a homeomorphism. □

Sublaminations of Masur domain laminations will play an important role in the present paper. We prove

Proposition 3.6. *Let γ be a simple closed curve which is contained in a lamination $\lambda \in \mathcal{O}$. Then γ does not represent an element in $\pi_1(N)$ which is contained in a rank-2 abelian subgroup. Moreover, $\pi_1(N)$ does not split as a non-trivial free product or as an HNN-extension such that γ represents an element which can be conjugated into one of the factors.*

The first assertion follows from the third part of Lemma 3.4. By Theorem 1.32 in [Kap01] the second assertion holds if γ intersects every meridian on $\partial_e N$. In the case that N is small, it is an immediate consequence of the definition of $\mathcal{O}$ that a lamination with compressible complement is not contained in a Masur domain lamination. Suppose now that N is not small and, seeking for a contradiction, assume that γ misses a meridian. The following lemma shows that $\gamma \in \mathcal{M}'$, contradicting $i(\lambda, \gamma) = 0$. This finishes the proof of Proposition 3.6.

Lemma 3.7. *Let N be a compression body which is not small. If μ is a minimal lamination on $\partial_e N$ and $\partial_e N - \mu$ is compressible, then $\mu \in \mathscr{M}'$.*

Proof. Otal [Ota88, Lemma 1.3.2] proved that $\mu \in \mathscr{M}'$ if $\partial_e N - \mu$ contains either several meridians, or one non-separating meridian.

We reproduce Otal's argument: If m is a non-separating meridian, then cut $\partial_e N$ along m and join the two resulting boundary components by an embedded arc κ. The boundary of a regular neighborhood of $m \cup \kappa$ in $\partial_e N$ is a meridian. Since κ can be chosen as close to μ as wanted, we deduce that $\mu \in \mathscr{M}'$. In the case that $\partial_e N - \mu$ contains several meridians one finds two meridians m_1 and m_2 such that μ is contained in a component of $\partial_e N - (m_1 \cup m_2)$ with two boundary components. Then one proceeds as above.

It remains to consider the case that there is only one meridian m in $\partial_e N - \mu$ which separates $\partial_e N$. Then m bounds an embedded disk D in N such that one of the components of $N - D$ is a trivial interval bundle over a closed surface, and the lamination μ is contained in the compressible component X of $\partial_e N - m$. The set of all measured laminations in X which are minimal arational (in X) and uniquely ergodic is dense in the space $\mathscr{ML}(X)$ of measured laminations on X [Mas82]. So it suffices to show that every such lamination is in $\mathscr{M}'$ because $\mathscr{M}'$ is closed.

Assume that $\mu \in \mathscr{ML}(X)$ is minimal arational and uniquely ergodic and let (γ_i) be a sequence of simple closed curves in X which converge to μ in the Hausdorff-topology. If γ_i is a meridian for infinitely many i, then $\mu \in \mathscr{M}'$ and we are done. Assume that no γ_i is a meridian.

Denote by Δ the complementary region of μ which contains $\partial X = m$ and fix a boundary leaf l of Δ. Let $\tau : [-1,1] \to \partial_e N$ be a geodesic segment contained in the closure of X with $\tau(1) \in m$, $\tau(0) \in l$ and $\tau(0,1) \subset \Delta$. For all i define

$$t_i = \max\{t \in [-1,1] \mid \tau(t) \in \gamma_i\}.$$

We have $t_i \to 0$ when $i \to \infty$ because μ is the Hausdorff-limit of the sequence (γ_i). For all i let $\Sigma_i \subset \partial_e N$ be a regular neighborhood of the 1-complex $\gamma_i \cup m \cup \tau[t_i, 1]$. The surface Σ_i is a 3-holed sphere and m and γ_i are represented by two components of $\partial \Sigma_i$. Denote by $\bar{\gamma}_i$ the geodesic in the free homotopy class of the third component of $\partial \Sigma_i$. In other words, $\bar{\gamma}_i$ is obtained from γ_i by sliding along the disk D.

Since N is not a small compression body, the surface X is not a torus with a hole. We deduce that the curves γ_i and $\bar{\gamma}_i$ are not freely homotopic in X. On the other hand, they are freely homotopic in N. So $\gamma_i \cup \bar{\gamma}_i$ is the support of a measured lamination in $\mathscr{M}'$ (see Lemma 3.1). Then up to choice of a subsequence, these measured laminations converge to some $\mu_{\mathscr{M}} \in \mathscr{M}'$. Further, the sequence $(\gamma_i \cup \bar{\gamma}_i)$ converges in the

Hausdorff-topology to a lamination $\mu_{\mathcal{H}}$ which contains the support of μ. Since μ is minimal arational, the supports of μ and $\mu_{\mathcal{M}}$ agree. Unique ergodicity of μ implies that μ and $\mu_{\mathcal{M}}$ represent the same element in $\mathcal{PML}$. Hence $\mu \in \mathcal{M}'$ and we are done. □

It is often useful to study laminations on $\partial_e N$ by considering their lifts to the covering associated to N. For that, let $\rho_0 : \pi_1(N) \to \mathrm{PSL}_2\mathbb{C}$ be a geometrically finite and minimally parabolic representation uniformizing the compression body N. Let S' be the unique invariant component of the discontinuity domain of the action of $\rho_0(\pi_1(N))$ on $\hat{\mathbb{C}}$; S' is the covering of the exterior boundary $\partial_e N$ with deck transformation group $\rho_0(\pi_1(N))$. The limit set Λ_{ρ_0} of the action of $\rho_0(\pi_1(N))$ on $\hat{\mathbb{C}}$ coincides with the boundary of S' in $\hat{\mathbb{C}}$ [Mak88].

A leaf l of a lamination on $\partial_e N$ is called *homoclinic* if there are two sequences $x_i, y_i \in \mathbb{R}$ and a lift $l' : \mathbb{R} \to S'$ of l to S', parameterized by arc-length, such that $|x_i - y_i| \to \infty$ when i goes to ∞ but such that the distances between $l'(x_i)$ and $l'(y_i)$ are bounded in S'. In particular, every meridian is homoclinic. Moreover, it is due to Casson (see Lecuire [Lec02, Appendix B]) that every Hausdorff-limit of meridians contains a leaf which is homoclinic.

We deduce the following partial converse to this fact.

Proposition 3.8. *Let N be a compression body which is not small. If $\mu \subset \partial_e N$ is a lamination containing a homoclinic leaf, then every minimal component of μ is the support of a lamination $\mu' \in \mathcal{M}'$.*

Remark 3.9. Proposition 3.8 was proved in [KS02, Proposition 1] in the case that N is a handlebody. This restriction was due to the fact that we could only prove Lemma 3.7 [KS02, Lemma 6] in the case of the handlebody. All the arguments in [KS02, Section 4.2] but this one remain unchanged.

Proposition 3.8 implies the following corollary which is originally due to Otal [Ota88].

Corollary 3.10. *The complement $\partial_e N - \lambda$ of a Masur domain lamination λ does not contain a homoclinic leaf.*

Let now $\rho : \pi_1(N) \to \mathrm{PSL}_2\mathbb{C}$ be an arbitrary discrete and faithful representation and $i_e : \partial_e N \to N$ the inclusion of $\partial_e N$ into N. In the next section we will consider a subset of $\mathcal{O}$ which takes into account the simple closed curves γ on $\partial_e N$ with $\rho \circ i_{e*}(\gamma)$ parabolic.

Definition 3.11. $\mathcal{O}_P = \mathcal{O}_P(\rho)$ is the set of those laminations in $\mathcal{O}$ which have positive intersection with every simple closed curve γ on $\partial_e N$ such that $\rho \circ i_{e*}(\gamma)$ is parabolic.

It follows from Proposition 3.6 that $\mathcal{O}_P = \mathcal{O}$ if ρ is minimally parabolic.

We will not prove that $\mathcal{O}_P$ is open, we only show the following slightly weaker result.

Lemma 3.12. *Let λ be a maximal lamination in $\mathcal{O}_P$, γ the (possibly empty) collection of simple closed curves in λ and (γ_i) a sequence of multicurves converging in $\mathcal{PML}$ to λ. If $\gamma \subset \gamma_i$ for all i, then there is i_0 such that $\gamma_i \in \mathcal{O}_P$ for all $i \geq i_0$. Moreover, every minimal arational lamination in $\mathcal{O}$ is an interior point of $\mathcal{O}_P$.*

Proof. To begin with, remark that every curve in γ represents a hyperbolic element in $\rho(\pi_1(N))$ because $\lambda \in \mathcal{O}_P$. Let $\lambda^1, \ldots, \lambda^k$ be the minimal components of $\lambda - \gamma$ and let $S(\lambda^j)$ be the component of $\partial_e N - \gamma$ which contains λ^j; the lamination λ^j is minimal arational in $S(\lambda^j)$.

For every $j \in \{1, \ldots, k\}$ let γ_i^j be the union of those components of γ_i which are contained in $S(\lambda^j)$; the sequence (γ_i^j) converges to λ_j when $i \to \infty$. By openness of $\mathcal{O}$, the multicurve $\gamma_i = \gamma \cup \gamma_i^1 \cup \ldots \cup \gamma_i^k$ is in $\mathcal{O}$ for large i.

Seeking for a contradiction, assume $\gamma_i \notin \mathcal{O}_P$ for infinitely many i, say for all i. There are curves η_i which represent parabolic elements in $\rho(\pi_1(N))$ and do not intersect γ_i. Every curve η_i intersects λ, but it does not intersect the multicurve γ; thus it is contained in one of the surfaces $S(\lambda^1), \ldots, S(\lambda^k)$. After passing to a subsequence, we may assume that (η_i) converges to a projective measured lamination μ whose support is contained in $S(\lambda^1)$. Moreover, $i(\lambda^1, \mu) = \lim_{i \to \infty} i(\gamma_i^1, \eta_i) = 0$. Since λ^1 is minimal arational in $S(\lambda^1)$, we deduce that the supports of λ^1 and μ agree. Hence there is i_0 such that for all $i \geq i_0$ the multicurve

$$\tau_i = \gamma \cup \{\eta_i, \gamma_{i_0}^2, \ldots, \gamma_{i_0}^k\}$$

is in $\mathcal{O}$. So by Lemma 3.1, η_i represents a primitive parabolic element in $\rho(\pi_1(N))$ for all i and by Proposition 3.6 the corresponding maximal parabolic subgroup is cyclic. By Sullivan's finiteness theorem [Sul81], we may extract a subsequence such that for all $i \geq i_0$ the curves η_i and η_{i_0} are freely homotopic in N. Corollary 3.5 says that for all $i \geq i_0$ there is $\phi_i \in Mod_0(N)$ such that $\phi_i(\tau_{i_0}) = \tau_i$. In particular, the multicurves τ_i are contained in a single $Mod_0(N)$-orbit and they are bounded in $\mathcal{O}$, contradicting Theorem 3.3. This shows that the multicurves γ_i are in $\mathcal{O}_P$ for large i.

Let now $\lambda \in \mathcal{O}$ be minimal arational; by definition $\lambda \in \mathcal{O}_P$. Assume that there is a sequence (λ_i) in $\mathcal{PML}$ with $\lambda_i \notin \mathcal{O}_P$ converging to λ. For all i there is a simple closed curve η_i which represents a parabolic element in $\rho(\pi_1(N))$ and which does not intersect λ_i. The sequence (η_i) converges in $\mathcal{PML}$, up to choice of a subsequence, to some λ' with the same support as λ. As remarked above, λ' is in $\mathcal{O}$ and hence in $\mathcal{O}_P$. This contradicts the first claim of the lemma because $\eta_i \notin \mathcal{O}_P$ for all i. $\square$

4. Compactness theorem

Let N be a non-trivial compression body and $i_e : \partial_e N \to N$ the inclusion of $\partial_e N$ into N. We consider in this section a fixed discrete and faithful representation $\rho : \pi_1(N) \to PSL_2\mathbb{C}$. A (geodesic) lamination λ on the exterior boundary $\partial_e N$ is realized in $M_\rho = \mathbb{H}^3/\rho(\pi_1(N))$ if there is $S \in \mathscr{T}(\partial_e N)$ and a pleated map $f : S \to M_\rho$ with $f_* = \rho \circ i_{e*}$ which maps every leaf of λ to a geodesic in M_ρ. We say that a measured lamination is realized if its support is.

In this section we will prove a compactness theorem for pleated surfaces which realize laminations in $\mathscr{O}_P = \mathscr{O}_P(\rho)$. All the results of this section are due to Otal [Ota88] if $\rho(\pi_1(N))$ does not contain parabolic elements.

To begin with, we show that multicurves in $\mathscr{O}_P$ are realized.

Lemma 4.1. *Every multicurve $\lambda \in \mathscr{O}_P$ is realized in M_ρ.*

Proof. Let $\hat{\lambda}$ be a lamination whose recurrent part is the multicurve λ and such that every complementary region of $\hat{\lambda}$ is an ideal triangle; so $\hat{\lambda} - \lambda$ is a finite union of isolated non-compact leaves. We will realize $\hat{\lambda}$ in M_ρ; by definition, λ is then realized, too.

Fix an arbitrary map $f : \partial_e N \to M_\rho$ in the right homotopy class and let $f' : S' \to \mathbb{H}^3$ be a lift. Recall that S' is the cover of $\partial_e N$ associated to $\pi_1(N)$. We deduce from the arguments used in [CEG87, Thm. 5.3.6] that the lamination $\hat{\lambda}$ is realized if for every leaf l of $\hat{\lambda}$ and every lift l' of l to S', the image $f'(l')$ in $\mathbb{H}^3$ has two well-defined endpoints in the limit set Λ_ρ of the action of $\rho(\pi_1(N))$ on $\hat{\mathbb{C}}$. If l is a simple closed curve, then $f(l)$ represents a hyperbolic element in $\rho(\pi_1(N))$ and thus $f'(l')$ has two endpoints. Assume that l is non-recurrent; by construction, l accumulates on one or two compact leaves. Let $l' : \mathbb{R} \to S'$ be a lift of l. The points $f'(l'(t))$ converge for $t \to \pm\infty$ to fixed points $p_\pm \in \Lambda_\rho$ of hyperbolic isometries $g_\pm$ represented by closed curves in λ. If p_+ and p_- are equal, then g_+ and g_- are in the same cyclic subgroup. This implies that the leaf l is homoclinic, contradicting Corollary 3.10. We deduce that both endpoints of $f'(l')$ are different. As remarked above, this is sufficient to show that $\hat{\lambda}$, and hence λ, is realized in M_ρ. $\square$

Later we will need the following lemma.

Lemma 4.2. *If $\lambda \in \mathscr{O}_P$ is realized in M_ρ and if $\bar{\lambda}$ is the union of the support of λ and finitely many non-compact isolated leaves, then $\bar{\lambda}$ is realized in M_ρ, too.*

Proof. Let $f : S \to M_\rho$ be a pleated surface realizing λ, $f' : S' \to \mathbb{H}^3$ a lift of f and l' a lift of a leaf $l \subset \bar{\lambda} - \lambda$. As in the proof of Lemma 4.1 it suffices to show that $f'(l')$

has two different well-defined endpoints in $\hat{\mathbb{C}}$. As above, we only have to rule out that $f'(l')$ has only one endpoint; suppose that this is the case.

We show that l' has also only one endpoint in S': Let x_i, y_i be two arbitrary sequences of points in l' such that $d_{\mathbb{H}^3}(f'(x_i), f'(y_i))$ tends to 0. There are elements $g_i \in \pi_1(N)$ such that the sequence $(g_i(x_i))$ converges (up to subsequence) to some $x_\infty \in S'$. It follows that $d_{\mathbb{H}^3}(f'(g_i(y_i)), f'(x_\infty))$ is bounded. Since f' is a proper map, the points $g_i(y_i)$ remain in a compact subset of S'. So, $d_{S'}(x_i, y_i) = d_{S'}(g_i(y_i), g_i(x_i))$ is bounded above implying that l' has only one endpoint in S'.

We conclude that the projection l of l' to S is homoclinic. This contradicts Corollary 3.10 because l is contained in the complement of the Masur domain lamination λ. $\qquad\square$

The main result of this section is the following proposition.

Proposition 4.3. *A lamination $\lambda \in \mathcal{O}_P$ is realized in M_ρ either if λ is not minimal arational or if λ is minimal arational and there is a sequence (γ_i) of multicurves converging to λ and a compact set $K \subset M_\rho$ such that γ_i is realized by a pleated surface $f_i : X_i \to M_\rho$ with $f_i(X_i) \cap K \neq \emptyset$ for all i.*

Recall that every minimal arational Masur domain lamination is in $\mathcal{O}_P$. If M_ρ is geometrically finite, then there is a compact set which intersects every pleated surface in M_ρ. As a consequence we obtain

Corollary 4.4. *If $\rho : \pi_1(N) \to \mathrm{PSL}_2\mathbb{C}$ is a geometrically finite and minimally parabolic representation, then every lamination in $\mathcal{O}$ is realized in M_ρ.*

The rest of this section is devoted to the proof of Proposition 4.3. We denote by γ the collection of simple closed curves in λ. Remark that we may assume that the measured lamination λ is maximal. So γ is empty if and only if λ is minimal arational.

We first claim that in the case that λ is not minimal arational there is also a compact set $K \subset M_\rho$ and a sequence (γ_i) of multicurves which converge to λ and are realized by pleated surfaces $f_i : X_i \to M_\rho$ with $f_i(X_i) \cap K \neq \emptyset$ for all i. Indeed, approximate λ by multicurves γ_i with $\gamma \subset \gamma_i$ for all i. Lemma 3.12 and Lemma 4.1 show that $\gamma_i \in \mathcal{O}_P$ and that γ_i is realized in M_ρ by a pleated map $f_i : X_i \to M_\rho$ for large i, say for all i. Since every pleated surface f_i realizes the multicurve γ, there is a compact set K as claimed.

From now on, we do not distinguish whether λ is minimal arational or not.

As the rest of the proof of Proposition 4.3 is quite long we continue with a short outline. Let $f_i : X_i \to M_\rho$ be the pleated surfaces realizing γ_i. We will show that, up to choice of a subsequence, the surfaces X_i converge in Teichmüller space to some point X_∞. This implies that there is an upper bound for the diameter of the surfaces X_i. Thus by the above, the images of the maps f_i are contained in some compact set $K' \subset M_\rho$. Furthermore, there are L_i-bi-Lipschitz maps $\psi_i : X_\infty \to X_i$ in the right homotopy class with $L_i \to 1$ when $i \to \infty$. The Arzela-Ascoli theorem implies that a subsequence of $(f_i \circ \psi_i)_i$ converges to a pleated map $X_\infty \to M_\rho$ which maps every leaf of the Hausdorff limit $\lambda_{\mathcal{H}}$ of (γ_i) to a geodesic in M_ρ [CEG87, Theorem 5.2.2]. Since the support of λ is contained in $\lambda_{\mathcal{H}}$, the lamination λ is realized in M_ρ. We now fill in the details.

We must prove that the surfaces X_i converge in Teichmüller space. We first show that the sequence (X_i) is bounded in moduli space. This is true if there is a uniform lower bound for the length $l_{X_i}(\alpha)$ of every simple closed curve $\alpha \subset \partial_e N$ on X_i.

Simple closed curves on $\partial_e N$ can be divided into three types: meridians, curves η which represent a parabolic element in $\rho(\pi_1(N))$, i.e. $\rho \circ i_{e*}(\eta)$ is parabolic, and curves η which represent a hyperbolic element in $\rho(\pi_1(N))$, i.e. $\rho \circ i_{e*}(\eta)$ is hyperbolic. We will successively bound the lengths of these curves from below. This is the content of the next three lemmas.

Lemma 4.5. *There is a uniform constant $\varepsilon_M > 0$ such that $l_{X_i}(m) > \varepsilon_M$ for every meridian m on $\partial_e N$ and all i.*

Proof. Suppose that (m_i) is a sequence of meridians such that the lengths $l_{X_i}(m_i)$ tend to 0. Denote by μ a limit of the sequence (m_i) in the space of projective measured laminations on $\partial_e N$. The lamination μ is in $\mathcal{M}'$, hence λ intersects μ. We find a segment $\kappa = \kappa[0, 1]$ in λ which can be closed by a small segment I contained in μ and such that $\frac{d\kappa}{dt}(0)$ and $\frac{d\kappa}{dt}(1)$ are close. Then for all i, there are subsegments $\kappa_i \subset \lambda_i$ and $I_i \subset m_i$ in the reference surface $\partial_e N$ such that the curve $\kappa_i \cup I_i$ is freely homotopic to $\kappa \cup I$.

In the surface X_i, we denote the corresponding segments in the geodesic representatives of λ_i and m_i by κ_i and I_i as well. As $l_{X_i}(m_i)$ tends to 0, the arcs $\kappa_i \subset X_i$ become long and the vectors $\frac{d\kappa_i}{dt}(0)$ and $\frac{d\kappa_i}{dt}(1)$ are close.

The map f_i maps the segment $\kappa_i \subset X_i$ isometrically and the vectors $\frac{d\kappa_i}{dt}(0)$ and $\frac{d\kappa_i}{dt}(1)$ are mapped nearby in M_ρ (compare with [BO01, Sous-lemme 10]). This implies that the geodesic in M_ρ representing the homotopy class of $f_i(\kappa_i \cup I_i)$ has almost the same length as the segment $\kappa_i \subset X_i$, contradicting the fact that this homotopy class is fixed. $\square$

Lemma 4.6. *There is a uniform constant $\varepsilon_P > 0$ such that $l_{X_i}(\eta) > \varepsilon_P$ for all i and every curve η on $\partial_e N$ which represents a parabolic element in $\rho(\pi_1(N))$.*

Proof. Suppose that there is a sequence (η_i) of curves on X_i which represent parabolic elements in $\rho(\pi_1(N))$ and such that $l_{X_i}(\eta_i) \to 0$ when $i \to \infty$. The Margulis lemma implies that the curves η_i do not intersect the collection γ of simple closed curves in λ. (Recall that γ is empty if λ is minimal arational.) So for all i, the curve η_i is contained in one of the components of $\partial_e N - \gamma$. As in the proof of Lemma 3.12 we may assume that the sequence (η_i) converges in the Hausdorff-topology to a lamination $\mu_{\mathscr{H}} \subset \partial_e N - \gamma$. If $\lambda - \gamma$ intersects $\mu_{\mathscr{H}}$, then the arguments in the proof of Lemma 4.5 yield a contradiction, so we may assume that the support of a component of $\lambda - \gamma$ is contained in $\mu_{\mathscr{H}}$ because $\lambda - \gamma$ is minimal arational in $\partial_e N - \gamma$. We conclude as in the proof of Lemma 3.12. $\qquad\square$

Lemma 4.7. *There is a compact set $K' \subset M_\rho$ such that $f_i(X_i) \subset K'$ for all i. In particular, curves on $\partial_e N$ which represent hyperbolic elements in $\rho(\pi_1(N))$ have length uniformly bounded from below in X_i.*

Proof. To begin with, we observe that there is a constant $\varepsilon > 0$ depending on ε_M, ε_P and the genus of $\partial_e N$ with the property that the surfaces $f_i(X_i)$ do not intersect any unbounded component of the ε-thin part $M_\rho^{<\varepsilon}$ of M_ρ. Moreover, up to reducing ε, we may assume that ε is less than the Margulis constant and that any two components of the ε-thin part $M_\rho^{<\varepsilon}$ are at least at distance one.

For a curve $\alpha \subset M_\rho$, let $l_{\varepsilon, M_\rho}(\alpha)$ be the length of the intersection $\alpha \cap M_\rho^{\geq \varepsilon}$ of α and the ε-thick part of M_ρ. Define $l_{\varepsilon, X_i}(\alpha)$ in a similar way for a curve α in X_i. $\qquad\square$

Lemma 4.8 (Bounded diameter lemma [Bon86]). *For every $\varepsilon > 0$ and every $A > 0$ there is a constant $c_\varepsilon > 0$ such that any two points x, y on a complete hyperbolic surface S with area at most A can be joined by a curve $\alpha \subset S$ with $l_{\varepsilon, S}(\alpha) \leq c_\varepsilon$.*

Proof. We are going to apply this lemma to the surfaces X_i. First remark that Lemma 4.5 implies that $f_i(X_i^{<\varepsilon}) \subset M_\rho^{<\varepsilon}$ for all i because $\varepsilon < \varepsilon_M$. In particular we have $l_{\varepsilon, X_i}(\alpha) \geq l_{\varepsilon, M_\rho}(f_i(\alpha))$ for any curve α on X_i.

By assumption there is a compact set $K \subset M_\rho$ with $K \cap f_i(X_i) \neq \emptyset$ for all i. We conclude that for all i the surface $f_i(X_i)$ is contained in the set K' of all points in M_ρ which can be joined to K by a path α with $l_{\varepsilon, M_\rho}(\alpha) \leq c_\varepsilon$ and such that α does not intersect the unbounded components of $M_\rho^{<\varepsilon}$; the set K' is compact. Since the injectivity radius is bounded in K', there is a uniform lower bound for the lengths in X_i of closed curves on $\partial_e N$ which represent hyperbolic elements in $\rho(\pi_1(N))$. $\qquad\square$

By Lemmas 4.5, 4.6 and 4.7, the sequence (X_i) is bounded in moduli space. We claim that it is also bounded in Teichmüller space.

Up to passing to a subsequence, there is X_∞ in Teichmüller space and a sequence (ϕ_i) of mapping classes such that $\phi_i X_i \to X_\infty$ in $\mathcal{T}(\partial_e N)$. First we prove

Lemma 4.9. *Up to choice of a subsequence, there is $i_0 > 0$ such that for all $i \ge i_0$ the mapping class $\phi_i \circ \phi_{i_0}^{-1}$ is in $Mod_0(N)$.*

Proof. Assume that the surfaces X_i are marked by $\sigma_i : \partial_e N \to X_i$ for $i = 1, 2, \ldots, \infty$.

The sequence $(\phi_i X_i)_i$ converges in Teichmüller space to X_∞. This implies that there are L_i-bi-Lipschitz maps $\psi_i : X_\infty \to X_i$ with $L_i \to 1$ when $i \to \infty$ such that the following diagram commutes up to homotopy. The vertical arrows are σ_∞ and σ_i respectively.

$$
\begin{array}{ccc}
\partial_e N & \xrightarrow{\phi_i} & \partial_e N \\
\downarrow & & \downarrow \\
X_\infty & \xrightarrow{\psi_i} X_i & \xrightarrow{f_i} M_\rho
\end{array}
$$

By Lemma 4.7, the image $f_i(X_i)$ is contained in the compact set K' for all i. The Arzela-Ascoli theorem shows that the sequence $f_i \circ \psi_i : X_\infty \to M_\rho$ has a subsequence, say the whole sequence, which converges to a pleated map $F : X_\infty \to M_\rho$. This implies that there is i_0 such that for $i \ge i_0$ the maps $f_i \circ \psi_i$ and $f_{i_0} \circ \psi_{i_0}$ are homotopic. We deduce that the maps $f_i \circ \sigma_i \circ \phi_i$ and $f_{i_0} \circ \sigma_{i_0} \circ \phi_{i_0}$ are homotopic, hence $\phi_i \circ \phi_{i_0}^{-1}$ is in $Mod_0(N)$.   □

To save notation we replace X_∞ by $\phi_{i_0}^{-1}(X_\infty)$; therefore we may assume that the maps ϕ_i are in $Mod_0(N)$ for all i.

Seeking for a contradiction we assume that (X_i) converges to a lamination v in $\mathscr{PML}$. Suppose for the moment that

$$i(\lambda, v) > 0. \tag{4.1}$$

By assumption the multicurves γ_i, which are realized by the pleated surfaces $f_i : X_i \to M_\rho$, converge to λ. By convergence of X_i to v there is a sequence ε_i tending to 0 such that

$$0 < i(\lambda, v) = \lim_{i \to \infty} \varepsilon_i \frac{l_{X_i}(\gamma_i)}{l_{X_1}(\gamma_i)}.$$

As $l_{X_i}(\gamma_i) \le l_{X_1}(\gamma_i)$ this yields the desired contradiction to the convergence of the surfaces X_i to v. This implies that the sequence (X_i) is bounded in Teichmüller space.

We prove (4.1). First remark that the lamination v is also the limit of the sequence $(\phi_i^{-1}(X_\infty))_i$. Let m be any meridian. Up to choice of a subsequence we may assume that the meridians $m_i = \phi_i(m)$ converge to $\mu \in \mathcal{M}'$. Moreover, since the lengths of m_i in X_i are bounded, we obtain that $i(\mu, \nu) = 0$. Since $\mu \in \mathcal{M}'$ we deduce that $i(\lambda, \mu) > 0$. There are $A > 0$ [FLP79, p.58] and $A' > 0$ such that

$$i(\lambda, \mu) = \lim_i \frac{i(\gamma_i, m_i)}{l_{X_1}(\gamma_i) l_{X_1}(m_i)} \leq \lim_i \frac{A \, l_{X_i}(\gamma_i) l_{X_i}(m_i)}{l_{X_1}(\gamma_i) l_{X_1}(m_i)} \leq A' \lim_i \frac{1}{l_{X_1}(m_i)}.$$

We deduce that there is an upper bound for $l_{X_1}(m_i)$ and hence that $\mu = m_i$ for i large enough, hence the complement of ν in $\partial_e N$ is compressible. In the case that N is small, it follows directly from the definition of $\mathcal{O}$ that $i(\lambda, \nu) > 0$. If N is not small, we deduce from Lemma 3.7 that every component of ν is in $\mathcal{M}'$ and thus $i(\lambda, \nu) > 0$. This concludes the proof of Proposition 4.3.

5. Main results

In this section we prove Theorem 1.1 and Theorem 1.3 from the introduction.

Theorem 5.1. *1.1Let E be a compressible end of a complete oriented hyperbolic 3-manifold M whose fundamental group is finitely generated but not free. If a Masur domain lamination on ∂E is not realized in M, then the end E is tame.*

Proof. The surface ∂E is a compressible boundary component of a core C of M. So the relative compression body $B_E \subset C$ associated to E is non-trivial. Moreover, $\pi_1(B_E)$ is not a free group because $\pi_1(C) = \pi_1(M)$ is not free.

The end E, the surface ∂E and the compression body B_E lift to the cover $M_E = \mathbb{H}^3/\pi_1(B_E)$ of M. We denote the lifts by E, ∂E and B_E again. Note that B_E is a core of M_E. By assumption, there is a Masur domain lamination $\lambda \subset \partial E$ which is not realized in M. It follows that λ is not realized in M_E.

In the case that λ is not minimal arational it follows from Proposition 4.3 that λ is not in the set $\mathcal{O}_P$ of those Masur domain laminations which intersect every simple closed curve on ∂E representing a parabolic element in $\pi_1(M_E)$. Let $\gamma \subset \partial_e N$ be a simple closed curve which represents a parabolic element in $\pi_1(M_E)$ and does not intersect λ. Since λ is a Masur domain lamination, $\lambda \cup \gamma \in \mathcal{O}$. We deduce from Proposition 3.6 that there is no splitting of $\pi_1(M_E) = \pi_1(B_E)$ as a non-trivial free product or as an HNN-extension such that the element represented by γ can be conjugated into one of the factors. By Bonahon's theorem, M_E is tame; thus E is tame in the case that λ is not minimal arational.

From now on, we assume that λ is a minimal arational lamination in $\mathcal{O}$ which is not realized in M_E. Let (γ_i) be a sequence of simple closed curves in the Masur domain of ∂E which converges to λ. By Lemma 3.12, the minimal arational lamination λ is an interior point of $\mathscr{O}_P$; therefore the curves γ_i are in $\mathscr{O}_P$ for large i, say for all i. By Lemma 4.1, every curve $\gamma_i \in \mathscr{O}_P$ is realized in M_E by a pleated surface $f_i : X_i \to M_E$. Since the lamination λ is not realized in M_E, we deduce from Proposition 4.3 that the surfaces $f_i(X_i)$ exit every compact set in M_E.

In particular, for large i the surface $f_i(X_i)$ is contained in a component of $M_E - B_E$, and the fundamental group of this component surjects onto $\pi_1(M_E)$. It follows that the surfaces $f_i(X_i)$ converge to the end E when i goes to ∞. Moreover, by construction, $f_i(X_i)$ is homotopic to the surface ∂E in M_E for all i. Since $\pi_1(M_E) = \pi_1(B_E)$ is not free, we can apply the Tameness Criterion from Section 2.   □

Before launching the proof of Theorem 1.3 we recall the notation from the introduction.

Let N be a (non-trivial) compression body, $\rho_0 : \pi_1(N) \to \mathrm{PSL}_2\mathbb{C}$ a geometrically finite and minimally parabolic representation which uniformizes N and let $QH(\rho_0)$ be the space of conjugacy classes of quasi-conformal deformations of ρ_0. Denote by P the union of the toroidal components of ∂N. A sequence $(\rho_i) \subset QH(\rho_0)$ converges into the Masur domain if it can be parameterized under the Ahlfors-Bers covering

$$\mathscr{T}(\partial N - P) = \mathscr{T}(\partial_e N) \times \mathscr{T}(\partial_{int} N - P) \to QH(\rho_0)$$

by a sequence $(S_i^e, S_i^{int})_i$ such that $(S_i^e)_i$ converges to a projective measured lamination $\lambda \in \mathscr{O}$. We prove:

Theorem 5.2. *1.3 Let N be a compression body which is not a handlebody and let $(\rho_i) \subset QH(\rho_0)$ be a sequence which converges algebraically to $\rho : \pi_1(N) \to \mathrm{PSL}_2\mathbb{C}$. If (ρ_i) converges into the Masur domain, then $M_\rho = \mathbb{H}^3/\rho(\pi_1(N))$ is tame.*

Proof. Assume that (ρ_i) is parameterized under the Ahlfors-Bers covering by a sequence (S_i^e, S_i^{int}) such that S_i^e converges to the Masur domain lamination $\lambda \subset \partial_e N$. We first show

Lemma 5.3. *There is a lamination $\bar{\lambda} \subset \partial_e N$ which contains the support of λ and which is not realized in M_ρ.*

Proof. By convergence of S_i^e to $\lambda \in \mathscr{O}$ there is a sequence $(\gamma_i) \subset \mathscr{O}$ of simple closed curves converging in $\mathscr{PML}$ to λ and satisfying

$$l_{S_i^e}(\gamma_i)/l_{S_0^e}(\gamma_i) \to 0 \quad \text{when } i \to \infty. \tag{5.1}$$

The continuity of the intersection form implies that there is some $A > 0$ such that $i(m, \lambda) > A$ for all meridians m. It follows from the defining property of the Thurston compactification that for all $L > 0$ there is some i_L with $l_{S_i^e}(m) > L$ for all $i \geq i_L$ and all meridians m. Then by a theorem due to Canary [Can91] there is $K > 0$ such that for all $i \geq i_L$

$$l_{\rho_i}(\gamma_i) \leq K\, l_{S_i^e}(\gamma_i) \qquad (5.2)$$

where $l_{\rho_i}(\gamma_i)$ is the length of the geodesic freely homotopic to γ_i in $M_{\rho_i} = \mathbb{H}^3/\rho_i(\pi_1(N))$. Combining (5.1), (5.2) and the second part of Proposition 2.3 we deduce that there is a constant $k > 0$ such that

$$\frac{l_{\rho_i}(\gamma_i)}{l_{\rho_0}(\gamma_i)} \leq \frac{K\, l_{S_i^e}(\gamma_i)}{k\, l_{S_0^e}(\gamma_i)} \to 0 \quad \text{when } i \to \infty. \qquad (5.3)$$

Let $\lambda_{\mathcal{H}}$ be the Hausdorff-limit of the sequence (γ_i) and let $\bar{\lambda}$ be the union of the recurrent components in $\lambda_{\mathcal{H}}$. The support of λ is contained in $\bar{\lambda}$, so $\bar{\lambda}$ is the support of a Masur domain lamination.

Seeking for a contradiction we assume that $\bar{\lambda}$ is realized in M_ρ. Then $\lambda_{\mathcal{H}}$ is also realized in M_ρ by Lemma 4.2. It follows from the first part of Proposition 2.3 that there is a compact set $K \subset M_\rho$ such that for all i the curve γ_i is freely homotopic in M_ρ to a geodesic γ_i^* contained in K. In particular we have that $l_\rho(\gamma_i) \asymp l_{\rho_0}(\gamma_i)$, i.e. there is a constant $c > 1$ with

$$c^{-1} l_\rho(\gamma_i) \leq l_{\rho_0}(\gamma_i) \leq c\, l_\rho(\gamma_i)$$

for all i. Algebraic convergence of ρ_i to ρ implies that there are homotopy equivalences $h_i : M_\rho \to M_{\rho_i}$, compatible with markings, which tend C^∞ to local isometries on K. For large i, the curves $h_i(\gamma_i^*)$ have small geodesic curvature. We obtain

$$l_{\rho_i}(\gamma_i) \asymp l_\rho(\gamma_i) \asymp l_{\rho_0}(\gamma_i)$$

contradicting (5.3). So $\bar{\lambda}$ is not realized in M_ρ. $\qquad\square$

Note that we do not know whether a core C of M_ρ is homeomorphic to N, so we cannot directly apply Theorem 1.1.

We continue with the proof of Theorem 1.3. In the case that the lamination $\bar{\lambda}$ is not minimal arational, the same argument as in the proof of Theorem 1.1 provides a parabolic element in $\rho(\pi_1(N))$ which cannot be conjugated into a factor of any decomposition of $\pi_1(N) \simeq \pi_1(M_\rho)$ as a non-trivial free product or as an HNN-extension. So Bonahon's condition (*) is satisfied, and M_ρ is tame if $\bar{\lambda}$ is not minimal arational.

For the rest of the proof we assume that $\bar{\lambda}$ is a minimal arational lamination on $\partial_e N$. We show.

Lemma 5.4. *Let C be a core of M_ρ. Then C is homeomorphic to N by a homeomorphism $f : N \to C$ with $f_* = \rho$.*

Proof. As in the proof of Theorem 1.1 there is a sequence $(\gamma_i) \subset \mathcal{O}$ of simple closed curves on $\partial_e N$ which converge to $\bar{\lambda}$ and which are realized in M_ρ by pleated surfaces $f_i : X_i \to M_\rho$. The surfaces $f_i(X_i)$ exit every compact set in M_ρ because the lamination $\bar{\lambda}$ is not realized. In particular, there is i_0 such that the pleated surface $f_{i_0}(X_{i_0})$ does not intersect the core C; let U be the closure of the component of $M_\rho - C$ containing $f_{i_0}(X_{i_0})$. We see that the homomorphism $\pi_1(U) \to \pi_1(M_\rho)$ is surjective. The Seifert-van Kampen theorem implies that the homomorphism $\pi_1(U \cap C) \to \pi_1(M_\rho) = \pi_1(C)$ is also surjective. Since $U \cap C$ is a component of the boundary ∂C of C, we have proved that C is a compression body. The compression bodies C and N are homotopy equivalent, thus they are homeomorphic. □

We finish the proof of Theorem 1.3. It follows from Lemma 5.4 that the compact core C is a compression body; in particular, the manifold M_ρ has only one compressible end and it is represented by the exterior boundary $\partial_e C$ of C. The image of $\bar{\lambda}$ under the identification $\partial_e N \simeq \partial_e C$ is a Masur domain lamination on $\partial_e C$ which is not realized in M_ρ. Theorem 1.1 implies that the manifold M_ρ is tame. □

Remark 5.5. The proof shows that the manifold M_ρ is homeomorphic to the interior of the compression body N in the case that the lamination λ is minimal arational but the topological type of M_ρ is undetermined if λ is not minimal arational.

For the sake of completeness we prove the following.

Proposition 5.6. *For every compression body N and every Masur domain lamination μ on $\partial_e N$ there is a discrete and faithful representation $\rho : \pi_1(N) \to \mathrm{PSL}_2\mathbb{C}$ such that no component of μ is realized in M_ρ.*

Proof. Let $\rho_0 : \pi_1(N) \to \mathrm{PSL}_2\mathbb{C}$ be a geometrically finite and minimally parabolic representation which uniformizes N.

Let S^{int} be a fixed surface in the Teichmüller space $\mathcal{T}(\partial_{int}N - P)$ of the interior boundary of N and let S_i^e be a sequence in the Teichmüller space $\mathcal{T}(\partial_e N)$ such that the length $l_{S_i^e}(\mu)$ of the measured lamination μ on the surface S_i^e tends to 0 when i goes to ∞. (See [Bon86] for the definition of the length of a measured lamination.)

Denote by $\rho_i \in QH(\rho_0)$ the image of (S_i^e, S^{int}) under the Ahlfors-Bers covering. As in the proof of Lemma 5.3, there is $K > 0$ with

$$l_{\rho_i}(\mu) \leq K \, l_{S_i^e}(\mu) \text{ for all } i \tag{5.4}$$

The following theorem shows that a subsequence of (ρ_i), say the whole sequence, converges to a discrete and faithful representation ρ.

Theorem 5.7. *Let N be a compression body and μ a measured lamination in the Masur domain. A sequence (ρ_i) of conjugacy classes of discrete and faithful representations of $\pi_1(N)$ into $\mathrm{PSL}_2\mathbb{C}$ has a convergent subsequence if there is $c > 0$ with $l_{\rho_i}(\mu) \leq c$ for all i.*

Remark 5.8. The last theorem was proved in [KS02, Theorem 2] if N is a handlebody. We restricted to this case because we could only prove Proposition 3.8 [KS02, Proposition 1] in the case of the handlebody. All the arguments but this one remain unchanged.

We continue with the proof of Proposition 5.6. By (5.4) the length of every component of μ tends to zero. The arguments in the proof of Lemma 5.3 show that no component of μ is realized in M_ρ. $\square$

Remark 5.9. The referee pointed out that one can generalize Proposition 5.6 by quite the same arguments to the statement that any allowable collection of ending invariants (with no rank-1-cusps) on a compact manifold with boundary (other than the handlebody) occurs as the ending invariants of a tame hyperbolic 3-manifold.

References

[And98]   J. Anderson (1998). A brief survey of the deformation theory of Kleinian groups, *The Epstein birthday schrift, Geom. Topol. Monogr., 1, Geom. Topol.*, 23–49.

[BO01]   F. Bonahon and J.-P. Otal (2001). Laminations mesurées de plissage des variétés hyperboliques de dimension 3, *preprint*.

[Bon83]   F. Bonahon (1983). Cobordism of automorphisms of surfaces, *Ann. Sci. Ec. Norm. Super., IV. Ser.* **16**, 237–270.

[Bon86]   F. Bonahon (1986). Bouts des variétés hyperboliques de dimension 3, *Ann. Math.* **124**, 71–158.

[Bro00]   J. Brock (2000). Continuity of Thurston's length function, *Geom. and Funct. Anal.* **10**, no. 4, 741–797.

[BBES]   J. Brock, K. Bromberg, R. Evans and J. Souto (2002). Boundaries of deformation spaces and Ahlfors' measure conjecture, *preprint*.

[Can91]   R.D. Canary (1991). The Poincaré metric and a conformal version of a theorem of Thurston, *Duke Math. J.* **64**, 349–359.

[Can93a] R.D. Canary (1993). Algebraic convergence of Schottky groups, *Trans. Amer. Math. Soc.* **337**, 235–258.

[Can93b] R.D. Canary (1993). Ends of hyperbolic 3-manifolds, *J. Amer. Math. Soc.* **6**, 1–35.

[CEG87]   R.D. Canary, D.B.A. Epstein and P. Green (1987). Notes on notes of Thurston. In *Analytical and geometric aspects of hyperbolic space (LMS lecture notes* **111***)*, 3–92.

[FLP79]   A. Fathi, F. Laudenbach and V. Poenaru (1979). Travaux de Thurston sur les surfaces, *Astérisque* **66-67**, Soc. Math. de France.

[Jac80]   W. Jaco (1980). Lectures on Three-Manifold Topology, *CBMS Regional Conference Series in Mathematics* **43**, Amer. Math. Soc..

[Joh79]   K. Johannson (1979). *Homotopy equivalences of 3-manifolds with boundaries (Lec. Notes Math.* **761***)*, Springer-Verlag.

[Kap01]   M. Kapovich (2001). Hyperbolic manifolds and discrete groups, *Progr. Math.* **183**, Birkhäuser.

[Ker90]   S. Kerckhoff (1990). The measure of the limit set of the handlebody group, *Topology* **29**, no. 1, 27–40.

[KS02]   G. Kleineidam and J. Souto (2002). Algebraic convergence of function groups, *Comment. Math. Helv.* **77**, 244–269.

[Lec02]   C. Lecuire (2002). Plissage des variétés hyperboliques de dimension 3, *preprint*.

[Mak88]   B. Maskit (1988). *Kleinian groups (Grundlehren der Mathematischen Wissenschaften* **287***)*, Springer-Verlag.

[Mar74]   A. Marden (1974). The geometry of finitely generated Kleinian groups, *Ann. Math.* **99**, 383–462.

[Mas82]   H. Masur (1982). Interval exchange transformations and measured foliations, *Ann. Math.* **115**, 169–200.

[Mas86]   H. Masur (1986). Measured foliations and handlebodies, *Ergodic Theory Dyn. Syst.* **6**, 99–116.

[MMS85]   D. McCullough, A. Miller and G.A. Swarup (1985). Uniqueness of cores of noncompact 3-manifolds, *J. Lond. Math. Soc.* **52**, 548–556.

[MMi86]   D. McCullough and A. Miller (1986). *Homeomorphisms of 3-manifolds with compressible boundary (Mem. Amer. Math. Soc.* **344***)*, Amer. Math. Soc..

[Ohs]   K. Ohshika. Kleinian groups which are limits of geometrically finite groups, *preprint.*

[Ota88]   J.-P. Otal (1988). Courants géodésiques et produits libres, *Thèse d'Etat, Université Paris-Sud, Orsay.*

[Ota94]   J.-P. Otal (1994). Sur la dégénérescence des groupes de Schottky, *Duke Math. J.* **74**, 777–792.

[Ota96]   J.-P. Otal (1996). *Le théorème d'hyperbolisation pour les variétés fibrées de dimension trois*, Astérisque No. 235.

[Sco73b]   G.P. Scott (1973). Compact submanifolds of 3-manifolds, *J. Lond. Math. Soc.* **7**, 246–250.

[Sou02]   J. Souto (2002). A note on the tameness of hyperbolic manifolds, *preprint.*

[Sul81]   D. Sullivan (1981). A finiteness theorem for cusps, *Acta Math.* **147**, 289–299.

[Thu79]   W.P. Thurston (1979). The geometry and topology of 3-manifolds, *Princeton University Lecture Notes.*
http://www.msri.org/publications/books/gt3m

Gero Kleineidam

Mathematisches Institut
Universität Bonn
Germany

kleineid@math.uni-bonn.de

Juan Souto

Mathematisches Institut
Universität Bonn
Germany

souto@math.uni-bonn.de

AMS Classification: 30F40, 57M50, 57N10

Keywords: Kleinian groups, 3-dimensional topology

Kleinian Groups and Hyperbolic 3-Manifolds   Y. Komori, V. Markovic & C. Series (Eds.)
Lond. Math. Soc. Lec. Notes **299**, 131–144   Cambridge Univ. Press, 2003

Quasi-arcs in the limit set of a singly degenerate group with bounded geometry

Hideki Miyachi[1]

Dedicated to Professor Hiroshi Yamaguchi on the occasion of his 60th birthday.

Abstract

In this paper we show that two "interior points" of the limit set can be connected by a quasi-arc in the limit set.

1. Introduction

Let G be a singly degenerate group, that is, G is a finitely generated Kleinian group with a single simply connected invariant component. The limit set $\Lambda(G)$ of such a Kleinian group G is a continuum with no interior point, but its topological and geometrical properties are not fully understood. According to beautiful recent work of Y. Minsky [Min94], if G is torsion free and the quotient manifold $\mathbb{H}^3/G$ has bounded geometry in the sense that its injectivity radius is bounded below by some positive constant within its convex core, $\Lambda(G)$ has rich topological properties. More precisely, it is locally connected and its interior (i.e. the set of non end points) has the structure of an R-tree whose distance is equivariant with respect to the action of the Kleinian group (see also [Abi88]). Furthermore, $\Lambda(G)$ is identified with the leaf space of the ending lamination, that is, topologically it is obtained by collapsing all leaves and closures of component of complements leaves of the ending lamination of the group via Moore's theorem (see [Min94] and [Thu82]).

In this paper, we study the limit set of such a Kleinian group G from the geometrical and analytical point of view. In this direction, B.Maskit showed in [Mak75] that deleting the limit set of any Schottky subgroup from $\Lambda(G)$ does not destroy the connectivity (see also [Mat01]). Furthermore, D.Sullivan [Sul80] showed that the Hausdorff dimension of the $\Lambda(G)$ is two, the maximal possible dimension (see also Bishop-Jones [BJ97]), and C.McMullen [McM96] observed that if the ending lamination of

[1]The author is partially supported by Research Fellowships of the Japan Society for the Promotion of Science for Young Scientists.

the given group G is the stable lamination of some pseudo-Anosov mapping, $\Lambda(G)$ contains uncountably many deep points in his sense. From these results, the limit set of a totally degenerate group whose manifold has bounded geometry looks *thick* in the geometrical and analytical sense. On the contrary, we will show

Theorem 1.1. *Let G be a singly degenerate group with bounded geometry. Then any two points in the interior (in the sense of R-trees) of $\Lambda(G)$ are connected by a quasi-arc in the interior of $\Lambda(G)$.*

Theorem 1.2. *Let G be as above. Then fixed points of any loxodromic element in G are connected by a quasi-arc in $\Lambda(G)$*

Here by a *quasi-arc* we mean a closed sub-arc in a quasi-circle. By definition, a quasi-arc is a simple arc in the Riemann sphere.

To state our final result, we assume that G is torsion free and consider a lift of the support of a measured lamination v of $\Omega(G)/G$, where $\Omega(G)$ is the region of discontinuity of G. Note that, since $\Lambda(G)$ is locally connected, two ends of each leaf of the support $|v|$ of μ land at two distinct points in $\Lambda(G)$ if $|\mu|$ is not the same as the ending lamination of G, otherwise two ends of any leaf land at same point.

Theorem 1.3. *Suppose that G is torsion free. Let λ be the support of a measured lamination on $\Omega(G)/G$ and $\tilde\lambda$ be its lift to $\Omega(G)$. Then*

(1) *the landing points (or point if $\lambda = \lambda_e$) of a leaf of $\tilde\lambda$ are connected by a quasi-arc which is contained in $\Omega(G)$ except for its end points,[2] and*

(2) *if $\lambda \neq \lambda_e$, any two landing points of a leaf of $\tilde\lambda$ are connected by a quasi-arc in $\Lambda(G)$.*

Moreover, in the case where $\lambda \neq \lambda_e$, we can take two quasi-arcs β_1 taken as in (1), and β_2 taken as in (2) for a leaf of $\tilde\lambda$ such that $\beta_1 \cup \beta_2$ consists of a quasi-circle in the Riemann sphere. In the other case, we can take the quasi-arc in (1) so that the arc consists of a quasi-circle.

Since such landing points in Theorem 1.3 are end points of the limit set (Corollary 3.2), our results tell us that uncountable many essentially different maximal quasi-arcs are included in the limit set of a totally degenerate group whose manifold has bounded geometry (see. S3.4). Thus our theorems imply that the limit set *seems* to be *not so distorted* (*crumpled*) in the geometrical and analytical sense.

---

[2]One can construct a simply connected domain in the complex plane whose boundary is locally connected but contains a point which cannot be reached by any quasi-arcs in that domain.

A motivation for this work comes from the following; One of the basic problems in complex dynamics is to compare results in the two theories of rational maps and of finitely generated Kleinian groups (see [Sul85], and also [McM96] and S5 of [MNTU]). In each of the two theories, there are well-studied dynamical objects, so called, *geometrically finite* ones (cf. [McM99] and [McM00]). The question here is:

How comparable (or related) are the two geometrically finite theories? More precisely, which results for geometric finite objects in one theory correspond to ones in the other?

A dynamical object which raised this question is a polynomial $P(z) = z^2 - 1.54368\ldots$, which satisfies $P^{\circ 3}(0) = P^{\circ 4}(0)$. It belongs to the category of geometrically finite polynomial maps because its post-critical set contains at most 3 points. Hence the Hausdorff dimension of its Julia set is less than 2 since its Fatou set is not empty (cf. [McM00]). In the case of a Kleinian group with non-empty region of discontinuity, such a condition on the Hausdorff dimension of its limit set is one of the characterizations of geometrically finiteness ([BJ97]). However. unfortunately this property does not characterize geometrically finiteness in the case of rational maps ([McM98]).

On the other hand, because the parameter of P is a Misiurewicz point, its Julia set is a dendrite (cf. Figure 7.6 of [McM94]), which is topologically similar to the limit set of a singly degenerate group with bounded geometry. Furthermore, $P^{\circ 2}$ is renormalizable and hybrid equivalent to $z \mapsto z^2 - 2$. Since the Julia set of $z \mapsto z^2 - 2$ is the interval $[-2, 2]$, the Julia set of P (the limit set of the semigroup generated by P) contains a quasi-arc connecting two of four fixed points of a cyclic sub-semigroup $\langle P^{\circ 2} \rangle$, these are just geometric intervals in the real axis (Hence a similar observation is obtained in the case where the given quadratic polynomial P satisfies the condition that near the critical point, some iteration $P^{\circ n}$ is hybrid equivalent to $z \mapsto z^2 - 2$). This observation is similar to the result of Theorem 1.2. Thus our theorems indicate that a geometrically infinite Kleinian group might have some comparable properties to those of geometrically finite rational maps.

Meanwhile, from the comparison by C. McMullen [McM96], the following problem has arisen naturally: Does an infinitely renormalizable quadratic polynomial with bounded combinatorics and definite moduli satisfy similar assertions to those in our theorems ? For example, are any two interior points of the Julia set of such polynomial connected by a quasi-arc (by analogy to Theorem 1.1.)?

2. Preliminaries

2.1. Notation

In this note, a Kleinian group G is said to have *bounded geometry* if, except for elliptic elements, the translation length of any element of G is greater than some constant depending only on G. Notice that from this definition we see that any Kleinian group with bounded geometry cannot contain parabolic elements. If G is torsion free, this condition is equivalent to saying that the injectivity radius of the quotient manifold $\mathbb{H}^3/G$ is bounded away from zero. (compare with the definition in Minsky [Min01].)

A compact set $X \subset \mathbb{C}$ is called a *dendrite* if X is connected and locally connected without interior and if the complement $\mathbb{C} - X$ is connected. For a set $X \subset \mathbb{C}$, a point $x \in X$ is said to be an *end point* of X if there is no arc contained in X which contains x in its interior with respect to the relative topology. Any two points of a dendrite X are connected by exactly one simple arc in X. A point $x \in X$ is an end point if and only if $X - \{x\}$ is connected. A point in X which is not an end point is called an *interior point*. It is conjectured that the limit set of of any singly degenerate group is locally connected, and hence a dendrite (cf.[Abi88]). In [Min94], Y.Minsky gave an affirmative answer to this conjecture for torsion free singly degenerate groups with bounded geometry. Thus any pair of points in the limit set of a singly degenerate group with bounded geometry can be connected via only one simple path.

An *R-tree* is a metric space in which any two points are joined by a unique topological arc which is isometric to an interval of $\mathbb{R}$. In his paper [Abi88], W.Abikoff observed that the local connectedness of the limit set of a singly degenerate group implies that the interior (with respect to some finer topology) of its limit set is an R-tree whose distance is equivariant with respect to the Kleinian group action.[3] Recently K.Matsuzaki [Mat01] showed that any point of approximation of a singly degenerate group with locally connected limit set is an end point (he treats a more general situation).

A map F between two metric spaces (X, d_X) and (Y, d_Y) is called a (K, δ)-*quasi-isometry* if

$$K^{-1} d_X(p,q) - \delta \le d_Y(F(p), F(q)) \le K d_X(p,q) + \delta$$

for all $p, q \in X$. A quasi-isometric image of an interval of $\mathbb{R}$ is said to be a *quasi-geodesic*. It is known that any quasi-geodesic in $\mathbb{H}^3$ is contained in some bounded neighborhood of the geodesic connecting its end points, furthermore the radius of the neighborhood depends only on the quasi-isometric constant.

---

[3]In 1940s, R.H.Bing and E.E.Moise independently showed that *any* locally connected compact continuum admits a path-metric (cf. [Bin49], [Moi49]).

The following result is well-known (cf. [GH90]). However, we give a proof for the sake of completeness.

Lemma 2.1. *Let A be the image of a quasi-isometry from the Poincaré 2-disk to the Poincaré 3 space. Then ∂A is a quasi-circle contained in the (Gromov) boundary of the Poincaré 3-space.*

Proof. Let us consider a (K,δ)-quasi-isometry g from $\mathbb{H}^2$ to $\mathbb{H}^3$ and put $A = g(\mathbb{H}^2) \subset \mathbb{H}^3$. This g can be extended to an injective and continuous mapping between the corresponding boundaries (cf. Minsky [Min94]). Thus ∂A is a Jordan curve. We now assume that $g(\infty) = \infty$. Let $\{z_i\}_{i=1}^3 \subset \mathbb{R}$ be three consecutive points and put $\zeta_i = g(z_i)$ $(i = 1,2,3)$. We show that $|\zeta_1 - \zeta_2| \leq C_0 |\zeta_1 - \zeta_3|$ for some $C_0 = C_0(K,\delta) > 0$, according to Ahlfors' three point property of quasi-circles.

First of all, we assume that $|z_1 - z_2| \geq |z_2 - z_3|$. Consider the ideal triangles Δ_0 and Δ_1 (resp. Δ_0' and Δ_1') in $\mathbb{H}^2$ (resp. in $\mathbb{H}^3$) whose vertices are $\{\infty, z_1, z_2\}$ and $\{\infty, z_1, z_3\}$ (resp. $\{\infty, \zeta_1, \zeta_2\}$ and $\{\infty, \zeta_1, \zeta_3\}$). Let x_0 and x_1 (resp. x_0' and x_1') be the nearest point on the edges $\overline{\infty z_1}$ (resp. on the edge $\infty \zeta_1$) from the the center of gravity of Δ_0 and Δ_1 (resp. Δ_0' and Δ_1'). By the assumption $|z_1 - z_2| \geq |z_2 - z_3|$, it holds that $d_{\mathbb{H}^2}(x_0, x_1) \leq C_1$ for some universal constant $C_1 > 0$. Hence $d_{\mathbb{H}^3}(g(x_0), g(x_1)) \leq KC_1 + \delta$. Further, since x_i $(i = 0,1)$ is contained in the C_2-neighborhood of each edge of $\partial \Delta_i$, for some universal constant C_2, $g(x_i)$ is also contained in the C_3-neighborhood of each edge of $\partial \Delta_i'$, for some C_3 depending only on (K,δ), since each edge of Δ_i is mapped to a (K,δ)-quasi-geodesic ending at the two of vertices of Δ_i' $(i = 0,1)$. Therefore, $d_{\mathbb{H}^3}(x_i', g(x_i)) \leq C_4$ for a constant $C_4 = C_4(K,\delta)$, and hence $d_{\mathbb{H}^3}(x_0', x_1') \leq 2C_4$. This means that $|\zeta_1 - \zeta_2| \leq C_0 |\zeta_1 - \zeta_3|$ for $C_0 = C_0(K,\delta)$.

Finally, we assume that $|z_1 - z_2| \leq |z_2 - z_3|$. By the argument above, we have $|\zeta_3 - \zeta_2| \leq C_0 |\zeta_3 - \zeta_1|$. Hence

$$|\zeta_1 - \zeta_2| \leq |\zeta_1 - \zeta_3| + |\zeta_3 - \zeta_2| \leq (C_0 + 1)|\zeta_1 - \zeta_3|.$$

$\square$

2.2. Model manifolds

Let G be a torsion free singly degenerate group with bounded geometry. In this subsection, following the work of Y.Minsky [Min94] we will construct a model manifold of the Kleinian manifold $\mathbb{H}^3/G$. A similar construction is given in Cannon-Thurston [CT89].

Let S be a closed surface of the same type as $\Omega(G)/G$. Let λ_e be the ending lamination of G. Then the following are known.

Lemma 2.2 (Lemma 2.4 in [Min94]). *If $\lambda \neq \lambda_e$ is the support of a measured lamination then λ_e and λ fill up S, that is, the union of their geodesic representatives in any hyperbolic metric cuts S into a union of disks each of whose boundaries consists of finitely many compact arcs of λ_e and λ.*

This lemma is proved by using the fact that each complementary component of λ_e is simply connected.

Lemma 2.3 (Gardiner–Masur [GM91]). *A pair of two measured laminations (or foliations) are obtained from a holomorphic quadratic differential provided that they fill up S.*

Let μ be a measured lamination on S. By Lemma 2.2, λ_e and the support of μ fill up S. Hence there exists a complex structure $\sigma = \sigma_\mu$ on S and a holomorphic quadratic differential Φ on (S, σ) with norm one so that the horizontal and vertical foliations Φ_h and Φ_v of Φ represent the support of μ and λ_e, respectively. Readers may consult Levitt [Lev83] to confirm the relation between geodesic laminations and foliations. Away from the zeros of Φ, the Φ-metric has the representation

$$|\Phi(z)||dz|^2 = dx^2 + dy^2,$$

where dx and dy are the transverse measures of Φ_v and Φ_h respectively. On the manifold $S \times \mathbb{R}$, we introduce the (singular) metric ds^2 by

$$ds^2 = e^{|2t|}dx^2 + e^{-2t}dy^2 + dt^2.$$

away from the zeros of Φ, and $ds^2 = 0$ on the rest. The path metric is defined on $S \times \mathbb{R}$ via the metric ds^2. We denote by M_μ the path-metric space (manifold) $(S \times \mathbb{R}, ds^2)$.

In this setting, Y.Minsky proved the following.

Proposition 2.4. *(Theorem 5.1 and Corollary 5.10 of [Min94]) There exists a surjective quasi-isometry f from $M = M_\mu$ to $\mathbb{H}^3/G$ whose lift is also a quasi-isometry between the universal covering space $\tilde{M}$ of M and $\mathbb{H}^3$.*

Let us denote by $\tilde{f}$ the lift of the quasi-isometry $f : M \to \mathbb{H}^3/G$.

Proposition 2.5. *(Lemma 7.2 in [Min94]) For any leaf l of Φ_h or Φ_v, the lift of $l \times \mathbb{R}$ is a totally geodesic plane in $\tilde{M}$, and is bilipschitz equivalent to the Poincaré half plane.*

Identification of the boundary mapping The universal cover $\tilde{S}$ and the covering group of S are identified with the upper half plane $\mathbb{H}^2$ and a fuchsian group H_0 representing (S, σ_μ) respectively. Fix an identification ψ_0 from $\mathbb{H}^2$ to $\tilde{S}$ compatible with their covering groups.

Let $\mathbb{H}^2 \subset \mathbb{H}^3$ be the geodesic plane preserved by the fuchsian group H_0. For $x \in \mathbb{H}^3$ let $p(x)$ denote the nearest point in $\mathbb{H}^2$ to x, and let $t(x)$ the signed distance to $\mathbb{H}^2$. Then we obtain the coordinates

$$\psi : \mathbb{H}^3 \ni x \mapsto (p(x), t(x)) \in \mathbb{H}^2 \times \mathbb{R}.$$

We denote by Δ_∞ (resp. $\Delta_{-\infty}$) the component of $\partial \mathbb{H}^3 - \overline{\mathbb{H}^2 \times \{0\}}$ corresponding to ∞ (resp. $-\infty$) under the identification ψ_0. These are geometric disks in $\hat{\mathbb{C}} = \partial \mathbb{H}^3$. Notice that the part $\partial \mathbb{H} \times \mathbb{R}$ of $\partial \mathbb{H}^3$ corresponds to the geometric circle $\partial \Delta_\infty = \partial \Delta_{-\infty}$ (i.e. the $\mathbb{R}$-direction is collapsed) under the identification. Here, we can define two measured foliations $\tilde{\Phi}_h$ and $\tilde{\Phi}_v$ on $\mathbb{H}^2$ by pulling back via ψ_0, which defines measured foliations on $\Delta_{\pm\infty}$ by extending canonically from the product structure of $\mathbb{H}^3$ above. To simplify, we denote these foliations by the same symbols as $\tilde{\Phi}_h$ and $\tilde{\Phi}_h$ respectively.

Proposition 2.6. *(p.574 of [Min94]) The mapping $\tilde{f} \circ \psi$ can be extended to a continuous mapping between boundaries. Moreover, let π be the continuous extension of $\tilde{f} \circ \psi$ from $\hat{\mathbb{C}}$ onto itself. Then π is a homeomorphism on $\Delta_{-\infty}$ onto the invariant component of G and sends $\overline{\Delta_\infty}$ onto $\Lambda(G)$. For $x, y \in \overline{\Delta_\infty}$, $\pi(x) = \pi(y)$ if and only if x and y are in the closure of the same leaf of $\tilde{\Phi}_v$.*

3. Proof of theorems

We prove theorems stated in Introduction in the order, Theorem 1.3, Theorem 1.2, and then Theorem 1.1.

3.1. Proof of Theorem 1.3

Let us prove Theorem 1.3. We use the notation from Section 2.2 freely. Let μ be a measured lamination on S whose support does not coincides with λ_e.

Let $\tilde{\lambda}$ be the lift of the support of μ. Let l be a leaf of either $\tilde{\lambda}$ or the lift of λ_e. Let M_μ be the model manifold associated to μ. Let $\tilde{L}$ be the leaf of $\tilde{\Phi}_h$ or $\tilde{\Phi}_v$ associated to l. Then $\tilde{L} \times \mathbb{R}$ is totally geodesic in $\tilde{M}$ and bilipschitz equivalent to $\mathbb{H}^2$ by Proposition 2.5. Hence by composing $\tilde{f}$ and this quasi-isometry, we have a quasi-isometry g from $\mathbb{H}^2$ to $\tilde{f}(\tilde{L} \times \mathbb{R})$. Let L_∞ (resp. $L_{-\infty}$) be the closure of the leaf of Φ_h or Φ_v in Δ_∞ (resp. $\Delta_{-\infty}$) associated with $\tilde{L}$.

Suppose first that l is not a leaf of λ_e. Since $\partial(\tilde{L} \times \mathbb{R}) = L_\infty \cup L_{-\infty}$, by Lemma 2.1, we have that $\eta_\infty = \pi(L_\infty)$ and $\eta_{-\infty} = \pi(L_{-\infty})$ are quasi-arcs as in Proposition 2.6, because these are closed arcs contained in $\partial \tilde{f}(\tilde{L} \times \mathbb{R})$. By definition, the interior of η_∞ is contained in the region of discontinuity of G and $\eta_{-\infty} \subset \Lambda(G)$. Moreover, the end points of each of these arcs are the same points in $\Lambda(G)$ and these two arcs terminating at the end points of l. This proves (1) in Theorem 1.3.

Next we show (2) in Theorem 1.3. Assume that l is a leaf of λ_e. Then by Proposition 2.6, $\eta_{-\infty}$ consists of a point in $\Lambda(G)$, and hence η_{∞} is a quasi-arc whose both ends land on the same point $\eta_{-\infty}$.

The last part of the statement of Theorem 1.3 follows from the argument above.

3.2. Proof of Theorem 1.2

Theorem 1.2 is proved immediately by combining Theorem 1.3 and the following reduction to the case where G is torsion free and $g \in G$ corresponds to a simple closed curve on S.

We first assume that Theorem 1.2 holds for any torsion free singly degenerate group. Let H be a singly degenerate group with bounded geometry. By the Selberg lemma (cf. e.g. Matsuzaki-Taniguchi [MT98]), we can take a torsion free subgroup G of finite index. For any loxodromic $h \in H$, some power h^k is loxodromic and contained in G. Hence by assumption, the fixed points of h^k are connected by a quasi-arc, and therefore so are those of h.

Second of all, we assume that G is torsion free. Assume further that Theorem 1.2 holds for all simple closed curves. Let $g \in G$. Since $\pi_1(S) = G$ is residually finite by Scott's theorem [Sco78], there exists a finite covering space so that a lift of the curve corresponding to g is a simple closed curve. This means that there exists a finite index subgroup G' of G so that $g \in G'$ and g corresponds to a simple closed curve on a surface associated to G'. Since $\mathbb{H}^3/G$ has bounded geometry, so does $\mathbb{H}^3/G'$. Hence, by assumption, there exists a quasi-arc in $\Lambda(G')$ connecting the fixed points of g. Since G' has finite index in G, $\Lambda(G) = \Lambda(G')$. Thus Theorem 1.2 holds for all $g \in G$.

3.3. Proof of Theorem 1.1

We begin with the following lemma.

Lemma 3.1. *A point $x \in \Lambda(G)$ is an interior point if and only if its preimage via π in Proposition 2.6 contains an end point of a leaf of $\tilde{\Phi}_v$.*

Proof. This is essentially proved by W.Abikoff in [Abi88]. Let φ be the Riemann mapping from $\mathbb{H}^2$ to the region of discontinuity of G and let H be the fuchsian equivalent of G via φ. By virtue of Theorem 3 in [Abi88], there exists a geodesic lamination

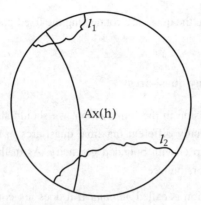

Figure 1: Axis intersecting each leaf

λ in $\mathbb{H}^2$ which is invariant under the action of H so that the interior of $\Lambda(G)$ is homeomorphic to the leaf space of λ (in the sense of R.L.Moore). More precisely, λ is constructed by taking a union of geodesics defined by either the connecting preimages $\varphi^{-1}(x)$ of $x \in \Lambda(G)$ if $^{\#}\varphi^{-1}(x) = 2$, or taking boundary edges of the convex hull of $\varphi^{-1}(x)$ of $x \in \Lambda(G)$ in $\mathbb{H}^2$ if $^{\#}\{\varphi^{-1}(x)\} \geq 3$ (see p.38–p.42 of [Abi88]).

Let l denote a leaf of λ and let $\{x,y\}$ denote the end points of l. Since $\pi|_{\mathbb{R}} = \varphi|_{\mathbb{R}}$ (see [Min94]), we have $\pi(x) = \pi(y)$ by the construction of λ. Therefore, by (3) of Proposition 2.6, we conclude that λ is contained in the lift of λ_e. On the other hand, since λ_e is minimal (see S3.4 for the definition), λ should agree with the lift of λ_e. Since $\tilde{\Phi}_v$ is associated with the lift of λ_e, we have proved the lemma. □

Corollary 3.2. *Let $\tilde{\lambda}$ be the lift of the support λ of a measured lamination in the region of discontinuity of G. If $\lambda \neq \lambda_e$, a landing point of an end of any leaf of $\tilde{\lambda}$ is an end point of $\Lambda(G)$.*

Proof. This follows from Lemmas 2.2, 3.1 and the minimality of λ_e. □

Proof of Theorem 1.1. It suffices to prove the theorem in the case where G is torsion free by a reduction similar to that in Theorem 1.2.

Take two interior points x,y in $\Lambda(G)$. By Lemma 3.1, there exist leaves l_1 and l_2 of $\tilde{\Phi}_v$ in Δ_∞ so that $\pi(l_1) = x$ and $\pi(l_2) = y$. Since non-singular leaves are dense in **B**, we may assume that each of l_1 and l_2 are non-singular. It is known that fixed points of loxodromic elements of H_0 are dense in $\partial\mathbf{B} \times \partial\mathbf{B}$ (cf. [GHe55]). Hence there exists an $h \in H_0$ so that l_j separates one of its fixed points from the other fixed point and the other leaf l_{3-j} for $j = 1,2$ (see Figure 1). Thus the axis of h passes through l_1 and l_2, and hence x and y lie on the quasi-arc connecting the fixed points of the element in G corresponding to h.

3.4. Uncountably many quasi-arcs

To complete the discussion in the Introduction, we should show that there exist uncountably many essentially different maximal quasi-arcs in the limit set $\Lambda(G)$ of a singly degenerate group G with bounded geometry. As in the previous section, we may assume that G is torsion free.

A geodesic lamination is called *minimal* if it does not contain a proper sublamination. A geodesic lamination is said to be *uniquely ergodic* if it supports only one transverse measure (up to scaling). We say that a measured lamination is minimal or uniquely ergodic if so is its support respectively. It is easy to see that unique ergodic measured lamination is minimal, because any geodesic lamination is decomposed into finitely many minimal laminations (see [Lev83]). It is known that there are uncountably many uniquely ergodic measured laminations on a compact hyperbolic surface and hence the same holds for minimal ones (cf. [KMS86]), This implies that, on any hyperbolic surface there are uncountably many minimal geodesic laminations each of which admits unique transverse measure. Therefore, we have the following proposition:

Proposition 3.3. *Let μ_1 and μ_2 be minimal measured laminations on S and let $l_1 \neq l_2$ be lifts of leaves of μ_1 and μ_2. Suppose that the support of μ_i is different from the ending lamination λ_e of G. Let ξ_i be a quasi-arc connecting the landing points of the ends of l_i in $\Lambda(G)$ $(i = 1, 2)$. Then $\xi_1 \neq \xi_2$ and the one does not contain the other.*

Proof. We first note the following:

(1) the points corresponding to the two end points of each l_i in $\Lambda(G)$ do not coincide since any half leaf of minimal geodesic lamination is dense in the lamination (cf. Lemma 2.4 of [Min94]), and the ending lamination is also minimal by Proposition 2.6, and

(2) each ξ_i is a maximal arc in $\Lambda(G)$, since each ξ_i connects two end points by Lemma 3.1.

Hence one of the quasi-arcs cannot contain the other. Furthermore, since each of the supports of μ_i is minimal, by hypothesis, no landing point of l_i coincides with that of any leaf of ending lamination. Hence for any landing point of l_1 or l_2, there is a sequence of leaves of $\tilde{\Phi}_\nu$ nesting down on it (see the argument of the last paragraph

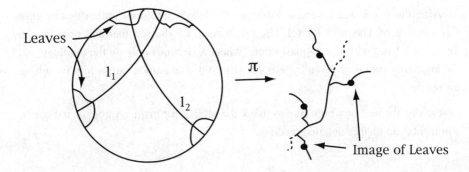

Figure 2: A landing point is separated from the other three.

in p.575 of [Min94]). Therefore there exist four leaves of $\tilde{\Phi}_\nu$ which separate one of the landing points of l_1 and l_2 from the other three points. Thus by Proposition 2.6, $\xi_1 \neq \xi_2$ (see Figure 2). □

Remarks 3.4.

1. The uncountability may also be proved by using the fact that there are uncountably many (minimal) laminations on a punctured (or a once holed) torus lying the surface S. The latter statement is proved using the fact that these laminations are in one-to-one correspondence with the set of irrational numbers.

2. The argument of this paper can be applied to the case of doubly degenerate groups with bounded geometry. In this case, the boundary of the lift of any leaf of the foliations (say $l \times \mathbb{R}$) in the given hyperbolic 3-manifold as in Section 2 is a quasi-circle in the Riemann sphere.

Acknowledgments   The author thanks the organizers for the invitation to give a talk. He would like to express his hearty gratitude to Professor Caroline Series for her interesting mathematical discussions and for her constant encouragement and advice, and to the University of Warwick for its hospitality during the author's visit from August 2001 to November 2001. This work was done during this visit. He is also very grateful to Professor Yair.N.Minsky for constant encouragements and advice during his visit SUNY from May 2000 to February 2001, when the essential part of this work was done.

The author also thanks Dr.Yusuke Okuyama for his useful comments from the complex dynamical point of view. Professor Toshihiro Nakanishi told the author the reference [GHe55] on the density of fixed points. He would like to thank him for his help.

After this paper was accepted, Professor B. Bowditch told the author that he gave a refinement of Theorem 1.1 (cf. [Bow]). Indeed, he showed that any quasi-arc in Theorem 1.1 is a part of a K-quasi-circle, where K depends only on the topology and the injectivity radius of given hyperbolic manifold. The author thanks him for telling his result.

Finally, the author would like to thank the referee for his/her valuable and useful comments and his/her detailed reading.

References

[Abi88]   W. Abikoff (1988). Kleinian groups – geometrically finite and geometrically perverse, *Contemp. Math.* **74**, 1–50.

[Bin49]   R.H. Bing (1949). Partitioning a set, *Bull. Amer. Math. Soc.* **55**, 1101–1110.

[BJ97]   C. Bishop and P. Jones (1997). Hausdorff dimension and Kleinian groups, *Acta Math.* **179**, 1–39.

[Bow]   B. Bowditch (2003). Hausdorff dimension and dendritic limit sets, *preprint*.

[CT89]   J.W. Cannon and W.P. Thurston (1989). Group invariant Peano curves, *preprint*.

[GH90]   E. Ghys and P. de la Harpe (eds.) (1990). *Sur les Groupes Hyperboliques d'après Mikhael Gromov*, Birkhäuser.

[GHe55]   W. Gottschalk and G. Hedlund (1955). *Topological dynamics*, Amer. Math. Soc..

[GM91]   F. Gardiner and H. Masur (1991). Extremal length geometry of Teichmüller space, *Complex Variables Theory Appl.* **16**, 209–237.

[KMS86]   S. Kerckhoff, H. Masur and J. Smillie (1986). Ergodicity of billiard flows and quadratic differentials, *Ann. Math.* **124**, no. 2, 293–311.

[Lev83]   G. Levitt (1983). Foliations and laminations on hyperbolic surfaces, *Topology* **22**, 119–135.

[Mak75]   B. Maskit (1975). A remarks on degenerate groups, *Math. Scand.* **36**, 17–20.

[Mat01]   K. Matsuzaki (2001). Locally connected, tree-like invariant continua under Kleinian groups, *Surikaiseki-kokyuroku* **1223**, 31–32.

[McM94] C. McMullen (1994). Complex dynamics and Renormalization, *Ann. Math. Studies* **135**, Princeton University Press.

[McM96] C. McMullen (1996). Renormalization and 3-manifolds which Fiber over the circle, *Ann. Math. Studies* **142**, Princeton University Press.

[McM98] C. McMullen (1998). Self-similarity of Siegel disks and the Hausdorff dimension of Julia sets, *Acta Math.* **120**, 691–721.

[McM99] C. McMullen (1999). Hausdorff dimension and conformal dynamics I: Kleinian groups and strong limits, *J. Diff. Geom.* **51**, 471–515.

[McM00] C. McMullen (2000). Hausdorff dimension and conformal dynamics II: Geometrically finite rational maps, *Comment. Math. Helv.* **75**, 535–593.

[Min94] Y. Minsky (1994). On Rigidity, Limit Sets and End Invariants of Hyperbolic 3-Manifolds, *J. Amer. Math. Soc.* **7**, (3), 539–588.

[Min01] Y. Minsky (2001). Bounded geometry for Kleinian groups, *Invent. Math.* **146**, 143-192.

[MNTU] S. Morosawa, Y. Nishimura, M. Taniguchi and T. Ueda (2000). *Holomorphic dynamics (Cambridge studies in advanced mathematics 66)*, Cambridge University Press.

[Moi49] E.E. Moise (1949). Grelle decomposition and convexification theorems for compact metric locally connected continua, *Bull. Amer. Math. Soc.* **55**, 1111–1121.

[MT98] K. Matsuzaki and M. Taniguchi (1998). *Hyperbolic manifolds and Kleinian groups*, Oxford Mathematical Monograph.

[Sco78] P. Scott (1978). Subgroups of surface groups are almost geometric, *J. Lond. Math. Soc.* **117**, 555–565.

[Sul80] D. Sullivan (1980). Growth of positive harmonic functions and Kleinian group limit sets of zero planar measure and Hausdorff dimension two. In *Geometry Symposium, Utrecht 1980 (Lec. Notes Math.* **894***)*, 127–144.

[Sul85] D. Sullivan (1985). Quasiconformal homeomorphisms and dynamics I: Solution of the Fatou-Julia problem on wandering domains, *Ann. Math.* **122**, 401–418.

[Thu82] W. Thurston (1982). Three-dimensional manifolds, Kleinian groups and hyperbolic geometry, *Bull. Amer. Math. Soc.* **6**, 357–381.

144   *H. Miyachi*

Hideki Miyachi

Department of Mathematics
Osaka City University
Sugimoto, Sumiyoshi-ku
Osaka, 558-8585
Japan

miyaji@sci.osaka-cu.ac.jp

AMS Classification: 28A99, 30F40, 37F50, 51N15

Keywords: Kleinian group, fractals

Kleinian Groups and Hyperbolic 3-Manifolds   Y. Komori, V. Markovic & C. Series (Eds.)
Lond. Math. Soc. Lec. Notes **299**, 145–163   Cambridge Univ. Press, 2003

On hyperbolic and spherical volumes for knot and link cone-manifolds

Alexander D. Mednykh[1]

Abstract

In the present paper links and knots are considered as singular subsets of geometric cone–manifolds with the three-sphere as an underlying space. Trigonometrical identities between lengths of singular components and cone angles for the figure eight knot, Whitehead link and Borromean rings are obtained. This gives a possibility to express the lengths in terms of cone angles. Then the Schläfli formula applies to find explicit formulae for hyperbolic and spherical volumes of these cone-manifolds.

1. Introduction

In 1975 R. Riley [Ril79] found examples of hyperbolic structures on some knot and link complements in the three-sphere. Later, in the spring of 1977, W. P. Thurston announced an existence theorem for Riemannian metrics of constant negative curvature on 3-manifolds. In particular, it turned out that knot complement of a simple knot (excepting torical and satellite) admits a hyperbolic structure. This fact allowed to consider knot theory as a part of geometry and Kleinian group theory. Starting from Alexander's works polynomial invariants became a convenient instrument for knot investigation. A lot of different kinds of such polynomials were investigated in the last twenty years. Among these there are Jones-, Kaufmann-, HOMFLY-, A–polynomials and others ([Kau88], [CCG+94], [HLM95a]). This relates the knot theory with algebra and algebraic geometry. Algebraic technique is used to find the most important geometrical characteristics of knots such as a volume, length of shortest geodesics and others.

This paper is a part of talk given by author on the conference " Kleinian Groups and Hyperbolic 3-manifolds" held at Warwick University, September 11-15th 2001. Mostly, it contains a survey of recent results obtained by author and his collaborators, but some new results are also given.

---
[1] Supported by the RFBR (grant 99-01-00630, 00-15-96165) and by DFG.

In the present paper links and knots are considered as singular subsets of the three-sphere endowed by Riemannian metric of constant curvature (negative, positive, or zero). More precisely, our aim is to investigate the structure of geometrical cone-manifolds whose underlying space is the three-sphere and the singular set is knot or link under consideration. Trigonometrical identities between lengths of singular components and cone angles for the figure eight knot, Whitehead link and Borromean rings are obtained. This gives a possibility to express the lengths in terms of cone angles. Then the Schläfli formula applies to find explicitly hyperbolic and spherical volumes for these cone-manifolds.

Section 2 contains a list of trigonometrical identities (Sine, Cosine, and Tangent rules) relating the lengths of singular geodesics of geometrical cone-manifolds with their cone–angles. Cone-manifolds are supposed to be hyperbolic, spherical, or Euclidean. Similar results are known for the right–angled hexagons in the hyperbolic 3-space which can be considered as triangles with complex lengths and angles [Fen89]. Related results can be also obtained for a class of knotted graphs. For example, they take place for the rational knots with bridges through their tunnels.

Section 3 is devoted to explicit calculation of volumes of the three above mentioned cone-manifolds in hyperbolic and spherical geometries. Partially, these results were obtained earlier in [HLM95b], [Koj98], [DM01], [MR], [MV95], and [MV01].

2. Trigonometrical identities for knots and links

2.1. Cone–manifolds, complex distances and lengths

We start with the definition of cone–manifold modeled in hyperbolic, spherical or Euclidean structure.

Definition 2.1. A 3–dimensional *hyperbolic cone–manifold* is a Riemannian 3–dimensional manifold of constant negative sectional curvature with cone-type singularity along simple closed geodesics. To each component of singular set we associate a real number $n \geq 1$ such that the cone-angle around the component is $\alpha = 2\pi/n$. The concept of the hyperbolic cone-manifold generalizes the hyperbolic manifold which appears in the partial case when all cone-angles are 2π. The hyperbolic cone-manifold is also a generalization of the hyperbolic 3–orbifold which arises when all associated numbers n are integers. Euclidean and spherical cone–manifolds are defined similarly.

In the present paper hyperbolic, spherical or Euclidean cone-manifolds C are considered whose underlying space is the three-dimensional sphere and the singular set

$\Sigma = \Sigma^1 \cup \Sigma^2 \cup \ldots \cup \Sigma^k$ is a link consisting of components $\Sigma^j = \Sigma_{\alpha_j}$, $j = 1, 2, \ldots, k$ with cone-angles $\alpha_1, \ldots, \alpha_k$ respectively.

Recall a few well-known facts from the hyperbolic geometry.

Let $\mathbb{H}^3 = \{(z, \xi) \in \mathbb{C} \times \mathbb{R} : \xi > 0\}$ be the upper half model of the 3 -dimensional hyperbolic space endowed by the Riemannian metric $ds^2 = \dfrac{dz\,d\bar{z} + d\xi^2}{\xi^2}$. We identify the group of orientation preserving isometries of $\mathbb{H}^3$ with the group $\mathrm{PSL}(2, \mathbb{C})$ consisting of linear fractional transformations

$$A : z \in \mathbb{C} \to \frac{az + b}{cz + d}.$$

By the canonical procedure the linear transformation A can be uniquely extended to the isometry of $\mathbb{H}^3$. We prefer to deal with the matrix $\widetilde{A} = \begin{pmatrix} a & b \\ c & d \end{pmatrix} \in \mathrm{SL}(2, \mathbb{C})$ rather than the element $A \in \mathrm{PSL}(2, \mathbb{C})$. The matrix $\widetilde{A}$ is uniquely determined by the element A up to a sign. If there will be no confusion we shall use the same letter A for both A and $\widetilde{A}$.

Let C be a hyperbolic cone–manifold with the singular set Σ. Then C defines a non-singular but incomplete hyperbolic manifold $N = C - \Sigma$. Denote by Φ the fundamental group of the manifold N.

The hyperbolic structure of N defines, up to conjugation in $\mathrm{PSL}(2, \mathbb{C})$, a holonomy representation

$$\hat{h} : \Phi \to \mathrm{PSL}(2, \mathbb{C}).$$

It is shown in [Zho99] (see also [Cul86]) that the representation $\hat{h}$ can be lifted to a representation $h : \Phi \to \mathrm{SL}(2, \mathbb{C})$. The lifting h will be also called holonomy representation. Chose an orientation on the link $\Sigma = \Sigma^1 \cup \Sigma^2 \cup \ldots \cup \Sigma^k$ and fix a meridian-longitude pair $\{m_j, l_j\}$ for each component $\Sigma_j = \Sigma_{\alpha_j}$. Then the matrices $M_j = h(m_j)$ and $L_j = h(l_j)$ satisfy the following properties:

$$\mathrm{tr}(M_j) = 2\cos(\alpha_j / 2), \quad M_j L_j = L_j M_j, \ j = 1, 2, \ldots, k.$$

Definition 2.2. A *complex length* γ_j of the singular component Σ^j of the cone-manifold C is defined as displacement of the isometry L_j of $\mathbb{H}^3$, where $L_j = h(l_j)$ is represented by the longitude l_j of Σ^j.

Immediately from the definition we get [Fen89, p.46]

$$2\cosh \gamma_j = \mathrm{tr}\,(L_j^2) \tag{2.1}$$

We note [BZ85, p.38] that the meridian-longitude pair $\{m_j, l_j\}$ of the oriented link is uniquely determined up to a common conjugating element of the group Φ. Hence, the complex length $\gamma_j = l_j + i\,\varphi_j$ is uniquely determined up to a sign and ($\mod 2\pi i$) by the above definition.

We need two conventions to choose correctly real and imaginary parts of γ_j. The first convention is the following. Since Σ^j does not shrink to a point, $l_j \neq 0$. Hence, we choose γ_j in such a way that $l_j = \operatorname{Re}\gamma_j > 0$. The second convention is concerned with the imaginary part $\varphi_j = \operatorname{Im}\gamma_j$. We want to choose φ_j such that the following identity holds

$$\cosh\frac{\gamma_j}{2} = -\frac{1}{2}\operatorname{tr}(L_j) \qquad (2.2)$$

By virtue of identity $\operatorname{tr}(L_j)^2 - 2 = \operatorname{tr}(L_j^2)$ equality (2.1) is a consequence of (2.2). The converse, in general, is true only up to a sign. Under the second convention (2.1) and (2.2) are equivalent. The two above conventions lead to convenient analytic formulas for calculation of γ_j and l_j. More precisely, there are simple relations between these numbers and eigenvalues of matrix L_j. Recall that $\det L_j = 1$. Since matrix L_j is loxodromic it has two eigenvalues f_j and $1/f_j$. We choose f_j so that $|f_j| > 1$. The case $|f_j| = 1$ is impossible because in this case the matrix L_j is elliptic and $l_j = 0$. Hence

$$f_j = -e^{\frac{\gamma_j}{2}}, \ |f_j| = e^{\frac{l_j}{2}}. \qquad (2.3)$$

2.2. Whitehead link cone–manifold

Denote by $W(\alpha,\beta)$ the cone-manifold whose underlying space is the 3-sphere and whose singular set consists of two components of the Whitehead link with cone angles $\alpha = 2\pi/m$ and $\beta = 2\pi/n$ (see Fig.1). It follows from Thurston's theorem that $W(\alpha,\beta)$ admits a hyperbolic structure for all sufficiently small α and β. The region of hyperbolicity of $W(\alpha,\beta)$ was investigated in [HLM95a] and [KM99]. In particular, this cone–manifold is hyperbolic for $m,n > 2.507$. The following theorems have been obtained in [Med00].

Theorem 2.3. (The Tangent Rule) *Suppose that cone–manifold $W(\alpha,\beta)$ is hyperbolic. Denote by γ_α and γ_β complex lengths of the singular geodesics of $W(\alpha,\beta)$ with cone angles α and β respectively. Then*

$$\frac{\tanh\frac{\gamma_\alpha}{4}}{\tanh\frac{\gamma_\beta}{4}} = \frac{\tan\frac{\alpha}{2}}{\tan\frac{\beta}{2}}.$$

Theorem 2.4. (The Sine Rule) *Let $\gamma_\alpha = l_\alpha + i\,\varphi_\alpha$ (resp. γ_β) be a complex length of the singular geodesic of a hyperbolic cone-manifold $W(\alpha,\beta)$ with cone angle α (resp. β). Then*

$$\frac{\sin\frac{\varphi_\alpha}{2}}{\sinh\frac{l_\alpha}{2}} = \frac{\sin\frac{\varphi_\beta}{2}}{\sinh\frac{l_\beta}{2}}.$$

Moreover, it was shown in [M, p. 300] that the following relations hold:

$$iB\coth\frac{\gamma_\alpha}{4} = iA\coth\frac{\gamma_\beta}{4} = u, \tag{2.4}$$

where $A = \cot\frac{\alpha}{2}$, $B = \cot\frac{\beta}{2}$, and u, $\text{Im}(u) > 0$, is a root of the cubic equation

$$u^3 - ABu^2 + \frac{1}{2}(A^2B^2 + A^2 + B^2 - 1)u + AB = 0 \tag{2.5}$$

This gives us a practical way to calculate the real length l_α it terms of cone angles α and β. Indeed, we have from (2.4) for a suitable choice of analytical branches

$$l_\alpha = \frac{\gamma_\alpha}{2} + \frac{\overline{\gamma_\alpha}}{2} = 2i\arctan\frac{\overline{u}}{B} - 2i\arctan\frac{u}{B} = 2i\arctan\frac{A}{z} - 2i\arctan\frac{A}{\overline{z}},$$

where $z = \dfrac{AB}{\overline{u}}$, $\text{Im}(z) > 0$ satisfy the following equation

$$z^3 + \frac{1}{2}(A^2B^2 + A^2 + B^2 - 1)z^2 - A^2B^2z + A^2B^2 = 0 \tag{2.6}$$

Thus, we have proved the following

Proposition 2.5. *Let $W(\alpha,\beta)$ be a hyperbolic Whitehead link cone–manifold. Denote by l_α and l_β real lengths of the singular geodesics of $W(\alpha,\beta)$ with cone angles α and β respectively. Then*

$$l_\alpha = 2i\arctan\frac{A}{z} - 2i\arctan\frac{A}{\overline{z}},$$

$$l_\beta = 2i\arctan\frac{B}{z} - 2i\arctan\frac{B}{\overline{z}},$$

where z, $\text{Im}(z) > 0$ is a root of equation (2.6), $A = \cot\frac{\alpha}{2}$, and $B = \cot\frac{\beta}{2}$.

In Section 3 we will apply this result to calculate the volume of $W(\alpha,\beta)$ via Schläfli formula.

Euclidean versions of Theorems 2.3 and 2.4 were obtained by R.N. Shmatkov [Sh]. The similar results take a place also for spherical cone–manifold $W(\alpha,\beta)$.

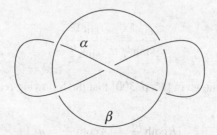

Figure 1: The Whitehead link cone–manifold $W(\alpha, \beta)$.

2.3. The Borromean cone–manifold

In this subsection we investigate geometric properties of a cone–manifold $B(\alpha, \beta, \gamma)$ with singular set the Borromean rings (Fig. 2). The cone angles of three components of the singular set are α, β, γ. As above, the corresponding lengths of the singular set components will be denoted by l_α, l_β, and l_γ.

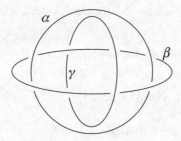

Figure 2: The Borromean cone–manifold $B(\alpha, \beta, \gamma)$.

It is well-known fact that $B(\alpha, \beta, \gamma)$ can be obtained by gluing together eight copies of the Lambert cube $Q(\alpha/2, \beta/2, \gamma/2)$ with essential dihedral angles $\alpha/2, \beta/2, \gamma/2$ (Fig. 3). See [Thu79] and [HLM92] for details. The Lambert cube $Q(\alpha/2, \beta/2, \gamma/2)$ (and hence $B(\alpha, \beta, \gamma)$) is hyperbolic ([Thu79], [HLM92]) if $0 \leq \alpha, \beta, \gamma < \pi$, Euclidean if $\alpha = \beta = \gamma = \pi$, and spherical [Dia99] if $\pi < \alpha, \beta, \gamma < 2\pi$.

Moreover, if $L_\alpha, L_\beta, L_\gamma$ denote the edge lengths of $Q(\alpha/2, \beta/2, \gamma/2)$ with dihedral angles $\alpha/2, \beta/2, \gamma/2$ we get

$$L_\alpha = \frac{l_\alpha}{4}, \quad L_\beta = \frac{l_\beta}{4}, \quad L_\gamma = \frac{l_\gamma}{4}. \tag{2.7}$$

As in the case of cone–manifolds $W(\alpha, \beta)$ there are simple trigonometrical identities relating the lengths $l_\alpha, l_\beta, l_\gamma$ of $B(\alpha, \beta, \gamma)$ with its cone angles α, β, γ. We start

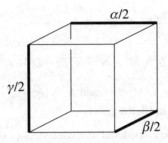

Figure 3: The Lambert cube $Q(\alpha/2, \beta/2, \gamma/2)$.

with the following

Theorem 2.6. (The Tangent Rule). *Let $B(\alpha, \beta, \gamma)$ be a hyperbolic Borromean cone–manifold with cone angles $0 < \alpha, \beta, \gamma < \pi$ and the singular geodesic lengths $l_\alpha, l_\beta, l_\gamma$. Then*

$$\frac{\tan \frac{\alpha}{2}}{\tanh \frac{l_\alpha}{4}} = \frac{\tan \frac{\beta}{2}}{\tanh \frac{l_\beta}{4}} = \frac{\tan \frac{\gamma}{2}}{\tanh \frac{l_\gamma}{4}} = T,$$

where T is a positive number defined by $T^2 = K + \sqrt{K^2 + L^2 M^2 N^2}$, $L = \tan \frac{\alpha}{2}$, $M = \tan \frac{\beta}{2}$, $N = \tan \frac{\gamma}{2}$, and $K = (L^2 + M^2 + N^2 + 1)/2$.

Proof. We prefer to deal with the Lambert cube $Q(\alpha/2, \beta/2, \gamma/2)$ rather then cone–manifold $B(\alpha, \beta, \gamma)$. It follows from the result of [Kel89] that the edge lengths L_α, L_β and L_γ are related with its angles by

$$\frac{\tan \frac{\alpha}{2}}{\tanh L_\alpha} = \frac{\tan \frac{\beta}{2}}{\tanh L_\beta} = \frac{\tan \frac{\gamma}{2}}{\tanh L_\gamma} = T, \tag{2.8}$$

where $T = \tan \theta$ for some angle θ such that $\alpha, \beta, \gamma \leq 2\theta \leq \pi$. The simple proof of this formula by means of Gram matrix techniques can be find also in [V]. The following equation for T was obtained in ([K], p.564, eq. (II)) and ([HLM1], eq. (A.2)) in slightly different terms

$$T^2 = \frac{T^2 - L^2}{1 + L^2} \frac{T^2 - M^2}{1 + M^2} \frac{T^2 - N^2}{1 + N^2}, \tag{2.9}$$

The last equation is equivalent to

$$(T^2 + 1)(T^4 - (L^2 + M^2 + N^2 + 1)T^2 - L^2 M^2 N^2) = 0.$$

Since T is a positive number we get

$$T^4 - (L^2 + M^2 + N^2 + 1)T^2 - L^2 M^2 N^2 = 0. \tag{2.10}$$

Hence $T^2 = K + \sqrt{K^2 + L^2 M^2 N^2}$, and $K = (L^2 + M^2 + N^2 + 1)/2$. Taking into account (2.7) and (2.8) we finish the proof.

The next theorem can be considered as a consequence of the Tangent Rule .

Theorem 2.7. (The Sine-Cosine Rule). *Let $B(\alpha, \beta, \gamma)$ be a hyperbolic Borromean cone–manifold with cone angles $0 < \alpha, \beta, \gamma < \pi$ and the singular geodesic lengths $l_\alpha, l_\beta, l_\gamma$. Then*

$$\frac{\sin \frac{\alpha}{2}}{\sinh \frac{l_\alpha}{4}} \frac{\sin \frac{\beta}{2}}{\sinh \frac{l_\beta}{4}} \frac{\cos \frac{\gamma}{2}}{\cosh \frac{l_\gamma}{4}} = 1.$$

Proof. We rewrite the statement of the Tangent Rule in the form

$$\sinh^2 L_\alpha = \frac{L^2}{T^2 - L^2}, \quad \sinh^2 L_\beta = \frac{M^2}{T^2 - M^2}, \quad \cosh^2 L_\gamma = \frac{T^2}{T^2 - N^2}. \tag{2.11}$$

We get also

$$\sin^2 \frac{\alpha}{2} = \frac{L^2}{1 + L^2}, \quad \sin^2 \frac{\beta}{2} = \frac{M^2}{1 + M^2}, \quad \cos^2 \frac{\gamma}{2} = \frac{1}{1 + N^2}. \tag{2.12}$$

By virtue of (2.9) we have from (2.11) and (2.12)

$$\frac{\sin^2 \frac{\alpha}{2}}{\sinh^2 L_\alpha} \frac{\sin^2 \frac{\beta}{2}}{\sinh^2 L_\beta} \frac{\sin^2 \frac{\gamma}{2}}{\sinh^2 L_\gamma} = \frac{T^2 - L^2}{1 + L^2} \frac{T^2 - M^2}{1 + M^2} \frac{T^2 - N^2}{1 + N^2} \frac{1}{T^2} = 1.$$

By taking the square root we obtain the statement of the theorem.

By similar arguments the following spherical analogs of the above two theorems can be obtained [DM].

Theorem 2.8. (The Tangent Rule). *Let $B(\alpha, \beta, \gamma)$ be a spherical Borromean cone–manifold with cone angles $\pi < \alpha, \beta, \gamma < 2\pi$ and the singular geodesic lengths $l_\alpha, l_\beta, l_\gamma$. Then*

$$\frac{\tan \frac{\alpha}{2}}{\tan \frac{l_\alpha}{4}} = \frac{\tan \frac{\beta}{2}}{\tan \frac{l_\beta}{4}} = \frac{\tan \frac{\gamma}{2}}{\tan \frac{l_\gamma}{4}} = T,$$

where T is a negative number defined by $T^2 = -K + \sqrt{K^2 + L^2 M^2 N^2}$, $L = \tan \frac{\alpha}{2}$, $M = \tan \frac{\beta}{2}$, $N = \tan \frac{\gamma}{2}$, and $K = (L^2 + M^2 + N^2 + 1)/2$.

Theorem 2.9. (The Sine-Cosine Rule). *Let* $B(\alpha,\beta,\gamma)$ *be a spherical Borromean cone–manifold with cone angles* $\pi < \alpha,\beta,\gamma < 2\pi$ *and the singular geodesic lengths* $l_\alpha, l_\beta, l_\gamma$. *Then*

$$\frac{\sin\frac{\alpha}{2}\;\sin\frac{\beta}{2}\;\cos\frac{\gamma}{2}}{\sin\frac{l_\alpha}{4}\;\sin\frac{l_\beta}{4}\;\cos\frac{l_\gamma}{4}} = -1$$

Remark 2.10. Up to cyclic permutation of angles α, β, and γ, the Sine-Cosine Rule contains three independent relations. They are sufficient to determine l_α, l_β, and l_γ through α, β, and γ. In particular, the Tangent Rule is a consequence of these three relations.

3. Explicit volume calculation

3.1. The Schläfli formula

In this section we will obtain explicit formulas for volume of some knot and link cone–manifolds in the hyperbolic and spherical geometries. In the case of complete hyperbolic structure on the simplest knot and link complements such formulas in terms of Lobachevsky function are well-known and widely represented in [Thu79]. In general situation, a hyperbolic cone–manifold can be obtained by completion of non-complete hyperbolic structure on a suitable knot or link complement. If the cone–manifold is compact explicit formulas are know just in a few cases ([Hod], [HLM95b], [MV95], [Koj98]). In all these cases the starting point for the volume calculation is the Schläfli formula (see, for example [Hod]).

Theorem 3.1. (The Schläfli volume formula). *Suppose that C_t is a smooth 1–parameter family of (curvature K) cone–manifold structures on a n-manifold, with singular locus Σ of a fixed topological type. Then the derivative of volume of C_t satisfies*

$$(n-1)KdV(C_t) = \sum_\sigma V_{n-2}(\sigma)d\theta(\sigma)$$

where the sum is over all components σ of the singular locus Σ, and $\theta(\sigma)$ is the cone angle along σ.

In the present paper we will deal mostly with three-dimensional cone–manifold structures of negative constant curvature $K = -1$, or positive constant curvature $K = 1$. The Schläfli formula in this case reduces to

$$KdV = \frac{1}{2}\sum_i l_{\theta_i} d\theta_i,$$ (3.1)

where the sum is taken over all components of the singular set Σ with lengths l_{θ_i} and cone angles θ_i.

We want to obtain the volume formulae for cone–manifolds $W(\alpha,\beta)$, $B(\alpha,\beta,\gamma)$ described in the above section and, also, for the figure eight cone–manifold $4_1(\alpha)$.

3.2. Volume of the Whitehead link cone–manifold

First of all we consider the case of the hyperbolic Whitehead link with one complete cusp.

Theorem 3.2. *Let $W(\alpha,0)$ be a hyperbolic Whitehead link cone–manifold with a complete hyperbolic structure on one cusp and cone angle α, $0 \le \alpha < \pi$ on the another. Then the volume of $W(\alpha,0)$ is given by the formula*

$$Vol\,W(\alpha,0) = 2\int_\alpha^\pi \operatorname{arcsinh}\left(\sin\frac{t}{2}\right)dt.$$

Proof. By [KM99] cone–manifold $W(\alpha,0)$ is hyperbolic for all $0 \le \alpha < \pi$. Denote by V_α the hyperbolic volume of $W(\alpha,0)$. By Schläfli formula we have $dV_\alpha = -\frac{1}{2}l_\alpha d\alpha$. By calculation produced in [Med00] we obtain

$$\cosh l_\alpha = \frac{M^4 + 10M^2 + 17}{(M^2+1)^2},$$ (3.2)

where $M = \cot\frac{\alpha}{2}$. Simplifying (3.2) we get $\cosh l_\alpha = 8 - 8\cos\alpha + \cos 2\alpha$ and hence

$$\sinh\frac{l_\alpha}{4} = \sin\frac{\alpha}{2}.$$ (3.3)

By integrating the Schläfli formula we have

$$V_\alpha = -2\int_\theta^\alpha \operatorname{arcsinh}\left(\sin\frac{t}{2}\right)dt + V_\theta,$$ (3.4)

for an arbitrary θ, $0 \le \theta < \pi$.

It was noted by Caroline Series (Warwick, February 2001) that the fundamental polyhedron $\mathscr{C}(\alpha)$ (Fig. 4) for cone-manifold $W(\alpha,0)$ in $\mathbb{H}^3$ coincides with a convex Nielsen hull for two generated quasifuchsian group described in Section 6 of [PS]. Two dihedral angles of the polyhedron are equal to α and all others are equal to $\frac{\pi}{2}$.

$\mathscr{C}(\alpha)$ has four proper vertices and four cusps shown by small circles. In particular, it follows from this consideration that $V_\alpha = Vol(\mathscr{C}(\alpha)) \to 0$ as $\alpha \to \pi$. Going over to the limit we immediately get from (3.4) the statement of the theorem.

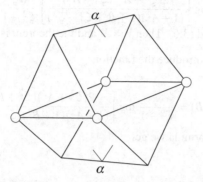

Figure 4: The fundamental polyhedron $\mathscr{C}(\alpha)$ for cone–manifold $W(\alpha,0)$.

In the case of closed cone–manifold $W(\alpha,\beta)$ the volume function becomes more complicated and can be expressed in terms of roots of a cubic equation. More precisely, the following theorem takes a place [MV01].

Theorem 3.3. *Let $W(\alpha,\beta)$ be a hyperbolic Whitehead link cone–manifold with cone angles α and β. Then the volume of $W(\alpha,\beta)$ is given by the formula*

$$Vol\, W(\alpha,\beta) = i \int_{\zeta_1}^{\zeta_2} \log\left[\frac{2(\zeta^2+A^2)(\zeta^2+B^2)}{(1+A^2)(1+B^2)(\zeta^2-\zeta^3)} \right] \frac{d\zeta}{\zeta^2-1},$$

where $A = \cot\frac{\alpha}{2}$, $B = \cot\frac{\beta}{2}$, $\zeta_1 = \bar{z}$, $\zeta_2 = z$, $\mathrm{Im}(z) > 0$ and z is a root of the cubic equation

$$z^3 + \frac{1}{2}(A^2B^2 + A^2 + B^2 - 1)z^2 - A^2B^2z + A^2B^2 = 0.$$

Proof of the theorem is based on the following arguments. Denote by $V = Vol\, W(\alpha,\beta)$ the hyperbolic volume of $W(\alpha,\beta)$. Then by virtue of the Schläfli formula we have

$$\frac{\partial V}{\partial \alpha} = -\frac{l_\alpha}{2}, \quad \frac{\partial V}{\partial \beta} = -\frac{l_\beta}{2}, \tag{3.5}$$

where l_α and l_α are lengths of singular geodesics corresponding to cone angles α and β respectively. Moreover, by Theorem 3.2 we obtain

$$V \to 0 \text{ as } \alpha \to \pi \text{ and } \beta \to 0. \qquad (3.6)$$

We set $\tilde{V} = i \int_{\zeta_1}^{\zeta_2} \log \left[\dfrac{2(\zeta^2 + A^2)(\zeta^2 + B^2)}{(1 + A^2)(1 + B^2)(\zeta^2 - \zeta^3)} \right] \dfrac{d\zeta}{\zeta^2 - 1}$ and show that $\tilde{V}$ satisfies conditions (3.5) and (3.6). Then $\tilde{V} = V$ and the theorem is proven.

To verify (3.5) we introduce the function

$$F(\zeta, A, B) = \frac{i}{\zeta^2 - 1} \log \left[\frac{2(\zeta^2 + A^2)(\zeta^2 + B^2)}{(1 + A^2)(1 + B^2)(\zeta^2 - \zeta^3)} \right].$$

Then by the Leibnitz formula we get

$$\frac{\partial \tilde{V}}{\partial \alpha} = F(\zeta_2, A, B) \frac{\partial \zeta_2}{\partial \alpha} - F(\zeta_1, A, B) \frac{\partial \zeta_1}{\partial \alpha} + \int_{\zeta_1}^{\zeta_2} \frac{\partial F(\zeta, A, B)}{\partial A} \frac{\partial A}{\partial \alpha} d\zeta. \qquad (3.7)$$

We note that $F(\zeta_1, A, B) = F(\zeta_2, A, B) = 0$ if ζ_1, ζ_2, A, and B are the same as in the statement of the theorem. Moreover, since $\alpha = 2 \operatorname{arccot} A$ we have $\dfrac{\partial A}{\partial \alpha} = -\dfrac{1 + A^2}{2}$ and

$$\frac{\partial F(\zeta, A, B)}{\partial A} \frac{\partial A}{\partial \alpha} = \frac{iA}{\zeta^2 + A^2}.$$

Hence, by Proposition 2.5 we obtain from (3.7)

$$\frac{\partial \tilde{V}}{\partial \alpha} = i \int_{\zeta_1}^{\zeta_2} \frac{A \, d\zeta}{\zeta^2 + A^2} = -i \arctan \frac{A}{\zeta_2} + i \arctan \frac{A}{\zeta_1} = -\frac{l_\alpha}{2}.$$

The equation $\frac{\partial \tilde{V}}{\partial \beta} = -\frac{l_\beta}{2}$ can be obtained in the same way. The boundary condition (3.6) for the function $\tilde{V}$ follows from the integral formula.

Similar arguments can be apply to find the spherical volume of the Whitehead link cone–manifold $W(\alpha, \beta)$. See [HLM4] for relationship between hyperbolic and spherical cases. In particular, in the spherical case, all roots of the cubic equation

$$z^3 + \frac{1}{2}(A^2 B^2 + A^2 + B^2 - 1)z^2 - A^2 B^2 z + A^2 B^2 = 0$$

are real and lengths of the singular geodesics of $W(\alpha, \beta)$ are given by the formulas

$$l_\alpha = 2 \arctan \frac{A}{\zeta_1} - 2 \arctan \frac{A}{\zeta_2},$$

$$l_\beta = 2\arctan\frac{B}{\zeta_1} - 2\arctan\frac{B}{\zeta_2},$$

where $\zeta_1, \zeta_2, 0 \le \zeta_1 < \zeta_2$ are non-negative roots of the equation. As a result we have

Theorem 3.4. *Let $W(\alpha,\beta)$ be a spherical Whitehead link cone–manifold with cone angles α and $\beta, 0 < \alpha, \beta \le \pi$. Then the volume of $W(\alpha,\beta)$ is given by the formula*

$$Vol\,W(\alpha,\beta) = \int_{\zeta_1}^{\zeta_2} \log\left[\frac{2(\zeta^2+A^2)(\zeta^2+B^2)}{(1+A^2)(1+B^2)(\zeta^2-\zeta^3)}\right]\frac{d\zeta}{\zeta^2-1},$$

where $A = \cot\frac{\alpha}{2}, B = \cot\frac{\beta}{2}, \zeta_1, \zeta_2, 0 \le \zeta_1 < \zeta_2$ are roots of the cubic equation

$$z^3 + \frac{1}{2}(A^2B^2 + A^2 + B^2 - 1)z^2 - A^2B^2z + A^2B^2 = 0.$$

3.3. Volume of the Borromean rings cone–manifold

It was noted in the Subsection 2.3 that the volume of the hyperbolic cone–manifold $B(\alpha,\beta,\gamma), 0 \le \alpha,\beta,\gamma < \pi$ is eight times the volume of the Lambert cube $L(\frac{\alpha}{2},\frac{\beta}{2},\frac{\gamma}{2})$. Hence, according to [Kel89], $Vol\,B(\alpha,\beta,\gamma)$ can be obtain as a linear combination of eight Lobachevsky functions. We indicate here a slightly different from [Kel89], but equivalent approach for volume calculation. It will be based on trigonometrical identities (Tangent and Sine–Cosine Rules) and the Schläfli formula and can be apply in both hyperbolic and spherical geometries. We have to use Theorem 2.8 and its spherical analog to find the lengths of singular geodesics of cone–manifold $B(\alpha,\beta,\gamma)$. The proof of the following proposition [DM01] is similar to the proof of Theorem 3.3.

Proposition 3.5. *The volume of a spherical Borromean rings cone–manifold $B(\alpha,\beta,\gamma)$ $\pi < \alpha,\beta,\gamma < 2\pi$ is given by the formula*

$$Vol\,(B(\alpha,\beta,\gamma)) = 2\int_{-\infty}^{T} \log\frac{(t^2+L^2)(t^2+M^2)(t^2+N^2)}{(1+L^2)(1+M^2)(1+N^2)t^2}\frac{dt}{t^2-1},$$

where T is a negative root of the equation

$$T^4 + (L^2+M^2+N^2+1)T^2 - L^2M^2N^2 = 0,$$
$$L = \tan\frac{\alpha}{2}, M = \tan\frac{\beta}{2}, and\ N = \tan\frac{\gamma}{2}.$$

As an immediately consequence we obtain

Theorem 3.6. *The volume of a spherical Borromean rings cone–manifold* $B(\alpha, \beta, \gamma)$, $\pi < \alpha, \beta, \gamma < 2\pi$ *is given by the formula*

$$Vol(B(\alpha, \beta, \gamma)) = 2\left(\delta\left(\frac{\alpha}{2}, \theta\right) + \delta\left(\frac{\beta}{2}, \theta\right) + \delta\left(\frac{\gamma}{2}, \theta\right) - 2\delta\left(\frac{\pi}{2}, \theta\right) - \delta(0, \theta)\right),$$

where $\delta(\xi, \theta) = \int\limits_{\theta}^{\frac{\pi}{2}} \log(1 - \cos 2\xi \cos 2\tau)\frac{d\tau}{\cos 2\tau}$ *and* θ, $\frac{\pi}{2} < \theta < \pi$ *is a principal parameter defined by conditions*

$$\tan\theta = T, \ T^4 + (L^2 + M^2 + N^2 + 1)T^2 - L^2M^2N^2 = 0.$$
$$L = \tan\frac{\alpha}{2}, M = \tan\frac{\beta}{2}, \text{ and } N = \tan\frac{\gamma}{2}.$$

Proof. From Proposition 3.5 we have

$$Vol(B(\alpha, \beta, \gamma)) = 2\int\limits_{-\infty}^{T}\log\left(\frac{t^2 + L^2}{1 + L^2}\frac{t^2 + M^2}{1 + M^2}\frac{t^2 + N^2}{1 + N^2} : \frac{t^2 + 0^2}{1 + 0^2}\right)\frac{dt}{t^2 - 1}$$

$$= 2\,(I(L, T) + I(M, T) + I(N, T) - I(0, T)),$$

where $I(K, T) = \int\limits_{-\infty}^{T}\log\left(\frac{t^2 + K^2}{1 + K^2}\right)\frac{dt}{t^2 - 1}$. Let $T = \tan\theta$, $K = \tan\xi$. Then under substitution $t = \tan\tau$ we obtain

$$I(K, T) = \int\limits_{\frac{\pi}{2}}^{\theta}\log\left(\frac{\tan^2\tau + \tan^2\xi}{1 + \tan^2\xi}\right)\frac{d\tau}{\cos^2\tau(\tan^2\tau - 1)}$$

$$= \left(\int\limits_{\theta}^{\frac{\pi}{2}}\log(1 - \cos 2\tau \cos 2\xi)\frac{d\tau}{\cos 2\tau}\right) - \left(\int\limits_{\theta}^{\frac{\pi}{2}}\log(1 + \cos 2\tau)\frac{d\tau}{\cos 2\tau}\right)$$

$$= \delta(\xi, \theta) - \delta\left(\frac{\pi}{2}, \theta\right).$$

Hence

$$Vol(B(\alpha, \beta, \gamma)) = 2\,(I(L, T) + I(M, T) + I(N, T) - I(0, T))$$

$$= 2\left(\delta\left(\frac{\alpha}{2}, \theta\right) + \delta\left(\frac{\beta}{2}, \theta\right) + \delta\left(\frac{\gamma}{2}, \theta\right) - 2\delta\left(\frac{\pi}{2}, \theta\right) - \delta(0, \theta)\right).$$

We note that the function $\delta(\alpha, \theta)$ can be considered as a spherical analog of the function $\Delta(\alpha, \theta) = \Lambda(\alpha + \theta) - \Lambda(\alpha - \theta)$, where $\Lambda(x) = -\int\limits_{0}^{x}\log|2\sin t|\,dt$ is the

Lobachevsky function. The hyperbolic volumes of many knots, orbifolds and cone manifolds (see [Thu79], [MV95], [Vin92], [Kel89]) can be expressed in terms of $\Delta(\alpha, \theta)$. To compare the above result with the hyperbolic case we rewrite the main theorem of [Kel89] in the following form.

Theorem 3.7. *The volume of a hyperbolic Borromean rings cone–manifold $B(\alpha, \beta, \gamma)$, $0 < \alpha, \beta, \gamma < \pi$ is given by the formula*

$$Vol(B(\alpha, \beta, \gamma)) = 2\left(\Delta\left(\frac{\alpha}{2}, \theta\right) + \Delta\left(\frac{\beta}{2}, \theta\right) + \Delta\left(\frac{\gamma}{2}, \theta\right) - 2\Delta\left(\frac{\pi}{2}, \theta\right) - \Delta(0, \theta)\right),$$

where θ, $0 < \theta < \frac{\pi}{2}$ is a principal parameter defined by conditions

$$\tan \theta = T, \quad T^4 - (L^2 + M^2 + N^2 + 1)T^2 - L^2 M^2 N^2 = 0.$$

$$L = \tan\frac{\alpha}{2}, M = \tan\frac{\beta}{2}, \text{ and } N = \tan\frac{\gamma}{2}.$$

3.4. Volume of the figure eight knot cone–manifold

Denote by $4_1(\alpha)$ the figure eight knot cone-manifold with cone angle α (Fig. 5). We start with the existence theorem for geometrical structure on $4_1(\alpha)$. Some particular cases of this theorem were obtained by many authors ([Thu79], [HLM95b], [HKM98], [Koj98], [MV95], [MR]). For example, in [Thu79] and [MV95] the Dehn surgery arguments have been used to obtain the hyperbolic structure on $4_1(\alpha)$, in [HKM98] non-convex fundamental set for the orbifold $4_1(\frac{2\pi}{n})$, $n \in \mathbb{N}$ in the hyperbolic space has been constructed, in [HLM95b] an explicit construction of the Dirichlet polyhedra in the hyperbolic and spherical spaces have been done. In [MR] a new method to create the Delaunay tessellation (which is dual to Dirichlet one) for two-bridge cone-manifolds in the hyperbolic, spherical and Euclidean spaces was suggested. This approach [MR99] allowed to construct non-convex fundamental set for the figure eight knot cone-manifold consisting of ten tetrahedra gluing around a common edge. The combinatorial type of these polyhedra remains the same in all three constant curvature geometries. Their geometrical parameters can be described in terms of quadratic equation whose coefficients are integer polynomials in $\cot\frac{\alpha}{2}$.

The following three theorems were obtained in [MR99].

Theorem 3.8. *The figure eight knot cone–manifold $4_1(\alpha)$ is hyperbolic for $0 \le \alpha < \frac{2\pi}{3}$, Euclidean for $\alpha = \frac{2\pi}{3}$, and spherical for $\frac{2\pi}{3} \le \alpha < \frac{4\pi}{3}$.*

Earlier [HLM95b] the existence of the spherical structure on $4_1(\alpha)$ was established only for $\frac{2\pi}{3} < \alpha \le \pi$. The cone–manifold $4_1(0)$ is the figure eight compliment with

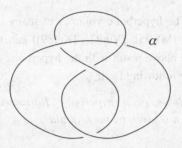

Figure 5: The figure eight knot cone–manifold $4_1(\alpha)$.

a complete hyperbolic structure. It can be also shown (see [MR99] for details) that the geometric limit of $4_1(\alpha)$ in the Hausdorff-Gromov topology as $\alpha \to \frac{4\pi}{3}$ coincides with a virtual knot cone–manifold whose underlying space is the three-sphere and the singular set is formed by graph having two vertices and four edges with cone angles $\frac{4\pi}{3}$. So, in some sense, $0 \le \alpha < \frac{4\pi}{3}$ is the natural interval for existence of constant curvature metric on the figure–eight knot cone–manifold $4_1(\alpha)$.

Theorem 3.9. *The volume of a hyperbolic figure–eight knot cone–manifold $4_1(\alpha)$, $0 \le \alpha < \frac{2\pi}{3}$ is given by the formula*

$$Vol\,4_1(\alpha) = \int_{\alpha}^{\frac{2\pi}{3}} arcosh\,(1+\cos t - \cos 2t)dt.$$

We remark that equivalent but more complicated formulas for hyperbolic $Vol\,4_1(\alpha)$ were obtained in [HLM95b], [MV95], and [Koj98].

Theorem 3.10. *The volume of a spherical figure–eight knot cone–manifold $4_1(\alpha)$ is given by the formula*

$$Vol\,4_1(\alpha) = \int_{\frac{2\pi}{3}}^{\alpha} arccos\,(1+\cos t - \cos 2t)dt, \frac{2\pi}{3} < \alpha \le \pi$$

and

$$Vol\,4_1(\alpha) = 2\pi(\alpha - 0.9\pi) - \int_{\pi}^{\alpha} arccos\,(1+\cos t - \cos 2t)dt, \pi < \alpha < \frac{4\pi}{3}.$$

The author is grateful to referee for helpful remarks and suggestions.

References

[BZ85]   G. Burde and H. Zieschang (1985). *Knots*, De Gruyter Studies in Mathematics.

[CCG+94] D. Cooper, M. Culler, H. Gillet, D.D. Long and P.B. Shalen (1994). Plane curves associated to character varieties of 3-manifolds, *Invent. Math.* **118**, 47–84.

[Cul86] M. Culler (1986). Lifting Representations to Covering Groups, *Adv. Math.* **59**, 64–70.

[Dia99] R. Diaz (1999). A Characterization of Gram Matrices of Polytopes, *Discrete Comput. Geom.* **21**, 551–601.

[DM01] D. Derevnin and A. Mednykh (2001). On the volume of spherical Lambert cube, *preprint*.

[Fen89] W. Fenchel (1989). *Elementary Geometry in Hyperbolic Space*, Walter de Gruyter.

[Hod] C. D. Hodgson. Schläfli revisited: Variation of volume in constant curvature spaces, *preprint*.

[HKM95] H. Helling, A.C. Kim and J.L. Mennicke (1995). Some Honey-combs in the hyperbolic 3-space, *Comm. Algebra* **23**, 5169–5206.

[HKM98] H. Helling, A.C. Kim and J.L. Mennicke (1998). A geometric study of Fibonacci groups, *Journal of Lie Theory* **8**, 1–23.

[HKW86] P. de la Harpe, M. Kervaire and C. Weber (1986). On the Jones Polynomial, *L'Enseignement Mathématique* **32**, 271–335.

[HLM92] H.M. Hilden, M.T. Lozano and J.M. Montesinos-Amilibia (1992). On the Borromean orbifolds: geometry and arithmetic. In *Topology'90, edited by B.A. Apanasov, W. Newmann, A. Reid and L. Siebenmann*, de Gruyter, 133–167.

[HLM95a] H.M. Hilden, M.T. Lozano and J.M. Montesinos-Amilibia (1995). On the arithmetic 2-bridge knots and link orbifolds and a new knot invariant, *Journal of Knots and its Ramification* **4**, no. 1, 81-114.

[HLM95b] H.M. Hilden, M.T. Lozano and J.M. Montesinos-Amilibia (1995). On a remarkable polyhedron geometrizing the figure eight knot cone manifolds, *J. Math. Sci. Univ. Tokyo* **2**, 501–561.

[HLM96] H.M. Hilden, M.T. Lozano and J.M. Montesinos-Amilibia (1996). Volumes and Chern-Simons invariants of cyclic covering over rational knots.

In *Proceedings of the 37th Taniguchi Symposium on Topology and Te-ichmüller Spaces held in Finland, July 1995, edited by Sadayoshi Kojima et al.*, 31–35.

[Kel89]   R. Kellerhals (1989). On the volume of hyperbolic polyhedra, *Math. Ann.* **28**, 245–270.

[Kau88]   L. Kauffman (1988). New Invariants in the Theory of Knots, *Amer. Math. Monthly* **95**, no. 3, 195-242.

[KM99]   A.-Ch. Kim and A. Mednykh (1999). On the hyperbolic structure on the Whitehead link cone–manifold, *preprint*.

[Koj98]   S. Kojima (1998). Deformation of hyperbolic 3-cone-manifolds, *J. Diff. Geom.* **49**, 469–516.

[Med00]   A. Mednykh (2000). On the remarkable properties of the hyperbolic Whitehead link cone–manifold, *Series Knots and Everything, Singapure et al., World Scientific* **24**, 290-305.

[MR]   A. Mednykh and A. Rasskazov. On the structure of the canonical funda-mental set for the 2-bridge link orbifolds, *Universität Bielefeld, Sonder-forschungsbereich 343, Discrete Structuren in der Mathematik, Preprint,* 98–062.

[MR99]   A. Mednykh and A. Rasskazov (1999). On the degenaration of geometric structures on the figure eight knot cone–manifold, *preprint*.

[MV95]   A. Mednykh and A. Vesnin (1995). Hyperbolic volumes of Fibonacci manifolds, *Siberian Math. J.* **36**, 235–245.

[MV01]   A. Mednykh and A. Vesnin (2001). On the volume of hyperbolic White-head link cone–manifolds, *preprint*.

[PSe95]   A.R. Parker and C. Series (1995). Bending formulae for convex hull boundaries, *J. Anal. Math.* **67**, 165–198.

[Ril79]   R. Riley (1979). An elliptical path from parabolic representations to hy-perbolic structure. In *Topology of Low-Dimension manifolds (Lec. Notes Math.* **722**), Springer-Verlag, 99-133.

[Shm01]   R.N. Shmatkov (2001). Euclidean structure on the Whitehead link cone-manifold, *Manuskripte der Forschergruppe Atithmetik, Universität Hei-delberg, Preprint Nr. 5, 1-26.*

[Thu79]   W.P. Thurston (1979). *The geometry and topology of three-manifolds*, Princeton University Lecture Notes.
http://www.msri.org/publication/books/gt3m/

[Vin92]   E.B. Vinberg (1992). Geometry II, *Encyclopeadia of Mathematical Sciences* **29**, Springer-Verlag.

[Zho99]   Qing Zhou (1999). The Moduli Space of Hyperbolic Cone Structures, *J. Diff. Geom.* **51**, 517–550.

Alexander Mednykh

Sobolev Institute of Mathematics
Novosibirsk 630090
Russia

mednykh@math.nsc.ru

AMS Classification: 30F40, 57M50, 57M25

Keywords: cone manifold, complex length, hyperbolic volume, Tangent rule, Borromean rings

Kleinian Groups and Hyperbolic 3-Manifolds
Lond. Math. Soc. Lec. Notes **299**, 165–179

Y. Komori, V. Markovic & C. Series (Eds.)
Cambridge Univ. Press, 2003

Remarks on the curve complex: classification of surface homeomorphisms

W. J. Harvey

1. Introduction and summary

The purpose of this article is to describe a brief and elementary approach to the classification of mapping-classes for a surface with negative Euler characteristic, using rudimentary facts about the structure of the curve complex. Work of W.P. Thurston around 1975 [Thu88] deepened the analogy with the classical case of the torus, which involves the structure of the modular group $SL_2(\mathbb{Z})$ acting on the upper half plane $\mathcal{H}$. He constructed a completion of $\mathcal{L}$, the set of simple loops in a base surface viewed modulo isotopy, which plays the part of boundary sphere S_∞ for a proper geometric action of the mapping-class group on Teichmüller space, the natural analogue of $\mathcal{H}$. This produced a classification of surface mappings into types closely resembling the genus 1 situation, where there is the familiar division of elements of $SL_2(\mathbb{Z})$ into conjugacy classes according to their fixed point set in the closed topological disc $\mathcal{H} \cup S_\infty^1$.

The dichotomy between reducible and irreducible surface mappings and the existence of an invariant axis for each hyperbolic irreducible mapping, which lie at the heart of Thurston's approach, were later treated in the setting of Teichmüller theory by L. Bers [Ber78], using techniques of extremal quasiconformal mappings; for a more detailed summary, which enlarges on the relationship between the various approaches, see for instance [Hrv79]. For a thorough treatment, the reader may consult the book [Abi80] by W. Abikoff, which presents an outline of the requisite background from Teichmüller theory as part of a self-contained account of the Bers approach in [Ber78], and the Orsay seminar report [FLP79] which describes this aspect of Thurston's work in depth. In fact, the existence of a (unique) invariant Teichmüller axis within the Teichmüller space, which follows rather directly in Thurston's approach, was only obtained by Teichmüller-theoretic methods much later (by A. Marden and K. Strebel). The classification had earlier been studied in great depth (but without complete solution) by J. Nielsen in the period between 1930 and 1950 using his fixed point analysis of the mappings of the circle which arise from lifting a map to the universal covering disc and extending it to the boundary. His contribution was later brought into closer relation with Thurston's work using the notion of geodesic lamination [HT85].

Our goal here is modest: using a purely combinatorial approach, we show that surface mapping-classes are distinguishable by their operation on a certain connected graph $\mathcal{K}$, to be defined below, which describes the permutation action on the set $\mathcal{L}$ of homotopy classes of simple loops in S. By application of results on automorphisms of trees due to J. Tits, we show that the (nontorsion) mapping-classes which are not reducible to mappings of a proper sub-surface are precisely those with hyperbolic $\mathcal{K}$-action, preserving an axis which is an infinite geodesic line in $\mathcal{K}$. The structure of reducible homeomorphisms is described in S3, and a technique for constructing hyperbolic elements is given in the last section, which indicates how to produce an infinite geodesic axis in $\mathcal{K}$ for this type of element; the basic idea is close to Thurston's original method for constructing pseudo-Anosov surface mappings.

It should be noted that this paper deals only with primitive topological aspects of Thurston's work on surfaces: the beautiful geometry and dynamics underlying his classification is not accessible without the more detailed analysis provided by either the theory of measured foliations or extremal quasiconformal mappings. The original version of this paper was written in 1978, soon after the discovery of the curve complex $\mathscr{C}(S)$ in the context of the problem of completing the Riemann modular variety of S, $\mathcal{M}_g = T(S)/\Gamma(S)$, reported in [Hrv81]; it was motivated by the original report [Thu88] of Thurston's work, which required a very substantial amount of background theory, at that time unpublished, to reach the full implications of what amounts in essence to a very simple dichotomy. I have rewritten the introduction, clarified a few proofs and added some comments and further references, but the core of the paper remains virtually the same.

Finally, the most compelling reason for offering the present revision of this (now rather aged) article is to motivate the reader to study the striking applications of the curve complex to the geometric classification of Kleinian groups which have been given recently by Yair Minsky ([Min00],[Min01]), partly based on his results in collaboration with H. Masur [MM]; see also his article in the present volume. In particular, section 4 makes it clear in an elementary way that there are infinite geodesic lines in $\mathcal{K}$.

2. A simplicial action for mapping-class groups

We fix a reference surface S with g handles and $n \geq 0$ boundary components. The pair (g,n) is the *topological type* of S. If $n = 0$, we omit it from the notation. We restrict attention usually to surfaces with negative Euler characteristic, so that $2g - 2 + n > 0$.

The *mapping classes* of S are the isotopy classes, relative to the boundary ∂S, of homeomorphisms of S which fix pointwise each component of ∂S; they form a group which we denote by $\Gamma = \Gamma(S)$.

The mapping class group $\Gamma(S)$ has a natural permutation action on the set $\mathscr{L}(S)$ of free isotopy classes of simple loops in S. I introduced in previous work [Hrv81] a structure of simplicial complex based on $\mathscr{L}$, which renders the Γ-action more accessible in some respects. It is closely related to a discrete action of $\Gamma(S)$ on the Teichmüller moduli space of marked hyperbolic structures on the base surface S which has received much attention over the years: see for instance [Abi80] , [Ber78] or [Iva02].

A *partition* of S is a non-empty subset $\Lambda \subset \mathscr{L}$ such that any pair of loop classes in Λ have disjoint representatives. It is often helpful to view a partition as a way to dissect S into subsurfaces modulo isotopy or, equally, as defining a class of singular surface with nodes obtained by collapsing the loops in Λ to nodal points. Performing successive dissections or collapses corresponds to a partial ordering by inclusion on partitions; this provides a structure of abstract *simplicial complex* on the set $\mathscr{C}(S)$ of all partitions. For our purposes, it is natural to impose the restriction to partitions that contain no class of loop enclosing a single boundary component; henceforth we tacitly make this condition. A k-simplex of $\mathscr{C}(S)$, which represents a partition Λ with $k+1$ loops, is denoted σ_Λ.

The topological structure of the surface $S \setminus \Lambda$, obtained by cutting up S along a disjoint set of loops representing the partition Λ, is conveniently summarised in the *partition graph* K_Λ, which is defined as follows. Assign a vertex with label (g', n') to each connected component of S/Λ with type (g', n'), and put an edge between two vertices (possibly the same one) for each loop interconnecting the corresponding component(s) in S.

Each simple closed curve ℓ in S has an associated *Dehn twist map* τ_ℓ , which is a self-homeomorphism of S fixing ℓ pointwise: it is obtained by cutting S along ℓ, applying a full 360° twist in a clockwise direction to a small annular neighbourhood of one bank of the cut and then sewing the surface back together. In view of the fact that the free homotopy class of ℓ determines the isotopy class of τ_ℓ, and because twists about disjoint loops commute up to homotopy, we obtain for each partition Λ a free abelian *twist subgroup* of Γ, denoted $Tw(\Lambda)$, generated by the mapping classes of twists about the loops in Λ.

For a proper inclusion of surfaces $S_0 \subseteq S$, with $S \setminus S_0$ containing no discs or boundary annuli, we define the *relative partition complex* $\mathscr{C}(S, S_0)$ to be the subcomplex of $\mathscr{C}(S)$ comprising partitions with all loops carried in S_0. The boundary loops of S_0 define a partition Λ_0 of S and one may relate $\mathscr{C}(S, S_0)$ to the corresponding simplex σ_0 and the curve complexes of the components of S_0. When S_0 is disconnected, it is not hard to see that $\mathscr{C}(S, S_0)$ is isomorphic to the join of σ_0 and the complexes $\mathscr{C}(S')$, with S' running through all distinct components of S_0.

Example. If $S_0 = S \setminus \Lambda$ for some partition Λ, then $\mathscr{C}(S, S_0)$ is the *link* of σ_Λ in $\mathscr{C}(S)$.

The framework which we use to study $\Gamma(S)$ is the 1-skeleton of the first barycentric subdivision of $\mathscr{C}(S)$. This admits a direct definition, without prior reference to $\mathscr{C}(S)$, as the simplicial complex $\mathscr{K}(S)$ whose *vertices* are the partitions Λ, with an *edge* between Λ and Λ' corresponding to a proper inclusion $\Lambda \subset \Lambda'$. There is a *grading* by positive integers defined on the vertices of $\mathscr{K}$ by the rule $\Lambda \mapsto \mathrm{card}(\Lambda)$. It follows from consideration of Euler characteristic for partitions of a genus g surface with n boundary components that the maximum value of this grading is $3g - 3 + n$.

Proposition 2.1. *If S is not of type $(1,1)$, $\mathscr{K}(S)$ is a connected 1-complex on which $\Gamma(S)$ operates by simplicial automorphisms that preserve the grading. The quotient is a finite combinatorial graph $\mathscr{X}(S)$.*

Proof. The first statement follows from the fact, proved in [Hrv81], that $\mathscr{C}(S)$ is connected. Briefly, one applies the theorem of Dehn and Lickorish that $\Gamma(S)$ is generated by the set of Dehn twists about a finite collection of loops in S (see [Bir74] or [Lic64]), to construct a path in $\mathscr{K}$ between any given pair of vertices which represent single non-dividing loops ℓ, $\gamma(\ell)$ ($\gamma \in \Gamma$) respectively; from this the path connectedness follows easily. The permutation action on classes of loops induces naturally a simplicial action on the first barycentric subdivision of $\mathscr{C}$, which in turn gives rise to the action on $\mathscr{K}$. Since elements of Γ preserve the grading and adjacent vertices must have distinct grade, no inversion of an edge can occur.

As there are only finitely many non-isomorphic trivalent graphs which represent the distinct ways to dissect a surface of finite type into pairs of pants (three-holed spheres) up to topological equivalence, the remaining assertions follow from the simplicial nature of the Γ-action. □

We note in passing that much more is known about the topology of the curve complex of a surface: for instance, $\mathscr{C}(S_g)$ is homotopically a wedge of spheres of dimension $2g - 2$ – see [Har88], [Iva02], for different proofs of this.

It is worthwhile examining briefly some simple examples of curve complex for surfaces of genus 1 and 2.

Examples 2.2.

(a) In genus 0 with $n \le 6$ boundary components, all loops separate the surface and the patterns correspond to distinct types of bracketing of the n labels.

(b) If S has type $(1,1)$, then vertices of $\mathcal{K}$ correspond bijectively to the points in $\mathbb{Q} \cup \{\infty\} \subseteq \partial \mathcal{H}$, acted upon by the classical modular group, $SL_2(\mathbb{Z})/\pm Id$. The quotient is a single point. As Minsky's work on ending laminations has emphasized, this 0-dimensional picture needs to be enhanced by placing it within the subdivision of the hyperbolic plane given by the Farey tesselation.

(c) For S of type $(1,2)$, a torus with two boundary components, there are two kinds of (one loop) vertex, giving rise to partition graphs with labels $(1,1)$ and $(0,4)$ respectively, depending on whether the loop separates S or not. They form two disjoint orbits V_0 and V_1 under the action of $\Gamma(S)$ and the quotient $X(S)$ consists of these two vertices, with edges joining them for each combinatorial type of grade-2 partition, one from V_0 to V_1 which includes a loop of each type, and a second linking V_1 to itself, representing two homotopically distinct type-V_1 loops which together separate S.

(d) If S is a closed surface of genus 2, $\mathscr{C}$ has dimension 2. There are two types of 2-simplex, σ_1 with one separating loop and two non-separating loops and σ_2 with three (individually) non-separating loops, forming the two $\Gamma(S)$−orbits. Hence $\mathscr{X}(S)$ is the union of two subcomplexes $\Sigma_1 \cup \Sigma_2$, representing the quotients of each type of simplex by its automorphism group, cyclic C_2 for σ_1 and the symmetric group S_3 for σ_2; the intersection of the Σ_j is a single edge.

Note. In the last three cases, the action of Γ on $\mathcal{K}$ is not effective: the canonical elliptic (or hyperelliptic) involution of the surface induces a symmetry fixing the entire graph in the respective cases. In case (d), this involution amounts to a halfturn rotation of a model surface $S_{2,0}$ in $\mathbb{R}^3$ about an axis skewering three representative non-separating loops. The quotient surfaces arising are all spheres, with 4, 5 or 6 distinguished points.

It is also worth noting that for all these non-effective values of g, n, the exceptional elliptic or hyperelliptic involution fixes every point of the corresponding Teichmüller space.

On all surfaces whose curve complex has dimension at least 3, the modular action *is* effective. For the reader's convenience, we include a proof of this well-known fact.

Proposition 2.3. *If S has type (g,n) with $3g+n > 6$, then $\Gamma(S)$ acts effectively on $\mathcal{K}(S)$.*

Proof. . For any surface $S = S_{g,n}$ with $3g+n > 6$, if a homeomorphism h fixes up to isotopy all simple loops in a given maximal partition, then h must preserve (modulo isotopy) the space S' obtained by cutting S apart along the loops. The components of S' are spheres with three boundary curves, and it is well known that their

self-homeomorphisms are generated up to isotopy by Dehn twists about the bound-
ary loops (see for instance [Bir74]). Now each Dehn twist τ_ℓ acts effectively on $\mathcal{K}$,
because the set of loops intersecting ℓ nontrivially are permuted non-trivially by τ_ℓ.
It follows, in view of the independence of the various twists, that the full group $\Gamma(S)$
acts effectively. $\square$

3. Groups operating on graphs

The graph $\mathcal{K} = \mathcal{K}(S)$ is equipped with a $\Gamma(S)$-invariant metric $d(\ ,\ )$ determined
by assigning the standard Euclidean metric of length 1 on each edge. A *geodesic*
between two vertices in $\mathcal{K}$ is a combinatorial path not containing any segments of
the form ee^{-1} with e an edge. A *geodesic segment* is a closed subinterval of such a
geodesic between two vertices. When restricted to the *vertex set* $V(\mathcal{K})$, the metric d
is an integer-valued function with $d(v,w)$ the number of edges in a shortest geodesic
linking v with w. Of course, there is in general no unique geodesic between points in
a graph.

We begin by outlining some elementary results about automorphisms of trees. For
more details the reader may refer to Serre [Se77] or Tits [Ti70]. Throughout we restrict
attention to automorphisms which do not invert an edge.

Proposition 3.1. *Let G be a group operating without inversions on a tree T. The
following conditions are equivalent:*

  (i) *there is a vertex $v \in T$ with the orbit Gv bounded;*

  (ii) *there is a vertex of T fixed by G.*

Proof. (ii) $\Longrightarrow$ (i) is trivial. To show (i) $\Longrightarrow$ (ii), embed the set Gv in a bounded
subtree T_0 of T by connecting each pair of vertices $gv, g'v$ with a geodesic. Now T_0 is
G-invariant, and an induction argument on the diameter of T_0 shows that some vertex
(or larger subset) is fixed by G. $\square$

The key fact about automorphisms of trees is the following theorem due to J. Tits.

Theorem 3.2. *If an automorphism γ operates without inversion on a metric tree T,
then precisely one of two exclusive possibilities occurs:* either

  (i) *there is a vertex fixed by γ, or*

  (ii) *there is an infinite geodesic line in T on which γ acts as a translation.*

If case (ii) prevails, the line is called an *axis* of γ, in analogy with terminology from hyperbolic geometry, and the tree automorphism γ is called *hyperbolic*.

Note. The proof (see [Se77] or [Ti70]) proceeds by considering the set of geodesic segments between pairs of vertices v, $\gamma(v)$ in T.

This result will be applied to characterise the elements of Γ by their action on a universal covering tree $\mathscr{T} = \widetilde{\mathscr{K}}$ of $\mathscr{K}$. Of course, the nature of an element $\gamma \in \Gamma$ is not necessarily reflected in the action of an individual lift $\widetilde{\gamma}$. We introduce the following terminology for mapping-classes.

Definition 3.3. An element $\gamma \in \Gamma$ is called

(i) *reducible* if it fixes a vertex of $\mathscr{K}$,

(ii) *irreducible* if it fixes no vertex of $\mathscr{K}$.

The corresponding subsets of $\Gamma^* = \Gamma \setminus [Id]$ are denoted Γ_R and Γ_I.

As an immediate consequence of Theorem 3.2, we deduce the following characterisation of the reducible/irreducible dichotomy using lifts of a mapping class acting on $\mathscr{K}$ to an automorphism of $\mathscr{T}$.

Theorem 3.4. *A mapping-class γ is irreducible if and only if every lift $\widetilde{\gamma}$ is a hyperbolic automorphism of $\mathscr{T}$.*

Proof. We employ the standard construction of the universal covering (metric) tree $\mathscr{T} = \widetilde{\mathscr{K}}$ as homotopy classes of path in $\mathscr{K}$ with chosen vertex v as base point : of course, geodesic segments $\alpha : [0,1] \to \mathscr{K}$ with initial point $\alpha(0) = v$ represent these path classes. The covering projection $\pi : \widetilde{\mathscr{K}} \to \mathscr{K}$ is given by the map $\alpha \mapsto \alpha(1)$ sending path to terminal point.

First of all, we observe that any reducible mapping class γ has a lift which fixes some vertex of $\mathscr{T}$, because we can choose to employ a base point $v \in \mathscr{K}$ fixed by γ and then γ acting on $\mathscr{K}$ lifts to an automorphism $\widetilde{\gamma}$ of T with fixed point the vertex $\widetilde{v}$ corresponding to the trivial path class.

Conversely, the action on $\mathscr{K}$ of any lifted automorphism of $\mathscr{T}$ which fixes some vertex $\widetilde{v}$ is clearly reducible, fixing $v = \pi(\widetilde{v})$. It follows that no lift of an irreducible mapping class γ can fix a vertex of $\mathscr{T}$ and so in this case each lift $\widetilde{\gamma}$ is hyperbolic, preserving an axis in $\mathscr{T}$. $\qquad\square$

From this proof, we also infer that the action of an irreducible non-torsion mapping class on $\mathscr{K}$ has fixed axes there:

Corollary 3.5. *For each lift of an irreducible, nontorsion mapping class γ, the projection of the invariant axis $A(\tilde{\gamma})$ to $\mathcal{K}$ is an infinite geodesic axis in $\mathcal{K}$.*

Proof. Assume not: then the projection $\pi(A)$ is a closed simplicial loop, given by a finite sequence of vertices in $\mathcal{K}$ on which γ acts as a cyclic permutation. Also, the $\tilde{\gamma}$-orbit of any vertex in $\mathcal{T}$ is unbounded, since otherwise Proposition 3.1 would imply that γ is reducible. But γ is irreducible and of infinite order, so that the same holds for any power γ^n, i.e. no lift of any power can fix a vertex of $\mathcal{T}$. Thus, each lift acts hyperbolically on $\mathcal{T}$, and the projection of each invariant axis $A(\tilde{\gamma})$ determines an infinite geodesic in $\mathcal{K}$. □

A classification of nonidentity surface mapping-classes now ensues. First we expand our terminology for types of mapping-class to emphasize the analogy with isometries of the hyperbolic plane. We need to consider torsion (finite order) elements carefully as they may be in either I or R:

Definition 3.6. A mapping class $\gamma \in \Gamma$ is

  (i) *hyperbolic* if it has an invariant axis in $\mathcal{K}$;

  (ii) *parabolic* if it is reducible and not torsion,

  (iii) *elliptic* if it has finite order.

The corresponding subsets of Γ^* are called H, P and E.

Corollary 3.7. $\Gamma^* = H \cup P \cup E$, *where* $H = I \setminus (I \cap E) \neq I$ *and* $P = R \setminus (R \cap E) \neq R$.

Proof. The first part of the statement, leaving aside torsion elements, follows from the previous two results: by the definitions and Theorem 3.4, $\Gamma^* = I \cup R$ and $I \cap R$ is empty.

To understand the situation for torsion elements, it is helpful to recall that any finite order mapping class can be realised as an automorphism of some complex structure on S – this follows by the Nielsen-Fenchel resolution [Fe48] of the Hurwitz-Nielsen Realisation problem for finite cyclic groups. The quotient is then a compact orbifold Riemann surface $S_0 = S/\langle \gamma \rangle$ and an elliptic mapping class γ is irreducible if and only if this quotient has the property that any closed loop is homotopic to a cone point or a boundary point. For an irreducible torsion element, the lifts to $\mathcal{T}$ have axes which project to finite simplicial loops in $\mathcal{K}$ invariant under the action of that element. Torsion surface mappings of each type exist in all genera. □

Note 3.8. For a hyperbolic element, as we saw, *every* lift to T has an invariant axis, but there is no apparent natural choice of projection in $\mathcal{K}$; for instance there are many with the same minimal translation length. In [MM], a notion of 'quasi-geodesic' line between vertices are employed very effectively to study the coarse geometry of the curve complex $\mathcal{C}$. This aspect of the structure would repay further study.

In the next section we indicate how to refine the classification somewhat by further analysis of the class R of reducible maps.

4. Reducible mapping-classes

The standard example of a reducible homeomorphism is a Dehn twist about a single loop ℓ. More complicated examples arise from products of twists around loops in a partition, composed with a symmetry of the partition graph arising from a permutation of loops which respects the topological structure of the partition.

Another type of example comes from selection of irreducible homeomorphisms on several components of a partitioned subsurface that has large enough grading; this can be extended to a homeomorphism of S in various ways, for instance, by fixing all complementary parts or by twisting along the partition loops.

We next show, by examining the fixed point set in $\mathcal{C}$, that any reducible homeomorphism falls into one or other of these types of pattern.

Theorem 4.1. *If γ is a reducible mapping-class, then either it fixes a maximal partition, or there are subsurfaces S_h, S_e, S_p with union S such that*

(i) *$\mathcal{C}(S, S_p)$ is the fixed set of γ in $\mathcal{C}(S)$, and*

(ii) *the restriction of γ to $\mathcal{C}(S, S_e)$ $\left(\mathcal{C}(S, S_h)\right)$ is elliptic (resp. hyperbolic).*

Proof. Let $\sigma_\Lambda \subset \mathcal{C}$ be a simplex of maximal dimension in $\mathrm{Fix}(\gamma)$. It follows that if ℓ is any closed loop in $S \setminus \Lambda$, distinct from Λ, then ℓ is not fixed by γ, nor is ℓ part of any γ-invariant partition of S. This implies that on any part of $S \setminus \Lambda$ that is not a 3-holed sphere, the action of γ is irreducible, hence either hyperbolic or elliptic.

If Λ is a maximal partition, then γ belongs to a group extension

$$1 \longrightarrow \mathrm{Tw}(\Lambda) \longrightarrow \mathrm{Stab}(\Lambda) \longrightarrow \mathrm{Aut}(K_\Lambda) \longrightarrow 1,$$

because each component of $S \setminus \Lambda$ has no non-trivial homeomorphisms apart from products of twists about the three boundary loops, and so any homeomorphism preserving

Λ must induce an automorphism of the associated partition graph K_Λ. In this case, we may say $S_p = S$ by convention.

If Λ is not a maximal partition, then $S \setminus \Lambda$ is expressible as a disjoint union $S_1 \cup S_2$, with $S_1 = \bigcup \{S' \mid \chi(S') < -1\}$ and S_2 a union of three-holed spheres. On S_2, γ acts as some combination of permutations and boundary twists, so we set $S_2 = S_p$. It remains to decompose S_1 into hyperbolic and elliptic parts. Now γ is hyperbolic on S_1 if and only if every $\langle \gamma \rangle$-orbit in $\mathscr{K}(S, S_1)$ is unbounded. Therefore the elliptic subsurface $S_e \subseteq S_1$ may be characterised as the union of all component subsurfaces $S' \subseteq S_1$ such that the restriction of γ to $\mathscr{K}(S, S')$ has a bounded orbit.   □

Remarks 4.2.

(i) S_h is the largest subsurface on which γ is hyperbolic. The action may involve a cyclic permutation of the distinct components of S_h.

(ii) The decomposition may be regarded as a kind of analogue of the Jordan canonical form for matrices.

(iii) Surface mappings with $S_h \neq \emptyset$ are termed *pseudo-hyperbolic* by Bers [Ber78]. They belong to $R \setminus E$, of course.

The structure of the full stability group of any simplex σ_Λ of $\mathscr{C}$ can be analysed as follows in terms of $\mathrm{Tw}(\Lambda)$, the twist subgroup of the partition.

Proposition 4.3. *The stability group of σ_Λ is the normaliser in $\Gamma(S)$ of $\mathrm{Tw}(\Lambda)$.*

Proof. If f is a homeomorphism of S, then $f \circ \tau_\ell \circ f^{-1}$ is, up to isotopy, the twist about $f(\ell)$. Therefore any homeomorphism which normalises $\mathrm{Tw}(\Lambda)$ must permute the loops defining Λ, and conversely.   □

Referring to the exact sequence for the stabiliser of a simplex, we remark that the stabiliser of a simplex σ' in the *barycentric subdivision* $\mathscr{C}'$ is obtained by taking the kernel of the epimorphism from the stability group of the carrier simplex $\sigma_\Lambda \subset \mathscr{C}$ onto the automorphism group of the associated partition graph K_Λ. This group $\mathrm{Stab}(\sigma')$ in fact *centralises* the subgroup $\mathrm{Tw}(\Lambda)$.

5. Construction of hyperbolic elements

At first sight, it is by no means obvious that hyperbolic surface homeomorphisms exist, in the higher genus case at least. The purpose of this section is to indicate a simple method for constructing them from twists, based on the original examples given by Thurston in [Thu88].

A *Thurston decomposition* of a compact surface S is, by definition, a cell decomposition of S determined by two loops which *fill up* the surface, partitioning it into polygonal cells with an even number of edges, which are segments coming alternately from each loop. The original construction in [Thu88] produced a counterexample to a conjecture of J. Nielsen from the 1940s that an algebraically finite mapping class, i.e. one induced by a surface mapping for which all the integral homology eigenvalues are roots of unity, should be periodic. To construct a hyperbolic homeomorphism, one takes a certain composition of twists along a pair of null-homologous loops α, β that *fill up* the surface: the loops are in general position with no triple intersection points and so determine a Thurston decomposition with the property that the complementary regions of S are cells with at least four edges. It may be worth pointing out here that this cellulation of S is a *Grothendieck dessin* (see for instance [Hrv93]), with the property that all vertices (given a label $\star$) have valency 4, so that there is in fact a structure of algebraic curve definable over a number field implicit in this data on S.

A simple modification of the decomposition produces a triangulation by the following procedure. Place a vertex labelled ○ inside each 2-cell, a vertex with ● as barycenter of each segment of a loop and a vertex labelled $\star$ (as before) at each intersection point. An edge runs from each ○ to adjacent (edge) ● vertices and to each adjacent $\star$ vertex of the graph. This can also be turned into a kind of cell partition of the surface into quadrilaterals, each with two vertices labelled $\star$, which is closely related to the cell partition of S dual to the original one: omitting the edges containing ● vertices gives the standard dual subdivision into quadrilaterals.

The method used by Thurston to construct hyperbolic (pseudo-Anosov) homeomorphisms from this combinatorial structure is ingenious: because the surface S has been decomposed into quadrilaterals, it has a branched piecewise-Euclidean structure given by requiring each 2-cell to be a unit square. This determines a structure of compact Riemann surface too – by removal of the branch point singularities at all corners where the valency is not 4 – but the important point is the following fact:

Proposition 5.1. *In the singular Euclidean structure on S, the Dehn twist mappings corresponding to the loops α and β act as translation by an integer.*

Proof. To see this, we recall that the (right) *Dehn twist* along a simple loop ℓ, τ_ℓ acts on a tubular neighbourhood N of the loop, fixing the complement of N pointwise while twisting N through one complete turn by rotating the far edge to the right relative to the other in the orientation induced from the surface S. To be more precise, on our geometric surface we can choose local coordinates (x, y) for N, viewing it as the quotient of a universal covering $\tilde{N}$, a horizontal channel of height 1 with core the x–axis, by the group of integer translations, with ℓ the core loop represented by the points $(t, 0,)$, $0 \le t < 1$. Now the right Dehn twist τ_ℓ is defined on this cylinder by the (rightward) horizontal shear $(x, y) \mapsto (x + y, y)$.

Note that one can perform the (re)construction of the surface S from the (unit) square tile pieces by first joining them in sequence following the path α to obtain the cylinder N and then making the necessary identifications of edge segments on ∂N to record the crossing pattern of β; in the local coordinate of the resulting geometric structure, a twist τ_α along the core of the cylinder is then clearly an integral translation. For the β loop, one similarly obtains an integer translation in the orthogonal direction, for instance by re-assembling the tiling of S along β first.   □

The group $G(\tau_\alpha, \tau_\beta)$ generated by the two twists is therefore isomorphic to a non-abelian subgroup (of finite index) in $SL(2, \mathbb{Z})$. Furthermore, it is clear that any word in the two twist maps that produces a modular group element with two real eigenvalues (hyperbolic) will define a hyperbolic mapping class for the surface. If one chooses α and β more carefully, to be null homologous on S, the mappings will act trivially on homology (see [Thu88] for more details).

Now we observe that in the context of the action of the subgroup G on $\mathscr{C}(S)$, hyperbolic elements will have infinite axes in $\mathscr{K}$, which therefore has infinite diameter. In fact, it is possible by a more detailed study of how such twist elements operate on $\mathscr{C}(S)$ to verify directly that this process yields hyperbolic elements in the sense of section 2. Each twist fixes a subcomplex of $\mathscr{C}$ containing the corresponding loop vertex, σ_α or σ_β, and these two fixed sets are easily seen to be disjoint. It can then be shown, for instance, that the geodesic segment linking σ_α and σ_β extends to an invariant axis for all positive products of twists $\tau_\alpha^m \tau_\beta^n$.

Notes 5.2.

(i) With reference to this construction, Masur [Mas81] showed, using the trajectory structure of the Jenkins-Strebel quadratic differentials associated to the loops α, β in S, that the extremal dilatations for the (hyperbolic) mapping classes $\tau_\alpha^n \tau_\beta^{-n}$ $(n = 1, 2, \ldots)$ are given by

$$\frac{2 + n^2 v^2 + v\sqrt{n^2 v^2 + 4}}{2}$$

where v is the intersection number of ℓ with m. These elements are involved in his proof that there are Teichmüller geodesic axes whose projection is dense in the quotient moduli space, and this fact in turn plays an important part in Ivanov's proof [Iva97] of Royden's Theorem, which states that the automorphism group of the Teichmüller space $T(S)$ is $\Gamma(S)$ if S is a closed surface of genus $g \geq 2$.

(ii) An analogous procedure on appropriate subsurfaces produces pseudo-hyperbolic elements with prescribed type.

(iii) Not all hyperbolic elements arise in this fashion. For instance, since the constructed maps have even translation distance on $\mathcal{K}(S)$, one expects that these elements are not always primitive.

It is possible to use the methods of this paper to provide simple proofs of other results about subgroups of mapping class groups. Along these lines, one might hope to find more elementary geometric proofs of other difficult results, such as the resolution of the conjugacy problem and its refinement by L. Mosher [Mo95]. The desire to strengthen Mosher's results was part of the motivation for the work of Masur and Minsky on the curve complex.

References

[Abi80]   W. Abikoff (1980). *Topics in the real analytic theory of Teichmüller space (Lec. Notes Math. 820)*, Springer-Verlag.

[Ber78]   L. Bers (1978). An extremal problem for quasiconformal mappings and a theorem by Thurston, *Acta Math.* **141**, 73–98.

[Bir74]   J.S. Birman (1974). *Braids, Links and Mapping-class Groups (Ann. Math. Studies 82)*, Princeton University Press..

[Fe48]   W. Fenchel (1948). Estensioni di gruppi e transformazioni periodiche delle superficie, *Atti Accad. Lincei* **5**, 326–329.

[FLP79]   A. Fathi, F. Laudenbach and V. Poenaru (1979). *Travaux de Thurston sur les surfaces (Astérisque 66-67)*, Soc. Math. de France.

[Har88]   J. Harer (1988). *The cohomology of the moduli space of curves (Lec. Notes Math. 1337)*, Springer-Verlag, 138–221.

[Hrv79]   W.J. Harvey (1979). Geometric structure of surface mapping-class groups. In *Homological Methods in Group Theory, edited by C.T.C. Wall (LMS Lecture Notes 36)*, Cambridge University Press, 255–269.

[Hrv81] W.J. Harvey (1981). Boundary structure of the modular group. In *Riemann Surfaces and Related Topics, (Stony Brook, N.Y.), Ann. Math. Studies* **97**, 245–251.

[Hrv93] W.J. Harvey (1993). On certain families of compact Riemann surfaces. In *Mapping Class Groups and Moduli Spaces, Contemp. Math.* **150**, Amer. Math. Soc., 137–148.

[HT85] M. Handel and W.P. Thurston (1985). New proofs of some results of Nielsen, *Adv. Math.* **56**, 173–191.

[Iva97] N.V. Ivanov (1997). Automorphisms of complexes of curves and of Teichmüller spaces, *Internat. Math. Res. Notices* **14**, 651–666.

[Iva02] N.V. Ivanov (2002). Mapping Class Groups (Chapter 12). In *Handbook of Geometric Topology, edited by R.J. Daverman and R.B. Sher*, Elsevier Science B.V..

[Lic64] W.B.R. Lickorish (1964). A finite set of generators for the homeotopy group of a 2-manifold, *Proc. Camb. Phil. Soc.* **60**, 769–778.

[Mas81] H. Masur (1981). Dense geodesics in moduli space. In *Riemann Surfaces and Related Topics (Stony Brook Conference), Ann. Math. Studies* **97**, 417–438.

[Min00] Y.N. Minsky (2000). Kleinian groups and the complex of curves, *Geometry and Topology* **4**, 117–147.

[Min01] Y.N. Minsky (2001). Bounded geometry for Kleinian groups, *Invent. Math.* **146**, 143–192.

[MM] H.A. Masur and Y.N. Minsky. Geometry of the complex of curves.
(1999). I: Hyperbolicity, *Inv. Math.* **138**, 103–149.
(2000). II: Hierarchichal structure, *Geom. and Funct. Anal.* **10**, 902–974.

[Mo95] L. Mosher (1995). Mapping class groups are automatic, *Ann. Math.* **142**, 303–384.

[Se77] J.P. Serre (1977). *Arbres, Amalgames et SL_2 (Astérisque **46**)*, Soc. Math. de France.

[Ti70] J. Tits (1970). Sur le groupe des automorphismes d'un arbre, *Essais sur la Topologie*, Springer-Verlag, 188–211.

[Thu79]   W.P. Thurston (1979). The geometry and topology of three-manifolds, *Princeton University Lecture Notes*.
http://www.msri.org/publications/books/gt3m/

[Thu88]   W.P. Thurston (1988). On the geometry and dynamics of diffeomorphisms of surfaces I, preprint 1976, revised and published, *Bull. Amer. Math. Soc.* **19**, 417–431.

W. J. Harvey

Department of Mathematics
King's College London
Strand, London, WC2R 2LS
U.K.

bill.harvey@kcl.ac.uk

AMS Classification: 20H10, 30F10, 30F60, 57M60

Keywords: curve complex, mapping clas group

Part II

Once-punctured tori

Kleinian Groups and Hyperbolic 3-Manifolds   Y. Komori, V. Markovic & C. Series (Eds.)
Lond. Math. Soc. Lec. Notes 299, 183–207   Cambridge Univ. Press, 2003

On pairs of once-punctured tori

Troels Jørgensen[1]

Abstract

This work is a detailed study of the space of quasifuchsian once punctured torus
groups in terms of their Ford (isometric) fundamental polyhedra. The key is a
detailed analysis of how the pattern of isometric planes bounding the polyhedra
change as one varies the group.

1. Introduction

One possible approach to "Kleinian groups" is to ask: "How do they look?" It makes
sense when the groups have been associated with natural fundamental polyhedrons.

In the case of Fuchsian groups, satisfactory answers were known to Fricke [FK26],
but generally the situation becomes rather complicated.

It is natural to restrict the considerations to finitely generated groups and, hence-
forward, we shall do so, since, in many respects, the class of groups which cannot be
generated by a finite number of Möbius transformations seems to be too extensive for
general studies – see for instance the examples of Abikoff [Abi71], [Abi73].

In preparation for an intuitive treatment of Kleinian groups, Ahlfors' finiteness
paper [Ahl65a] contains much of the ground material. Also, it indicates one direction
in which the above question might be specified, for instance, in order to attack the
problems related to the characteristics of the limit sets, namely, whether the set of
limit points situated on the boundary of the Dirichlet fundamental polyhedron is finite.
Ahlfors proved that it has zero area [Ahl65b].

Contributing techniques from 3-dimensional topology, Marden [Mar74] described
those groups which have finite sided polyhedrons and observed that they are stable in
the sense of Bers [Ber70b].

---

[1]*Editor's note.* This paper first appeared in preprint form around 1975. Although not widely circulated
it became hugely influential, inspiring many recent developments in 3-dimensional hyperbolic geometry,
including some of Thurston's original insights about geometrization of 3-manifolds. Despite its importance,
the paper was never published and has long been difficult to obtain. We are therefore delighted to have
Jørgensen's permission to print it, unmodified, for the first time.

It is the groups with infinite polyhedrons which present the main difficulties. An important result is due to Scott [Sco73a]: Every finitely generated discrete group of Möbius transformations has a finite presentation.

Unlike the polyhedron, the Ford polygon [For51] is always finite-sided [BJ75]; this is another formulation of the finiteness theorem, introduced by J. R. McMillan. Hence, as to the visual understanding one would like at least to some extent to be able to recognize the small deformations of a group by their effects on its polygon. However, it demands that the infinite polyhedrons possess a certain rigidity near the limit points on their boundaries which has not yet been proved.

One knows that the set of non-elementary discrete groups is closed in regard to continuous deformations and that each of its connected components consists of isomorphic groups [Jør76].

This paper is about the space of quasi-fuchsian groups, freely generated by two loxodromic Möbius transformations X and Y with parabolic commutator

$$K = XYX^{-1}Y^{-1}.$$

It is one of the simplest spaces within the scope of the above remarks, not large enough to illustrate all the troubles which occur in the deformation theory, but it is concrete which, I find, compensates for the lack of generality.

The pioneering papers of Bers [Ber70a] and Maskit [Mak70] about Teichmüller spaces and their boundaries were one of the challenges leading to this study.

2. Matrices and traces

To begin with, we shall mention some elementary facts about Lie products of Möbius transformations which are worth knowing, although they are of minor importance as to our main subject.

In $\mathrm{SL}(2,\mathbb{C})$, most elements have square roots. If E denotes the unit element and τ is the trace function, then we have

$$A + A^{-1} = \tau(A)E$$

or, equivalently,

$$A^2 + E = \tau(A)A.$$

Therefore, an element B whose trace is different from -2 has the square roots

$$\pm(2 + \tau(B))^{-\frac{1}{2}}(B + E).$$

The elements with traces equal to 0 are square roots of $-E$, and $-E$ is the only element with trace -2 for which a square root exists.

The determinant of $AB - BA$ is equal to $2 - \tau(ABA^{-1}B^{-1})$. Hence, if $\tau(ABA^{-1}B^{-1})$ is different from 2, then $AB - BA$ determines a Möbius transformation $\varphi(A,B)$ which is elliptic of order 2.

If A and B represent two Möbius transformations which have no common fixed point in the complex plane, then φ is well defined and we have

$$A^{-1} = \varphi A \varphi^{-1} \quad \text{and} \quad B^{-1} = \varphi B \varphi^{-1}.$$

It follows easily that the axis of φ in the hyperbolic space is the common perpendicular on the axis of A and the axis of B, suitably interpreted in case of A or B being parabolic. Equivalently, in the complex plane the fixed points of φ are the pair of points which is harmonic with the fixed points of A and with the fixed points of B. One may observe, that

$$\varphi = \varphi(A,B) = (-ABA^{-1}B^{-1})^{\frac{1}{2}}BA.$$

Notice that every Möbius transformation has a square root; in general it is not unique.

Proposition 2.1. *If G is a quasi-fuchsian group freely generated by two loxodromic transformations A and B with parabolic commutator, then the two fixed points of $\varphi(A,B)$ are separated by the Jordan curve carrying the limit set of G.*

Proof. If G is Fuchsian, then it is easy to see that the fixed points of the parabolic transformations $ABA^{-1}B^{-1}$, $BA^{-1}B^{-1}A$, $A^{-1}B^{-1}AB$ and $B^{-1}ABA^{-1}$ lie in the given cyclic order on the invariant circle of G. Hence, in succession, they are the points of tangency of four circles a_+, b_+, a_-, and b_- which are perpendicular on the invariant circle and paired in the sense that A maps a_- onto a_+ and B maps b_- onto b_+.

More generally, G is conjugate to a Fuchsian group by a quasi-conformal homeomorphism of the extended complex plane. It follows that G has a fundamental set Ω with two connected components and bounded by four quasi-circles a_+, b_+, a_- and b_- which are the images of four defining circles of a Fuchsian group as described above.

Each element of $G \setminus \{\text{id}\}$ is represented as a finite product of powers of A and B by turns. Denote by $F(a_+)$, $F(b_+)$, $F(a_-)$ and $F(b_-)$ the sets of attractive fixed points of elements "beginning with" a positive power of A, B, A^{-1} and B^{-1}, respectively. The fixed points of parabolic elements may be considered as being attractive. Each of the four sets lies in the quasi-disc exterior to Ω which is bounded by the curve marking it. Together, they are dense in the limit set of G.

By conjugation, φ defines an automorphism of G. Explicitly, it is given by

$$\prod_j A^{m_j} B^{n_j} \rightarrow \prod_j A^{-m_j} B^{-n_j}.$$

Since φ maps attractive fixed points onto attractive fixed points, it interchanges $F(a_+)$ and $F(a_-)$ and, also, it interchanges $F(b_+)$ and $F(b_-)$. It follows easily that the fixed points of φ cannot be limit points of G, and since they lie in separate components if G is Fuchsian, it must be so in general, by continuity. $\qquad\Box$

Next, the aim is in a suitable fashion to represent the groups to be considered as subgroups of $\mathrm{SL}(2,\mathbb{C})$.

Let X and Y be two loxodromic Möbius transformations. Certain choices make the group $\langle X,Y \rangle$ generated by X and Y a Schottky group. Such a group may be deformed into a quasi-fuchsian group of the first kind by making the commutator $K = XYX^{-1}Y^{-1}$ parabolic. All groups of the latter type lie on the boundary of the Schottky space of genus 2! They have finite polyhedrons with one double-sided cusp which then can be opened.

Now, we represent X and Y by matrices in $\mathrm{SL}(2,\mathbb{C})$ and let the elements of the group $\langle X,Y \rangle$, which is free, be represented by the corresponding matrix products. An easy investigation shows that K, being parabolic, must have trace $\tau(K) = -2$; otherwise X and Y have a common fixed point and $\langle X,Y \rangle$ cannot be discrete.

By conjugation, it can be achieved that ∞ is the fixed point of K. We take $K(z) = z+2$. As generators A and B, we shall only consider such pairs for which $ABA^{-1}B^{-1} = K$; they will be spoken of as generator pairs and $\{A, AB, B\}$ is then called a generator triple. The matrices of a generator pair $\{A,B\}$ always have the form

$$A = \begin{pmatrix} * & * \\ \tau(A) & * \end{pmatrix} \quad \text{and} \quad B = \begin{pmatrix} * & * \\ -\tau(B) & * \end{pmatrix}, \tag{2.1}$$

which shows that the difference between the pole of a generator and the pole of its inverse is equal to plus or minus 1.

One further conjugation leads to a normalization; taking 0 as pole of XY (and of YX), the matrices may be given by

$$K = \begin{pmatrix} -1 & -2 \\ 0 & -1 \end{pmatrix}$$

$$X = \begin{pmatrix} \tau(X) - \tau(Y)\tau(XY)^{-1} & \tau(X)\tau(XY)^{-2} \\ \tau(X) & \tau(Y)\tau(XY)^{-1} \end{pmatrix}$$

$$Y = \begin{pmatrix} \tau(Y) - \tau(X)\tau(XY)^{-1} & -\tau(Y)\tau(XY)^{-2} \\ -\tau(Y) & \tau(X)\tau(XY)^{-1} \end{pmatrix}$$

$$XY = \begin{pmatrix} \tau(XY) & -\tau(XY)^{-1} \\ \tau(XY) & 0 \end{pmatrix}$$

where the traces satisfy the identity

$$\tau(X)\tau(Y)\tau(XY) = \tau(X)^2 + \tau(Y)^2 + \tau(XY)^2.$$

We denote by T the connected space of quasi-fuchsian groups arising from the above representation. The traces may serve as parameters and to make their signs unambiguous, we demand that the Fuchsian groups in T be defined by three positive numbers. The formula

$$\tau(A)\tau(B) = \tau(AB) + \tau(AB^{-1})$$

is useful as to change of parameters, and the essential structure on the set of generators in a free group of rank 2 may be pictured by an infinite triangle-graph in which the traces of each pair of generators appear as opposite end-points of one edge. Locally, the graph looks as sketched:

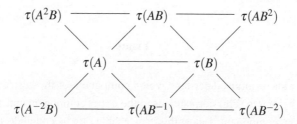

To obtain fundamental regions for the groups in T, we shall make use of isometric circles and hemispheres. For a Möbius transformation g which does not fix ∞, the equation $|dg(z)| = |dz|$ determines a circle which is called the isometric circle of g. Certain incidence relations between isometric circles of elements of a group follow directly from the chain rule for differentiating composite functions.

Geometrically, it is easy to see that in a discrete group containing the transformation $z \rightarrow z+1$, no element has an isometric circle whose radius exceeds 1.

If one extends the groups in T by adjoining the square root of K or, what amounts to the same thing, the Lie products of pairs of generators, then the discreteness is preserved. Thus, referring to (1), it follows in particular that all generators have traces of absolute value greater than or equal to 1.

Since isometric circles of elements close to the identity are large, one may also conclude that every group which can be approximated by groups in T or subgroups thereof is itself discrete and free.

If $\{A, B\}$ is a generator pair for a group in T, then the Lie product of A and B has the same isometric circle as AB and its two fixed points in the plane are

$$(AB)^{-1}(\infty) \pm i \, \tau(AB)^{-1}.$$

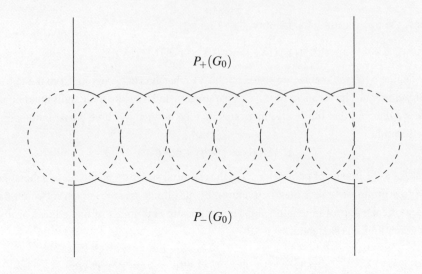

$P_+(G_0)$

$P_-(G_0)$

Figure 1

The signs plus or minus determine which component of the ordinary set of the group the points belong to. Plus corresponds to the upper component and minus corresponds to the lower component. Later, this fact enables us to conclude that if a generator appears as accidental parabolic transformation for a group on the boundary of T, then it has trace $+2$.

3. Polygons and polyhedrons

The group G_o is given by $\tau(X) = \tau(Y) = \tau(XY) = 3$. It is a Fuchsian group with two once-punctured tori as quotient surfaces. A fundamental polygon $P(G_o) = P_+(G_o) \cup P_-(G_o)$ for G_o is sketched in Figure 1. From the left, the seven circles are the isometric circles of

$$X^{-1}Y^{-1}, Y^{-1}, X, \left\{ \begin{array}{c} YX \\ XY \end{array} \right\}, Y, X^{-1}, \text{ and } Y^{-1}X^{-1}.$$

By $K = XYX^{-1}Y^{-1}$, the straight line to the left is mapped onto the straight line to the right. The two lines and the seven circles each determine a hyperbolic plane in the upper half space H. The points in H which lie between the two half planes and exterior to each of the seven isometric hemispheres constitute a fundamental polyhedron for G_o with respect to the action on H. This polyhedron is denoted by $Ph(G_o)$.

It is easy to check the above assertions. A polyhedron with boundary identifications as $Ph(G_o)$ is a fundamental polyhedron for the group generated by the side-pairing

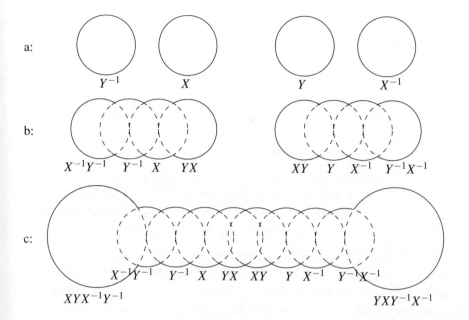

Figure 2

transformations. This follows from a theorem which goes back to Poincaré [Mak71]. Moreover, since X and Y both preserve the upper half plane, they generate a Fuchsian group. The signatures of the quotient surfaces can be read off from the polygon.

As a curiosity, one may observe that the same polygon as $P(G_o)$, but with other identifications on the boundary, arises from the group generated by

$$g = \begin{pmatrix} i & -\frac{1}{3}i \\ 3i & -2i \end{pmatrix} \quad \text{and} \quad h = \begin{pmatrix} i & -\frac{1}{3}i \\ -3i & -2i \end{pmatrix}.$$

In this case $g^{-2}h^2(z) = z + 2$ and the seven isometric circles belong to

$$h^2, h, h^{-1}, \left\{ \begin{matrix} h^{-2} \\ g^{-2} \end{matrix} \right\}, g^{-1}, g, g^2.$$

Again the limit set is the real axis including ∞, but the quotient merely consists of one twice-punctured torus.

Figure 2 indicates how G_o can be approximated by Schottky groups, each having ∞ as an ordinary point, in such a way that $P(G_o)$ becomes the limit of their Ford polygons. For a moment, let X and Y denote hyperbolic Möbius transformations preserving the real axis. To begin with, assume that the isometric circles $I(X), I(X^{-1}), I(Y)$ and

$I(Y^{-1})$ lie as sketched on Figure 2a. Then X and Y generate a Schottky group of genus 2; its Ford polygon consists of the points lying exterior to each of the four circles. By pushing $I(X)$ and $I(Y^{-1})$ together and similarly $I(X^{-1})$ and $I(Y)$, this group can be deformed into a group with a Ford polygon as sketched on Figure 2b. Finally the two blocks of circles may be pushed together. Then $I(YX)$ and $I(XY)$ meet, $I(K)$ breaks out through $I(X^{-1}Y^{-1})$ and $I(K^{-1})$ through $I(Y^{-1}X^{-1})$: Figure 2c shows a Ford polygon of this type. The groups remain Schottky groups until $I(YX)$ and $I(XY)$ coincide and K becomes parabolic with ∞ as its fixed point.

Returning to the groups in T, isometric circles and hemispheres have exactly those elements of G which do not belong to the cyclic subgroup $\{K^n\}_{n\in Z}$ stabilizing ∞. The isometric circle $I(g)$ of an element $g \in G$ coincides with $I(K^n g)$ for each $n \in Z$, while $I(gK^n)$ is the image of $I(g)$ under the translation $z \to z - 2n \, (= K^{-n}(z))$.

Denote by $\widetilde{Ph}(G)$ the subset of the upper half space which consists of all points lying exterior to each of the isometric hemispheres defined by G. The points of $\widetilde{Ph}(G)$ which lie between an euclidean plane perpendicular on the real axis and its image under K form a fundamental polyhedron $Ph(G)$ for G. Among the two-dimensional facets of the boundary of $Ph(G)$ or $\widetilde{Ph}(G)$, those which are situated on an isometric hemisphere will be spoken of as faces. A face or an edge is said to be outer or inner according as it has points in common with the complex plane or not.

Denote by $\widetilde{P}(G)$ the subset of the complex plane which consists of all points lying exterior to each of the isometric circles defined by G. The intersection of $\widetilde{P}(G)$ with a strip of breadth 2 which is bounded by lines perpendicular on the real axis constitutes a fundamental polygon $P(G)$ for G.

The two quotient surfaces defined by G correspond to an upper polygon $P_+(G)$ and a lower polygon $P_-(G)$, respectively, and $P(G)$ is the union of these two parts. Accordingly, $\widetilde{P}(G)$ is considered as the union of two parts $\widetilde{P_+}(G)$ and $\widetilde{P_-}(G)$. The upper and the lower polygons play symmetric roles and the symbols $P_*(G)$ and $\widetilde{P}_*(G)$ are used when distinction is unessential.

The aim is to describe how $\widetilde{Ph}(G)$ looks and how it changes under deformation of G.

Theorem 3.1. *For each group in T, the polyhedron has the following characteristics:*

(i) *In H, each point belonging to the boundary of $\widetilde{Ph}$ lies on the isometric hemi-sphere of a generator.*

(ii) *In H, each vertex of $\widetilde{Ph}$ is the common point of exactly three different faces of $\widetilde{Ph}$. The three faces meeting at such a vertex are situated on isometric hemi-spheres defined by a generator triple $\{A,AB,B\}$.*

(iii) Each vertex of $\widetilde{Ph}$ belonging to the complex plane lies on the boundary of $\widetilde{P}$.

(iv) $\widetilde{P}_*$ is connected and simply connected.

(v) The boundary of $\widetilde{P}_*$ looks as described in a or in b, i.e. two possibilities occur:

> *a) There is a generator triple $\{A, AB, B\}$ such that a characteristic part of the boundary of $\widetilde{P}_*$ is made up by sides lying in the following succession on the isometric circles of*
>
> $$\ldots, A^{-1}B^{-1}, B^{-1}, A, \left\{ \begin{matrix} BA \\ AB \end{matrix} \right\}, B, A^{-1}, B^{-1}A^{-1}, \ldots$$
>
> *From these, the full collection of sides arises by translations with powers of K.*

> *b) There is a generator pair $\{A, B\}$ such that a characteristic part of the boundary of $\widetilde{P}_*$ is made up by sides lying in the following succession on the isometric circles of*
>
> $$\ldots, \quad B^{-1}, \quad A, \quad B, \quad A^{-1}, \quad \ldots$$
> $$(A^{-1}B^{-1})\ (AB^{-1})\ \begin{pmatrix} BA \\ AB \end{pmatrix}\ (BA^{-1})\ (B^{-1}A^{-1})$$
>
> *As indicated, each vertex lies on the isometric circles defined by a generator triple. Translations with powers of K yield the full collection of sides.*

The proof of Theorem 1 is based on the method of geometric continuity. The crucial observation is that polyhedrons which look as described above can only be deformed into polyhedrons having the same characteristics as long as the groups defining them remain inside T. In order to establish it and, thereby, prove the theorem, certain geometrical investigations will be made.

First notice that if $I(AB)$ and $I(B)$ intersect, then also $I(A)$ and $I(B^{-1})$ intersect and $I(A^{-1})$ and $I(B^{-1}A^{-1})$ intersect. The sum of the exterior angles determined by these three pairs of circles is 2π. Assuming that A, AB and B form a generator triple, it follows from (1) in Section 2 that $I(A^{-1})$ and $I(B^{-1}A^{-1})$ are images of $I(A)$ and $I(AB)$ by the translation $z \to z + 1$. Hence, also $I(A)$ and $I(AB)$ intersect and the sum of the angles between $I(AB)$ and $I(B)$ and $I(AB)$ and $I(A)$ is greater than π. Therefore, if the three isometric circles defined by a generator triple mutually cut each other, then they make with each other angles whose sum exceeds π and, hence, the intersection of the finite open discs they bound is not void. This fact makes 3 a rather easy consequence of 2 and, also, it is essential as to the proof of 4.

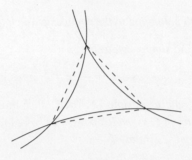

Figure 3

In the following it is used that the isometric circle of (a transformation) g is tangent from the inside to the isometric circle of h and x is the common point if and only if the isometric circles of h^{-1} and gh^{-1} are externally tangent with $h(x)$ as the common point.

Consider a polygon $\widetilde{P}_*$ (of a group in T) which looks as described in 4 and 5a. Such a polygon is said to be of type $[A,AB,B]$. One may say as well that it is of type $[AB,ABA^{-1},A^{-1}]$; different generator triples determining the same polygon correspond to a fixed trace triangle. The vertices of $\widetilde{P}_*$ fall into two equivalence classes, each side joining inequivalent points. The angles at three consecutive equivalent vertices add up to 2π. Using (1), one observes that the isometric circles of BA, B, A^{-1} and $B^{-1}A^{-1}$ are the images of the isometric circles of $A^{-1}B^{-1}$, B^{-1}, A and AB under the translation $z \rightarrow z + 1$. This periodicity shows that the angles at any three consecutive vertices of $\widetilde{P}_*$ add up to 2π. Combining it with the fact that the sum of the angles in a triangle bounded by circular arcs as shown on Figure 3 is less than π, an easy inspection shows that if the polygon (and the group) is subject to a small deformation, then no two sides of $\widetilde{P}_*$ can become externally tangent. It implies that no new side of $\widetilde{P}_*$ breaks out through an old side under deformation and that no new connected component of $\widetilde{P}_*$ results from cutting off a region of the original polygon.

A connected polygon $\widetilde{P}_*$ is said to be of type $[A,B]$ if it looks as described in 4 and 5b. One may say as well that such a polygon is of type $[ABA^{-1},A^{-1}]$; different generator pairs determining the same polygon correspond to a fixed edge in the trace graph. Four consecutive vertices of a polygon of type $[A,B]$ form a cycle as vertices of a fundamental polygon P_* and, because of periodicity, the angles at any two neighbouring vertices have sum π. It is easy to verify that no two sides of such a polygon can become externally tangent and, therefore, that no new side breaks out through an old side when the group is deformed a little. In particular, no new connected component of $\widetilde{P}_*$ results from cutting of a region of the original polygon.

The closure of P_* makes up a connected surface if equivalent points on the boundary of P_* are identified. Therefore, in order to destroy the connectedness by deforming a polygon $\widetilde{P}_*$ which looks as described in Theorem 1 new sides must break out through the vertices (since not through the sides) and be paired to sides bounding new components. If one assumes that the first three assertions of Theorem 1 hold, it follows that no such new component can arise; it would have to be contained in a triangle like the one sketched on Figure 3 determined by the isometric circles of a generator triple, and it was shown that such a configuration cannot occur.

Once it is known that the polygons remain connected, it is rather easy to see what can happen under deformation.

A polygon $\widetilde{P}_*$ of type $[A, AB, B]$ is determined by three consecutive sides. By deformation, it may happen that one among these disappears (together with its images under the cyclic group generated by $z \to z + 1$, the square root of K). According as the side on $I(A)$, $I(AB)$ or $I(B)$ disappears, the new polygon becomes of type $[AB, B]$, $[A, B]$ or $[A, BA]$. If two among three defining sides disappear simultaneously or if one of two defining sides of a polygon which looks as described in 5b disappears, then the side-pairing generator corresponding to the remaining side becomes parabolic and, hence, the new group does not belong to T, but to the boundary of T. If a polygon of type $[A, B]$ is deformed into a polygon of different type, then there are two possibilities: either the new type is $[A, AB, B]$ or it is $[AB^{-1}, A, B]$. In the former case $I(AB)$ carries a new side of $\widetilde{P}_*$ while $I(AB^{-1})$ has moved away from $\widetilde{P}_*$ and in the latter case it is $I(AB^{-1})$ which carries a side of $\widetilde{P}_*$ and $I(AB)$ which lies exterior to $\widetilde{P}_*$.

The previous description indicates that classification by types of polygons supplies the deformation space of the punctured torus with a tessellation which looks like the one coming from the triangle graph in the plane, the vertices, however, corresponding to boundary points. Modulo the assumption of connectivity, the above description exhausts the different possibilities, because additional identifications on the boundary of $\widetilde{P}_*$ would imply the existence of elliptic fixed points or parabolic fixed points, inequivalent to ∞ on the boundary, and it is impossible.

Mainly, the purpose in giving the next argument is to explain things from a slightly different point of view; as to the proof of Theorem 1, one can do without.

Every polygon P_* of a group in T defines a torus. By projection of the boundary of the polygon, one obtains a graph on the torus whose edges come from pairs of equivalent sides of P_* and whose vertices come from cycles of equivalent vertices of P_*. The fixed point at infinity constitutes one cycle. All other cycles contain at least three distinct points! It is no restriction to assume that P_* is finite sided. If $2s$ is the

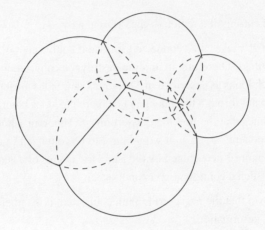

Figure 4

number of sides of P_* and c is the number of cycles, then one gets

$$2s + 2 \geq 3c,$$

using that the number of sides is equal to the number of vertices. Assume further that P_* is connected and simply connected. The first assumption can be satisfied if $\widetilde{P}_*$ is connected. The latter assumption is easily justified, because the limit set of the group is connected (a Jordan curve) and each isometric circle defined by an element of the group contains a limit point. Under these assumptions one deduces that

$$s - 1 = c,$$

since the graph has s edges and c vertices and the genus is 1. It follows that $c \leq 4$. In the opposite direction, it is clear that $c > 1$. An examination shows that $\widetilde{P}_*$ must look as described in 5. If $c = 2$, then $\widetilde{P}_*$ is of type $[A, B]$ for some pair of generators. If $c = 3$, then in most cases $\widetilde{P}_*$ is of type $[A, B]$, but the type may be $[A, AB, B]$. If $c = 4$, then $\widetilde{P}_*$ is of type $[A, AB, B]$.

After these preliminary studies of the polygons we shall turn the attention to the polyhedrons and investigate the portion of their boundaries which lies above the complex plane. Vertices, edges and faces may be pictured by their images under the orthogonal projections of H onto $\mathbb{C}$. Angles between the faces can be read off in the plane as exterior angles between the corresponding isometric circles, and the images by the projection are completely and easily determined by these circles (see Figure 4). Each facet is mapped onto a convex set!

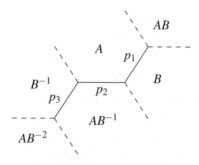

Figure 5

Figure 5 is supposed to originate in a polyhedron $\widetilde{Ph}$ which looks as described in Theorem 1. The aim is to explain that under deformation of such a polyhedron, the only essential replacements of its boundary take place at the boundary of the corresponding polygon $\widetilde{P}$, i.e. in the complex plane. It means that inner faces or edges do not degenerate as long as the groups belong to T. Only as the types of polygons change, inner faces may become outer and outer faces may disappear or, in the opposite direction, new outer faces may arise and outer faces become inner. In particular, it becomes clear that inner faces can only arise from outer faces being "pushed in", because the polyhedrons obtained as described above possess the properties which make Poincaré's theorem about fundamental polyhedrons applicable. More directly, one can show that additional identifications on the boundaries would imply the existence of elliptic fixed points contrary to the reality. It follows that the first assertion in Theorem 1 holds, i.e. isometric hemispheres of generators determine $\widetilde{Ph}$.

The segments p_1, p_2 and p_3 in Figure 5 picture the projections of three edges e_1, e_2 and e_3 which join vertices of $\widetilde{Ph}$ lying above the complex plane. By observing how the edges arise, it becomes clear that every edge of $\widetilde{Ph}$ having two points in H as its extremities belongs to a triple like e_1, e_2 and e_3, where e_1 and e_3 are equivalent and e_2 is equivalent to the image of e_1 under the square root of K. Since the polyhedron is isometric, the euclidean length of e_1, e_2 and e_3 are equal. Therefore, if one edge degenerates, then all three edges collapse into one point, and if the groups converge, then a generator becomes parabolic with this point as its fixed point and the new group can not belong to T.

Let F be a face of $\widetilde{Ph}$ situated on the isometric hemisphere of a generator A. In regard to hyperbolic geometry, F is convex and the inductive procedure for proving Theorem 1, furthermore, permits us to assume that it is symmetric relatively to the axis carrying the fixed points of $K^{-\frac{1}{2}}A$ (which is elliptic of order 2 and has the same isometric circles as A). Any two vertices of F which belong to H and are opposite as to the symmetry will be equivalent under the group in question.

Now, assume that F degenerates under deformation by which the corresponding groups converge to a group inside T. Then, at least from a certain stage, F must be an outer face, either with three or four vertices all of which belong to the complex plane or with three vertices among which one and only one lies above the plane.

To verify this assertion, first consider the possibility of both $\widetilde{P_+}$ and $\widetilde{P_-}$ having points in common with F. Hence, F necessarily shrinks to its axis of symmetry and it can not posses any vertices in H since pairs of opposite vertices then would collapse into a fixed point of a generator, contrary to the assumption that the groups remain inside T. In all other cases which have to be excluded, F would possess an edge having two points in H as its extremities and in the limit F had to become an edge, either situated on the axis of symmetry or being perpendicular on it. In the first case pairs of opposite vertices would converge towards a fixed point. To obtain a contradiction in the latter case too, one may observe that an adjacent face, situated on the isometric hemisphere of B, say, would degenerate together with F so that the mid-point of the resulting edge were fixed under both $B^{-1}K^{-\frac{1}{2}}$ and $K^{-\frac{1}{2}}A$, thus, by AB^{-1}. Of course, this is absurd.

Let us emphasize the geometrical facts attached to the symmetries which arise from the set L of Lie products of pairs of generators.

Theorem 3.2. *For each group in T, the polyhedron has the following characteristics:*

(vi) *In regard to hyperbolic geometry, each face of $\widetilde{Ph}$ is symmetric relatively to the axis of fixed points of one element from L.*

(vii) *Each vertex of $\widetilde{Ph}$ which lies above the complex plane is fixed by one element from L.*

For every Fuchsian group in T, the types of the upper polygon and the lower polygon are equal, and the properties 1,…,7 are easily verified. Also, for such a group it is clear what are the possible effects on $\widetilde{Ph}$ by deformations. Only if the types of $\widetilde{P_+}$ and $\widetilde{P_-}$ become different, then inner vertices of $\widetilde{Ph}$ arise.

We shall describe what happens when $\widetilde{Ph}$ gets new vertices inside H. Starting off from a group which has a polygon $\widetilde{P_*}$ of type $[A, AB, B]$ and for which no face of $\widetilde{Ph}$ lies on the isometric hemisphere of AB^{-1}, first the side of $\widetilde{P_*}$ which lies on $I(AB)$ shrinks to its midpoint so that $\widetilde{P_*}$ becomes of type $[A, B]$. In the critical situation all the vertices of $\widetilde{P_*}$ are equivalent. Thereafter, as $I(AB^{-1})$ breaks out making $\widetilde{P_*}$ of type $[AB^{-1}, A, B]$, this string of equivalent vertices gets lifted into H. One of the new vertices inside H

belongs to the isometric hemisphere of AB; it is a fixed point of $K^{-\frac{1}{2}}AB$. Another new vertex in H lies on the isometric hemisphere of AB^{-1} which carries an outer face with three edges; it is a fixed point of $K^{-\frac{1}{2}}AB^{-1}$ and so is the midpoint of the new side of $\widetilde{P}_*$ lying on $I(AB^{-1})$.

The general incidence relations between isometric circles make sure that the symmetries are still extant after further deformation. Hence, the relative stability of the boundary of Ph can be proved using that the symmetry axis cannot intersect each other since, for a group G in T, the product of any two Lie products from L belongs to $G \setminus \{K^n\}$ while the only elements in G having fixed points on the boundary of $\widetilde{Ph}$ are the powers of K.

Owing to the previous investigations, we may conclude that the asserted characteristics of $\widetilde{Ph}$ are preserved under continuous deformations. Since T is connected, this proves the theorems.

It is possible to describe $\widetilde{Ph}$ rather detailed provided the types of $\widetilde{P_+}$ and $\widetilde{P_-}$ are given.

Consider the triangle graph. To each pair Δ_+ and Δ_- of triangles, there corresponds a unique shortest chain of triangles, each having an edge in common with its successor, for which Δ_+ and Δ_- are the two extremities. These triangles are said to lie between Δ_+ and Δ_-. The set of triangles lying between Δ_+ and Δ_- is denoted by $[\Delta_+,\Delta_-]$.

Referring to Theorem 1, there is a natural mapping from the set of faces of $\widetilde{Ph}$ into the triangle graph, the image of a face simply being the vertex determined by the trace of the generator whose isometric hemisphere supports the face.

Theorem 3.3. *For a group in T whose upper polygon is of type $[A_+,A_+B_+,B_+]$ and whose lower polygon is of type $[A_-,A_-B_-,B_-]$, denote by Δ_+ and Δ_-, respectively, the trace triangles determined by the two generator triples. Then, concerning the boundary of $\widetilde{Ph}$, the following is true:*

(viii) *Inside H, the number of edges situated on the isometric hemisphere of a given generator is equal to twice the number of triangles in $[\Delta_+,\Delta_-]$ of which the trace of the generator is a vertex. In particular, there is a bijective correspondence via the hemispheres between on the one hand the faces and on the other hand the generators whose traces are vertices of the triangles between Δ_+ and Δ_-.*

(ix) *If and only if $\tau(A)$ and $\tau(B)$ determine a common edge of two different triangles in $[\Delta_+,\Delta_-]$, then both the A-face and the B^{-1}-face have a common edge and the A-face and the B-face have a common edge.*

Having seen how the more complex polyhedrons arise from the simpler polyhedrons, it is not difficult to verify Theorem 3 by the continuity method, extending the description to the cases where one or both of the polygons look as explained in 5b of Theorem 1.

4. Parameters and proportions

In order to characterize the groups in T and also the groups on the boundary of T, it is useful under various circumstances to know a little about the relative positions of the isometric circles of generators and about the orders of magnitude of these. As already mentioned, the radii do not exceed 1 or, equivalently, the absolute values of traces of generators are bounded from below by 1; this fact is proved in [Jør76]. It will be shown that in some cases the circles cannot be too small either.

Let $\{A, AB, B\}$ be a generator triple and consider the polygonal line which begins at the pole of B^{-1}, passes through the pole of A, next, the pole of AB and terminates at the pole of B. As an easy calculation shows, its segments are

$$A^{-1}(\infty) - B(\infty) = \tau(AB)\,\tau(A)^{-1}\,\tau(B)^{-1}$$
$$B^{-1}A^{-1}(\infty) - A^{-1}(\infty) = \tau(B)\,\tau(AB)^{-1}\,\tau(A)^{-1}$$
$$B^{-1}(\infty) - B^{-1}A^{-1}(\infty) = \tau(A)\,\tau(B)^{-1}\,\tau(AB)^{-1}.$$

Since the difference between the pole of B and the pole of B^{-1} is equal to 1, summation yields

$$1 = \tau(AB)\tau(A)^{-1}\tau(B)^{-1} + \tau(B)\tau(AB)^{-1}\tau(A)^{-1} + \tau(A)\tau(B)^{-1}\tau(AB)^{-1}.$$

Hence, the parameter equation

$$\tau(A)\tau(B)\tau(AB) = \tau(A)^2 + \tau(B)^2 + \tau(AB)^2$$

has been explained geometrically; it originates in the fact that the trace of the commutator is -2. As to the geometrical understanding, it may be convenient to consider triples of "complex probabilities" instead of the trace parameters:

$$1 = a_1 + a_2 + a_3.$$

The products $(a_i a_j)^{-1}$, $i \neq j$, reproduce the squares of the traces and the basic operation, changing parameters, it looks as follows:

$$a_i \to 1 - a_i$$
$$a_j \to a_j a_i (1 - a_i)^{-1}$$
$$a_k \to a_k a_i (1 - a_i)^{-1}.$$

After this insertion, consider again the parameter equation of the traces. Using that

$$\tau(A)\tau(B) = \tau(AB) + \tau(AB^{-1}),$$

one gets

$$\tau(AB)\tau(AB^{-1}) = \tau(A)^2 + \tau(B)^2.$$

This identity can be used to obtain another proof of the fact that if a side of a polygon P_* lies on $I(AB)$, say, then its midpoint is

$$(AB)^{-1}(\infty) * i\, \tau(AB)^{-1}.$$

First, one verifies the statement for the group G_0, given by $\tau(x) = \tau(y) = \tau(xy) = 3$; thereafter, observe that the property is preserved under change of sides.

A more important consequence of the parameter equation is that if one among the traces belonging to a generator triple has relatively small absolute value, but the absolute value of another is large, then also the third trace has relatively large absolute value. Otherwise the identity could not be satisfied, the square of the large element being too predominant. We formulate this fact as a statement about isometric circles.

Lemma 4.1. *If one of the isometric circles belonging to a generator triple is very small, then at least two of then are small. Under deformation, if one circle becomes infinitely small, then two circles become infinitely small.*

Considering the groups in T and on its boundary, it is geometrically clear that not all generators can have small isometric circles. As we shall see, the hemispheres supporting faces with many edges must cover a certain area in the plane separately. Although, formally, we do not yet know about the boundary groups, the following results extend by continuity to all groups in the closure of T.

Lemma 4.2. *Suppose $\{A,B\}$ is a generator pair of a group whose polyhedron has faces lying in succession on the hemispheres of B^{-1}, A, B and A^{-1}. Then we have*

$$|\tau(A)|^{-1} + |\tau(B)|^{-1} > \frac{1}{2}.$$

The proof is trivial. The edge between $Ih(B)$ and $Ih(A^{-1})$ is the image of the edge between $Ih(B^{-1})$ and $Ih(A)$ under the translation $z \rightarrow z + 1$. Also $Ih(A)$ and $Ih(B)$ possess a common edge and, therefore, the sum of the diameter of $I(A)$ and the diameter of $I(B)$ exceeds 1. It is true but unessential that the constant $\frac{1}{2}$ is best possible.

Lemma 4.3. *If τ is the trace of a generator whose hemisphere supports a face with at least 6 edges in H, then we have*

$$|\tau| < 4 + 2\sqrt{5}.$$

Proof. It follows from Theorem 3 and Lemma 2 that τ belongs to a trace triangle in which the two other vertices τ_1 and τ_2 satisfy

$$|\tau|^{-1} + |\tau_i|^{-1} > \frac{1}{2}, \quad i = 1, 2.$$

It is no restriction to assume that $|\tau| > 2$. Then we have

$$|\tau_i| \leq 2|\tau|(|\tau| - 2)^{-1}, \quad i = 1, 2.$$

Writing the parameter equation on the form

$$-\tau^2 = \tau_1^2 + \tau_2^2 - \tau\tau_1\tau_2,$$

we obtain

$$|\tau|^2 < 4|\tau|^2(2 + |\tau|)(|\tau| - 2)^{-2}.$$

Hence we have

$$|\tau|^2 - 8|\tau| - 4 < 0,$$

from which the assertion follows. □

Lemma 4.4. *Let $(A_n, B_n)_{n \in \mathbb{N}}$ be a convergent sequence of generator pairs of groups in T or on the boundary of T. Suppose that $m \in \mathbb{N}$ tends to infinity with n and that the sequence $|\tau(A_n^m, B_n)|_{n \in \mathbb{N}}$ is bounded. Then A_n converges to a parabolic transformation.*

Proof. The following general formula can be verified:

$$(\tau(A^m B A^{-m} B^{-1}) - 2)(\tau(A)^2 - 4) = (\tau(ABA^{-1}B^{-1}) - 2)(\tau(A^m)^2 - 4).$$

In the present case, omitting the indices, it gives

$$(\tau(A)^2 - 4)(\tau(A^m)^2 + \tau(B)^2 + \tau(A^m B)^2 - \tau(A^m)\tau(B)\tau(A^m B) - 4)$$
$$= -4(\tau(A^m)^2 - 4),$$

showing that if $\tau(A^m)$ become arbitrarily large, then $\tau(A)$ would accumulate at 0 contrary to the fact that $|\tau(A)| \geq 1$. It is used that also $|\tau(B)|$ is bounded. Hence, $\tau(A^m)$ stays inside a compact set as m tends to infinity, and since A cannot approach elliptic transformations, this is only possible if A in the limit becomes parabolic. □

Lemma 4.5. *Consider a convergent sequence of pairs (A, B) generating groups whose polyhedrons each have faces supported by the hemispheres of A and B. Assume that the faces situated on $Ih(A)$ get more and more edges. Then A converges towards a parabolic transformation.*

Proof. Attach to each group an interval of trace triangles as done at the end of Section 3. This is possible also for groups on the boundary of T, as we shall see later, but there the intervals may be infinite in one or both directions.

As the face supported by $Ih(A)$ gets more and more edges, the subinterval $[\Delta_+(A), \Delta_-(A)]$ consisting of the triangles with vertices $\tau(A)$ gets longer and longer. If m and n are integers so that $\Delta_+(A)$ is the triangle determined by $\{A, A^m B, A^{m+1}B\}$ and $\Delta_-(A)$ is the triangle determined by $\{A, A^n B, A^{n+1}B\}$, then we can assume without essential loss of generality that m tends to plus infinity and that $n \leq 0$, the latter since also $Ih(B)$ supports a face.

Consider a group whose interval passes by $\Delta_+(A)$ so that also $A^{m+1}BA^m B$ comes into play. In this case $\tau(A^m B)$ belongs to at least 3 different triangles of the interval. Thus, by Theorem 3, the face supported by $Ih(A^m B)$ will have at least 6 edges in H. Using Lemma 3 and Lemma 4, we see that if such polyhedrons occur more that finitely many times as m tends to infinity, then A becomes parabolic.

The alternative is a group with a polygon of type $[A, A^{m+1}B, A^m B]$. Suppose that this case occurs infinite often as m increases. If $|\tau(A^m B)|$ stays bounded, then again Lemma 4 tells us that A becomes parabolic. Otherwise, the isometric circle of $A^m B$ and, by Lemma 1, also the isometric circle of $A^{m+1}B$ will become infinitely small in the limit; the isometric circle of A cannot vanish. By geometry A must become parabolic; the resulting group has a polygon with semicircular sides lying on $I(A)$ and $I(A^{-1})$, tangent to each other at the fixed point of A. $\square$

Before making use of the above result, another set of geometrical quantities shall be introduced.

To a polygon of type $[A, AB, B]$ there exist three positive numbers $\theta(A)$, $\theta(AB)$ and $\theta(B)$ such that

$$B(\infty) * i\tau(B)^{-1}\exp[- * i\theta(B)] = A^{-1}(\infty) * i\tau(A)^{-1}\exp[* i\theta(A)]$$
$$A^{-1}(\infty) * i\tau(A)^{-1}\exp[- * i\theta(A)] = B^{-1}A^{-1}(\infty) * i\tau(AB)^{-1}\exp[* i\theta(AB)]$$
$$B^{-1}A^{-1}(\infty) * i\tau(AB)^{-1}\exp[- * i\theta(AB)] = B^{-1}(\infty) * i\tau(B)^{-1}\exp[* i\theta(B)],$$

where as usual the star stands for plus or minus according as the polygon is upper of lower. Each of these identities arises from expressing the common vertex of two adjoining sides by its position on each of the two circles, relatively to the mid-points of the sides. Thus, $\theta(A)$ measures half the angle from $A^{-1}(\infty)$ which the side on $I(A)$ determines. These angles may and will be assumed to lie in the interval between 0 and $\frac{\pi}{2}$. We shall allow one of the three numbers to be 0; then the polygons corresponding to "side-shifts" are included.

The equations may be rewritten as

$$* \, \tau(B) \, \exp[*i\theta(A)] = i \, \tau(AB) \, * \, \tau(A) \, \exp[- *i\theta(B)]$$
$$* \, \tau(A) \, \exp[*i\theta(AB)] = i \, \tau(B) \, * \, \tau(AB) \, \exp[- *i\theta(A)]$$
$$* \, \tau(AB) \, \exp[*i\theta(B)] = i \, \tau(A) \, * \, \tau(B) \, \exp[- *i\theta(AB)],$$

and it follows immediately that

$$\theta(A) + \theta(AB) + \theta(B) = \frac{\pi}{2},$$

which is one way of saying that each of the quotient surfaces of the groups has hyperbolic area 2π. Knowing this, the three equations are equivalent.

Every group in T has an upper and a lower polygon. To each of these corresponds a trace triangle and a set of side parameters, i.e. the θ's introduced above. Together, these characteristics form what will be called the signature of the group. Essentially, the signature is unique, the only overlappings being in cases of side shifts, i.e. when a side parameter is 0. In the opposite direction one verifies easily that each abstract signature gets realized by one group in T.

Theorem 4.6. *Each pair of trace triangles with associated side parameters* $\{\theta(A_*),$ $\theta(A_*B_*), \theta(B_*)\}$ *satisfying*

$$\theta(A_*) + \theta(A_*B_*) + \theta(B_*) = \frac{\pi}{2},$$

where $* \in \{+, -\}$ *and* $0 \le \theta < \frac{\pi}{2}$, *yields the signature of exactly one group in* T.

The proof is straightforward. As to the existence part, the continuity method fits; it is used that the obstructions which ensure that T has a boundary are few: only polygons with two vanishing side parameters must be evaded.

5. Regularity and degeneration

One way to reach groups on the boundary of T is by introducing accidental parabolic transformations; it corresponds to making two of the side parameters of a polygon

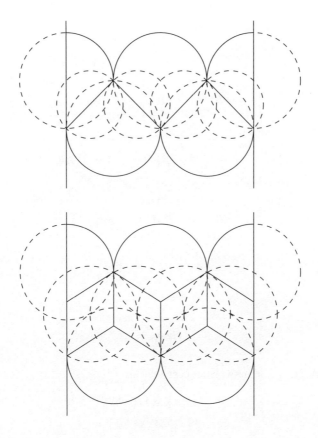

Figure 6

equal to 0. Then the quotient surface becomes a sphere with three punctures. Such a group is called a cusp, provided it still has two quotient surfaces, and it is said to be a double cusp if both of these are spheres.

Figure 6 indicates how two of the simplest kinds of double cusps look. The full-drawn lines and arcs show the projections of the edges of their fundamental polyhedrons. The drawing above arises by taking $\tau(X) = \tau(XY) = 2$ and $\tau(Y) = 2 - 2i$. The drawing below arises by taking $\tau(X) = \sqrt{3} - i$ and $\tau(Y) = \sqrt{3} + i$ so that $\tau(XY) = \tau(XY^{-1}) = 2$. Besides K and K^{-1}, the transformations pairing the faces are

$$X^{-1}Y^{-1}, YX, XY, Y^{-1}X^{-1}$$
$$X, X^{-1}$$

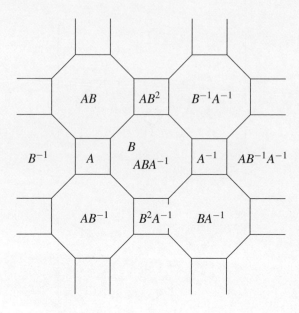

Figure 7

in the former case, while in the latter case they are

$$X^{-1}Y^{-1}, YX, XY, Y^{-1}X^{-1}$$
$$Y^{-1}, X, Y, X^{-1}$$
$$XY^{-1}, YX^{-1}.$$

Necessarily, a boundary group which is not a cusp has most one quotient surface left, i.e. at least one of the polygons has disappeared. The process leading to such a group is called degeneration. If both surfaces degenerate, then the resulting group is not Kleinian, but it is still discrete and one shows easily by means of the geometrical observations made in Section 3 that no accidental parabolic transformation will appear. Figure 7 shows a part of a projection originating in a group of this kind. If just one surface is missing, then the remaining surface can be either a torus with one puncture or a sphere with three punctures.

It is clear that T is not compact. For instance, the subvariety consisting of Fuchsian groups does not lead to groups on the boundary of T. Another fact is that one cannot obtain a double cusp by means of essentially only one accidental parabolic generator. If one tries, then the situation may look as sketched on Figure 8; both $\tau(X)$ and $\tau(Y)$ should be infinitely large in order that XY defined punctures on both surfaces. However this phenomenon is an exception. In particular, the following is true:

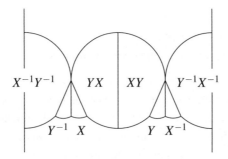

Figure 8

Theorem 5.1. *For each pair of distinct vertices τ_+ and τ_- of the triangle graph, there exists exactly one double cusp on the boundary of T whose accidental parabolic generators relatively to P_+ and P_- are the generators belonging to τ_+ and τ_-, respectively.*

As well as the groups inside T, the cusps are regular in the sense that their polyhedrons have only finitely many faces. Such groups are relatively easy to deal with in regard to deformations. Near a cusp a rather large portion of the space consists of groups in T. Therefore, it is not difficult to get to a cusp, but the various paths leading to it may "look" very different; for certain kinds of approach the polyhedrons do not converge to the polyhedron of the cusp. The most natural way to obtain the cusps is from Theorem 4 by a local compactification, i.e. by allowing the side parameters also to take the value $\frac{\pi}{2}$. It is then straightforward to figure out how their polyhedrons look; roughly, they can be characterized by saying that given the upper and the lower signatures, the set of generators whose isometric hemispheres support the faces is minimal.

It is natural to regard T as the product of the space of deformations of the upper polygon and the space of deformations of the lower polygon. For the study of the Kleinian groups on the boundary of T, it is no serious restriction to keep the signature of one of the polygons fixed and merely consider the deformations of the other.

Let $\{A, AB, B\}$ be an arbitrary but in the following fixed generator triple. Consider the set of groups whose upper polygon is of type $[A, AB, B]$ or of one of the adjacent types. For each upper signature $\theta = (\theta(A), \theta(AB), \theta(B))$, denote by $S(\theta)$ the slice consisting of all groups in the closure of T whose upper signature is the one given by θ. The complex dimension of $S(\theta)$ is 1 and a natural parameter will be the trace of one generator, for instance $\tau(A)$ or $\tau(B)$. By geometry, using Lemma 1, it follows that $S(\theta)$ is compact unless θ defines a sphere; in fact, the absolute values of $\tau(A)$ and $\tau(B)$ cannot be too large except when two of the side parameters given by θ are

0. In the latter case one can compactify $S(\theta)$ by adjoining one point; it corresponds to a generalized boundary group for which two three-punctured spheres with common accidental parabolic generator are the quotient surfaces.

For fixed θ, let $S(\theta)$ be identified with a subset of the sphere of complex numbers by means of a natural parameter. The various types of lower polygons yield an imbedding of the triangle graph so that $S(\theta)$ becomes the union of curvelinear "type triangles" whose vertices lie on the boundary and represent cusps, and of the rest of its boundary which consists of degenerated groups. It is obvious that $S(\theta)$ is connected and that its interior is simply connected.

References

[Abi71]   W. Abikoff (1971). Some remarks on Kleinian groups. In *Advances in the theory of Riemann surfaces, Ann. Math. Studies* **66**, 1–5.

[Abi73]   W. Abikoff (1973). Residual limit sets of Kleinian groups, *Acta Math.* **130**, 127–144.

[Ahl65a]   L.V. Ahlfors (1965). Finitely generated Kleinian groups, *Amer. J. Math.* **86**, 413–429 and **87**, 759.

[Ahl65b]   L.V. Ahlfors (1965). Some remarks on Kleinian groups. In *Proceedings Tulane conference on quasi-conformal mappings*, see *Lars V. Ahlfors: Collected Papers, Volume 2, 1954–1979*, Birkhäuser Boston, 1982, 316–319.

[Ber70a]   L. Bers (1970). On boundaries of Teichmüller spaces and on Kleinian groups: I, *Ann. Math.* **91**, 570–600.

[Ber70b]   L. Bers (1970). Spaces of Kleinian groups. In *Several Complex Variables I, Maryland 1970 (Lec. Notes Math.* **155***)*, Springer-Verlag, 9–34.

[BJ75]   A.F. Beardon and T. Jørgensen (1975). Fundamental domains for finitely generated Kleinian groups, *Math. Scand.* **36**, 21–26.

[Chu68]   V. Chuckrow (1968). On Schottky groups with applications to Kleinian groups, *Ann. Math.* **88**, 47–61.

[FK26]   R. Fricke and F. Klein (1926). *Vorlesungen über die Theorie der automorphen Funktionen*, B. G. Teubner.

[For51]   L.R. Ford (1951). *Automorphic Functions, 2nd edn.*, Chelsea, New York.

[Jør76] T. Jørgensen (1976). On discrete groups of Möbius transformations, *Amer. J. Math.* **98**, 739–749.

[Mak70] B. Maskit (1970). On boundaries of Teichmüller spaces and on Kleinian groups II, *Ann. Math.* **91**, 607–639.

[Mak71] B. Maskit (1971). On Poincaré's theorem for fundamental polygons, *Adv. Math.* **7**, 219–230.

[Mar74] A. Marden (1974). The geometry of finitely generated Kleinian groups, *Ann. Math.* **99**, 383–462.

[Sco73a] G.P. Scott (1973). Finitely generated 3-manifold groups are finitely presented, *J. Lond. Math. Soc.* **6**, 437–440.

Troels Jørgensen

Department of Mathematics
Columbia University
2990 Broadway
509 Mathematics Building
Mail Code: 4406
New York, NY 10027

tj@math.columbia.edu

AMS Classification: 30F40, 20H10, 37F30

Keywords: isometric circle, punctured torus, Kleinian group, quasifuchsian group, fundamental polyhedron

Kleinian Groups and Hyperbolic 3-Manifolds
Lond. Math. Soc. Lec. Notes **299**, 209–246

Y. Komori, V. Markovic & C. Series (Eds.)
Cambridge Univ. Press, 2003

Comparing two convex hull constructions for cusped hyperbolic manifolds

Hirotaka Akiyoshi and Makoto Sakuma

Dedicated to Professor Mitsuyoshi Kato on the occasion of his sixtieth birthday

Abstract

By imitating the Epstein–Penner convex hull construction in the Minkowski space
for cusped hyperbolic manifolds of finite volume, we define a natural decomposition of (a subspace of) the convex core of a cusped hyperbolic manifold of (possibly) infinite volume. The main purpose of this paper and its sequel [ASWY]
is to compare the decomposition with the "cellular structure" of the convex core
inherited from the cellular structure of its universal cover, i.e., the convex hull of
the limit set. We show that the underlying space of the decomposition is completely determined by the cellular structure of the convex core. In particular, for
any quasifuchsain punctured torus group, it is equal to the complement of the
bending lamination in the convex core. Moreover, we prove that, for a quasifuchsain punctured torus group, the restriction of the decomposition to the boundary
of the convex core is determined by the bending lamination.

1. Introduction

Let $M(\Gamma) = \mathbb{H}^d/\Gamma$ be a cusped hyperbolic manifold. Then we have the following two
convex hull constructions associated with Γ.

(i) The convex hull $\mathscr{C}(\Lambda)$ of the limit set Λ of Γ in the closure of the hyperbolic
space. If the dimension d is 3, then the boundary of $\mathscr{C}(\Lambda)$ has the structure of
a complete hyperbolic 2-manifold bent along a measured geodesic lamination,
which projects to the *bending measured lamination* $pl(\Gamma)$ of the boundary of the
convex core $M_0(\Gamma) = \mathscr{C}(\Lambda)/\Gamma$ of $M(\Gamma)$. (See Thurston [Thu02] and Epstein–
Marden [EM87].)

(ii) The convex hull $\mathscr{C}(\mathscr{B})$ in the Minkowski space of the Γ-invariant set $\mathscr{B}$ of
lightlike vectors which correspond to the horoballs projecting to cross sections
of the cusps of $M(\Gamma)$. If the volume of $M(\Gamma)$ is finite, then the cellular structure

of the boundary of $\mathscr{C}(\mathscr{B})$ descends to a finite ideal polyhedral decomposition, $\Delta(\Gamma)$, of $M(\Gamma)$. (See Epstein–Penner [EP88] and Weeks [Wee93]. See also Kojima [Koj90] and Frigerio–Petronio [FP] for analogous constructions for finite-volume hyperbolic manifolds with totally geodesic boundaries.)

The main purpose of this paper and its sequel [ASWY] is to study the relation between these two convex hull constructions. Our first task is to generalize the construction of the decomposition $\Delta(\Gamma)$ to the case where the volume of $M(\Gamma)$ is infinite (see Definitions 4.16 and 4.18). To explain a rough idea of the generalization, we give a more detailed description of the Epstein–Penner construction. Let T^d be the set of time-like vectors in the Minkowski space inside the positive light cone L^d. Then the following facts are proved by Epstein and Penner [EP88]. Suppose the volume of $M(\Gamma)$ is finite. Then $\partial \mathscr{C}(\mathscr{B}) \cap T^d$ is a locally finite, countable union of codimension one faces $\{F_i\}$ such that F_i is a convex hull of a finite subset of $\mathscr{B}$ and that the affine hull of F_i is Euclidean, i.e., the restriction of the Minkowski inner product to the hull is positive-definite. The cellular structure of $\partial \mathscr{C}(\mathscr{B}) \cap T^d$ projects homeomorphically to the projective model of $\mathbb{H}^d$ and gives a Γ-invariant locally finite tesselation. The interior of each cell of the tesselation injects into the hyperbolic manifold $M(\Gamma)$. Thus the tesselation descends to a finite ideal polyhedral decomposition of $M(\Gamma)$. Since each cell in the decomposition has a natural Euclidean structure, the decomposition induces a singular Euclidean structure on the hyperbolic manifold, and it is called a *Euclidean* decomposition. However, if we drop the condition that the volume of $M(\Gamma)$ is finite, then the following troublesome phenomena occur (See [Koj90] and [FP] for analogous phenomena for finite-volume hyperbolic manifolds with totally geodesic boundaries.)

(i) The cellular structure of $\partial \mathscr{C}(\mathscr{B})$ is not necessarily locally finite, and we must be careful about the definition of a face (see Remark 2.12). This forces us to introduce the notion of a *facet* by refining the notion of a face (see Definition 2.11).

(ii) Not all facets of $\partial \mathscr{C}(\mathscr{B})$ are Euclidean, that is, there may be a facet such that the restriction of the Minkowski inner product to the affine hull of the facet is not positive-definite.

(iii) The stabilizer of a facet of $\partial \mathscr{C}(\mathscr{B})$ with respect to Γ is not necessarily trivial, and hence an (open) piece of the decomposition may have a nontrivial fundamental group (see Corollary 5.4).

(iv) Some part of $\partial \mathscr{C}(\mathscr{B})$ may be "invisible" from the origin, that is, there may be a point of $\partial \mathscr{C}(\mathscr{B}) \cap T^d$ such that the line segment between the origin and the point contains some other points of $\partial \mathscr{C}(\mathscr{B})$. So, we need to consider only the *visible* facets of $\partial \mathscr{C}(\mathscr{B})$, i.e., those facets whose affine hulls do not contain the origin (see Definition 4.9 and Lemma 4.10).

(v) The image of $\partial \mathscr{C}(\mathscr{B}) \cap T^d$ in the projective model of $\mathbb{H}^d$ may be strictly smaller than $\mathbb{H}^d$. In fact, it is easy to see that the image is a subset of the convex hull $\mathscr{C}(\Lambda)$ of the limit set Λ.

We define $\Delta(\Gamma)$ to be the family of the images of visible open facets of $\mathscr{C}(\mathscr{B})$ in the hyperbolic manifold $M(\Gamma)$ and call it the *EPH-decomposition* of the convex core $M_0(\Gamma)$ (see Definition 4.18). The letters E, P and H, respectively, stand for Euclidean (or elliptic), parabolic and hyperbolic, and the naming reflects the fact that we may have parabolic or hyperbolic facets as well as Euclidean facets. Then $\Delta(\Gamma)$ gives a partition of a subset of the convex core $M_0(\Gamma)$ (see Proposition 4.15), and the underlying space $|\Delta(\Gamma)|$ contains the interior of the convex core (see Proposition 4.20). Moreover, the Ford domain with respect to a parabolic fixed point (of a hyperbolic manifold with only one cusp) can be regarded as a geometric dual to a certain subcomplex $\Delta_{\mathbb{E}}(\Gamma)$ of $\Delta(\Gamma)$ (see Section 10). For a more detailed description of this duality for 3-dimensional cusped hyperbolic manifolds and for an introduction to the EPH-decomposition from this view point, please see [ASWY, Section 2].

The Main Theorem 5.7 gives an explicit description of the underlying space $|\Delta(\Gamma)|$, of the EPH-decomposition, which arises from the convex hull construction in the Minkowski space, in terms of the cellular structure of the convex core $M_0(\Gamma)$, which arises from the convex hull construction in the hyperbolic space.

For a punctured torus group Γ, i.e., a Kleinian group freely generated by two isometries whose commutator is parabolic, we obtain the following corollary.

Corollary 1.1. *If Γ is a quasifuchsian punctured torus group, then we have*

$$|pl(\Gamma)| = M_0(\Gamma) - |\Delta(\Gamma)|.$$

Thus the bending lamination $|pl(\Gamma)|$ is determined by the decomposition $\Delta(\Gamma)$.

Since punctured torus groups are so special, we are led to the following problem.

Problem 1.2. Does $|pl(\Gamma)|$ determine the combinatorial structure of $\Delta(\Gamma)$ for a punctured torus group Γ?

Section 11 and the sequel to this paper [ASWY] are devoted to the study of this problem. We show that $|pl(\Gamma)|$ determines the restriction of $\Delta(\Gamma)$ to $\partial M_0(\Gamma)$ (see

212   H. Akiyoshi & M. Sakuma

Theorem 11.2). In [ASWY], we give a conjectural picture of $\Delta(\Gamma)$ in terms of $|pl(\Gamma)|$ and present some partial results and experimental results supporting the conjecture.

This paper is organized as follows. In Section 2, we recall basic properties of the convex hulls. Though most of the properties stated in this section seem to be already known, we tried to give proofs to all of them except those which we could find exactly the same statements in literature. In Section 3, we describe basic properties of the convex cores of hyperbolic manifolds. After giving an explicit definition of the EPH-decompositions in Section 4, we give the statements of the main results in Section 5. The proofs of the results are given in Sections 6–9. In Section 10, we explain the duality between the Ford domains and the EPH-decompositions. In the last Section 11, we study Problem 1.2 for punctured torus groups, and prove Theorem 11.2, which gives a partial answer to the problem.

Acknowledgment   The main bulk of this work was done when the authors were staying at the University of Warwick, and it was completed when they were visiting the University Paul Sabatier and the University of Geneva. They would like to thank these universities for their hospitality. They would also like to thank David Epstein, Caroline Series, Michel Boileau, Cam Van Quach Hongler, and Claude Weber for enlightening conversations and warm encouragement.

Notation

- $\mathscr{C}(V)$: the closed convex hull of V

- $\mathscr{C}(x_1,\ldots,x_n)$: the closed convex hull of the finite set $\{x_1,\ldots,x_n\}$

- $\mathbb{E}^{1,d}$: the $(d+1)$-dimensional Minkowski space

  - $\langle\cdot,\cdot\rangle$: the quadratic form on $\mathbb{E}^{1,d}$ of type $(1,d)$ defined by

  $$\langle x,y\rangle = -x_0y_0 + x_1y_1 + \cdots + x_dy_d$$

  - L^d: the positive light cone in $\mathbb{E}^{1,d}$
  - T^d: the set of time-like vectors of $\mathbb{E}^{1,d}$ inside the positive light cone L^d
  - $\mathbb{H}^d$: the d-dimensional hyperbolic space
  - $\partial\mathbb{H}^d$: the sphere at infinity
  - $\overline{\mathbb{H}^d} = \mathbb{H}^d \cup \partial\mathbb{H}^d$: the closure of the hyperbolic space
  - $\pi: L^d \cup T^d \to \overline{\mathbb{H}^d}$: the radial projection

- $\mathbb{S}^d = (\mathbb{E}^{1,d} - \{0\})/\sim$: the d-dimensional sphere, where $x \sim y$ if $y = \lambda x$ for some $\lambda > 0$

- cl(Z,Y): the closure of Z in Y

- int(Z,Y): the interior of Z in Y

- For a subset Z of $X = \mathbb{E}^{1,d}$ or $\overline{\mathbb{H}^d}$,

    - $\Pi(Z)$: the smallest plane containing Z in $\mathbb{E}^{1,d}$ or $\mathbb{S}^d$ (cf. Convention 2.2)
    - $\overline{Z}$: the closure of Z in $\Pi(Z)$, or equivalently, the closure of Z in X
    - int(Z): the interior of Z in $\Pi(Z)$
    - ∂Z: the frontier of Z in $\Pi(Z)$, i.e., $\partial Z = \overline{Z} - \text{int}(Z)$
    - $\delta(Z) = \overline{Z} - Z$

- $M = M(\Gamma) = \mathbb{H}^d/\Gamma$: a d-dimensional cusped hyperbolic manifold

    - Γ: the fundamental group $\pi_1(M) < \text{Isom}(\mathbb{H}^d)$
    - $\text{Stab}_\Gamma(\cdot)$: the stabilizer of $\cdot$ with respect to Γ
    - $p : \mathbb{H}^d \to M$: the universal covering projection
    - $\Lambda = \Lambda(\Gamma)$: the limit set of Γ
    - $\Lambda_p = \{v_n \,|\, n \in \mathbb{N}\}$: the set of parabolic fixed points of Γ
    - $\mathscr{C}(\Lambda)$: the closed convex hull of the limit set in $\overline{\mathbb{H}^d}$
    - $M_0 = M_0(\Gamma)$: the convex core $(\mathscr{C}(\Lambda) \cap \mathbb{H}^d)/\Gamma$ of M
    - $\mathscr{B}$: the Γ-invariant subset of L^d determined by cross sections of the cusps of M
    - $\mathscr{C}(\mathscr{B})$: the closed convex hull of $\mathscr{B}$ in $\mathbb{E}^{1,d}$
    - $\widetilde{\Delta} = \widetilde{\Delta}(\Gamma)$: the EPH-decomposition of $\mathscr{C}(\Lambda)$
    - $\Delta = \Delta(\Gamma)$: the EPH-decomposition of $M_0(\Gamma)$

2. Basic properties of convex hulls

Throughout this paper, X denotes either $\mathbb{E}^{1,d}$ or $\overline{\mathbb{H}^d}$.

Definition 2.1.

(i) A k-dimensional *plane* in $\mathbb{E}^{1,d}$ is a k-dimensional affine subspace of $\mathbb{E}^{1,d}$.

   (ii) A k-dimensional *plane* in $\overline{\mathbb{H}^d}$ is the closure of a k-dimensional totally geodesic subspace of $\mathbb{H}^d$. A singleton in $\partial\mathbb{H}^d$ is also regarded as a 0-dimensional plane in $\overline{\mathbb{H}^d}$.

  (iii) A *line segment* in X is a closed connected subset of a 1-dimensional plane in X.

  (iv) A *hyperplane* of X is a codimension one plane in X.

  (v) A *closed half space* of X is the closure of a component of $X - W$ for a hyperplane W of X.

Convention 2.2. When we mention the topology of subspaces of $\overline{\mathbb{H}^d}$, we sometimes need to regard $\overline{\mathbb{H}^d}$ as a subspace of $\mathbb{S}^d = (\mathbb{E}^{1,d} - \{0\})/\sim$, where $x \sim y$ if $y = \lambda x$ for some $\lambda > 0$. To be precise, we identify $\overline{\mathbb{H}^d}$ with the subset $(T^d \cup L^d)/\sim$ of $\mathbb{S}^d$, where L^d is the positive light cone and T^d is the set of time-like vectors inside L^d. By a k-dimensional *plane* in $\mathbb{S}^d$, we mean a subspace of $\mathbb{S}^d$ of the form $(W - \{0\})/\sim$, where W is a $(k+1)$-dimensional vector subspace of $\mathbb{E}^{1,d}$.

Definition 2.3. Let Z be a subset of X.

   (i) When $X = \mathbb{E}^{1,d}$, $\Pi(Z)$ denotes the smallest plane in $\mathbb{E}^{1,d}$ containing Z.

  (ii) When $X = \overline{\mathbb{H}^d}$, $\Pi(Z)$ denotes the smallest plane in $\mathbb{S}^d$ (not in $\overline{\mathbb{H}^d}$) containing Z.

 (iii) $\overline{Z}$ denotes the closure of Z in the plane $\Pi(Z)$, or equivalently, the closure of Z in X.

 (iv) $\text{int}(Z)$ denotes the interior of Z in the plane $\Pi(Z)$, i.e., $\text{int}(Z) = \text{int}(Z, \Pi(Z))$.

  (v) ∂Z denotes the frontier of Z in the plane $\Pi(Z)$, i.e., $\partial Z = \overline{Z} - \text{int}(Z)$.

Remark 2.4. If $X = \overline{\mathbb{H}^d}$, then $\Pi(Z) \cap \overline{\mathbb{H}^d}$ is the smallest plane in $\overline{\mathbb{H}^d}$ containing Z. Except when we consider $\text{int}(Z)$ or ∂Z, $\Pi(Z)$ may be regarded as $\Pi(Z) \cap \overline{\mathbb{H}^d}$.

Definition 2.5.

   (i) A subset $\mathscr{C}$ of X is *convex* if any two points in $\mathscr{C}$ are connected by a line segment contained in $\mathscr{C}$.

  (ii) The *dimension* of a convex set $\mathscr{C}$ is defined to be the dimension of the plane $\Pi(\mathscr{C})$. $\mathscr{C}$ is said to be *thick* if $\dim\mathscr{C} = \dim X$.

 (iii) A subset $\mathscr{C}$ of X is called an *open convex set* if it is convex and open in $\Pi(\mathscr{C})$.

 (iv) Given a subset V of X, we denote by $\mathscr{C}(V)$ the smallest closed convex set which contains V.

Definition 2.6. Let $\mathscr{C}$ be a closed subset of X. We call a closed half space H of X a *supporting half space* for $\mathscr{C}$ if $H \supset \mathscr{C}$ and $\partial H \cap \mathscr{C} \neq \emptyset$. A *support plane* at $x \in \mathscr{C}$ is a hyperplane of X which contains x and is the boundary of a supporting half space for $\mathscr{C}$.

Remark 2.7. If $\mathscr{C}$ is not thick, then there is a support plane W which is "inefficient" in the sense that $W \cap \mathscr{C} = \mathscr{C}$. However, every support plane W for $\mathscr{C}$ in $\Pi(\mathscr{C})$ is efficient, i.e., $W \cap \mathscr{C}$ is a proper subset of $\mathscr{C}$.

The following three lemmas are well-known (see [Ber87, Section 11.3 and Proposition 11.5.2], [EM87, Proposition 1.4.1 and Lemma 1.4.5]).

Lemma 2.8. *Let $\mathscr{C}$ be a closed convex set in X. Then for any point $z \in \partial\mathscr{C}$, there exists a support plane for $\mathscr{C}$ at z.*

Lemma 2.9. *A non-empty closed subset $\mathscr{C}$ of X is convex if and only if it is the intersection of all its supporting half spaces.*

Lemma 2.10. *Let $\mathscr{C}$ be a convex set in X of dimension d'. Then $\mathscr{C}$ is homeomorphic to $B^{d'} - L$ for some subset L of $\partial B^{d'}$, and $\mathrm{int}(\mathscr{C})$ is homeomorphic to a d'-dimensional open ball. If $\mathscr{C}$ is closed, then L is closed in $\partial B^{d'}$.*

Definition 2.11. Let $\mathscr{C}$ be a closed convex set in X.

(i) A subset F of $\mathscr{C}$ is called a *face* of $\mathscr{C}$ if there is a support plane W such that $F = W \cap \mathscr{C}$ and it is a proper subset of $\mathscr{C}$.

(ii) A subset F of $\mathscr{C}$ is called a *closed facet* (resp. an *open facet*) if there exists a sequence $\{F_i\}_{i=0}^{k}$ ($k \geq 0$) of subsets of $\mathscr{C}$ such that

    a. $F_0 = \mathscr{C}$,

    b. F_{i-1} is a closed convex set and F_i is a face of F_{i-1} for every $i \in \{1, \ldots, k\}$, and

    c. $F = F_k$ (resp. $F = \mathrm{int}(F_k)$, the interior of F_k in $\Pi(F_k)$).

We call $\{F_i\}_{i=0}^{k}$ a *face sequence* for F. The *dimension* of F is defined to be the dimension of the plane $\Pi(F)$.

(iii) By a *facet*, we mean a closed facet or an open facet.

Remark 2.12.

(i) If F is a d'-dimensional open facet, then F is homeomorphic to a d'-dimensional open ball by Lemma 2.10, and $\overline{F}$ is a d'-dimensional closed facet of $\mathscr{C}$. Conversely, if F is a d'-dimensional closed facet, then $\mathrm{int}(F)$ is a d'-dimensional open facet.

(ii) A facet of $\mathscr{C}$ is not necessarily a face of $\mathscr{C}$. For example, let $\mathscr{C}$ be the convex set in $\mathbb{E}^{1,1} = \mathbb{R}^2$ defined by

$$\mathscr{C} = \{(x,y) \mid x \le 1, y \le f(x)\},$$

where $f : (-\infty, 1] \to \mathbb{R}$ is the function defined by

$$f(x) = \begin{cases} 1 & (x \le 0), \\ \sqrt{1-x^2} & (0 < x \le 1). \end{cases}$$

Then $W := \{(x,y) \mid y = 1\}$ is a support plane for $\mathscr{C}$ and $F_1 := W \cap \mathscr{C} = \{(x,1) \mid x \le 0\}$ is a face of $\mathscr{C}$. So, $F_2 := \{(0,1)\}$ is a closed facet of $\mathscr{C}$. However, F_2 is not a face of $\mathscr{C}$, because W is the unique support plane for $\mathscr{C}$ containing F_2, and F_2 is strictly smaller than $F_1 = W \cap \mathscr{C}$.

Similar phenomena occur for 3-dimensional hyperbolic convex hulls whose boundaries are bent along irrational geodesic laminations (cf. Proposition 11.1 and [EM87, Definition 1.6.4]).

Lemma 2.13. *Let $\mathscr{C}$ be a closed convex set in X, and let F be a closed facet of $\mathscr{C}$. Then for any $x \in \partial F$, there is an open facet F' of $\mathscr{C}$ containing x. Moreover, any face sequence $\{F_i\}_{i=0}^k$ for F can be extended to a face sequence $\{F_i'\}_{i=0}^{k'}$ for F' so that $k' > k$ and $F_i' = F_i$ for every $i \in \{0, \dots, k\}$.*

Proof. Let x be a point in ∂F. Then, by Lemma 2.8, there is a support plane, W, at x for the closed convex set F in $\Pi(F)$. Set $F_{k+1} = F \cap W$. Then F_{k+1} is a closed facet of $\mathscr{C}$ and $\{F_i\}_{i=0}^{k+1}$ is a face sequence for F_{k+1} (see Remark 2.7). If $x \in \mathrm{int}(F_{k+1})$, then we obtain the conclusion. If $x \in \partial F_{k+1}$, then we can repeat the above argument. After repeating this argument at most $\dim F$ times, we obtain the conclusion. $\square$

Lemma 2.14. *Let $\mathscr{C}$ be a closed convex set in X and F an open facet of $\mathscr{C}$. Let D be an open convex set contained in $\mathscr{C}$ such that $D \cap F \ne \emptyset$. Then $D \subset F$.*

Proof. **Step 1** We first prove the assertion when $D = \mathrm{int}(l)$ for some line segment l. Let $\{F_i\}_{i=0}^k$ be a face sequence for F. We show $l \subset F_i$ for every $i \in \{0, \dots, k\}$. Suppose that this does not hold. Then, since $l \subset \mathscr{C} = F_0$, there exists $i_0 \in \{0, \dots, k-1\}$ such that $l \subset F_{i_0}$ and $l \not\subset F_{i_0+1}$. Let W be a support plane for F_{i_0} in $\Pi(F_{i_0})$ such that $F_{i_0+1} = W \cap F_{i_0}$. Pick a point, x_0, in $\mathrm{int}(l) \cap F$. Then $x_0 \in l \cap W$ and $l \not\subset W$. Hence the line segment l in $\Pi(F_{i_0})$ intersects W transversely at the interior point x_0. Since $l \subset F_{i_0}$, this contradicts the assumption that W is a support plane for F_{i_0} in $\Pi(F_{i_0})$. Hence $l \subset F_i$ for every $i \in \{0, \dots, k\}$. In particular, $l \subset F_k = \overline{F}$.

Let U be an open ball neighborhood of x_0 in F. Then, since F is a convex set in $\Pi(F)$, we have $\mathrm{int}(l) \subset \mathrm{int}(\mathscr{C}(l \cup U)) \subset F$.

Step 2 To prove the assertion for the general case, pick a point x_0 in $D \cap F$. Let x be a point in D. Then, since D is an open convex set, there is a line segment l such that $\text{int}(l)$ contains both x_0 and x. By Step 1, we see that $\text{int}(l) \subset F$, and hence $x \in F$. So, we have $D \subset F$. □

Lemma 2.15. *Let $\mathscr{C}$ be a closed convex set in X. Then any two distinct open facets of $\mathscr{C}$ are disjoint.*

Proof. Let F and F' be open facets of $\mathscr{C}$ such that $F \cap F' \neq \emptyset$. Since F' is an open convex set, we have $F' \subset F$ by Lemma 2.14. Similarly, we have $F \subset F'$, and hence $F = F'$. □

By Lemmas 2.13 and 2.15, we have the following proposition.

Proposition 2.16. *Let $\mathscr{C}$ be a closed convex set in X. Then any closed facet F of $\mathscr{C}$ is the disjoint union of the open facets of $\mathscr{C}$ contained in F. In particular, $\mathscr{C}$ is the disjoint union of the open facets of $\mathscr{C}$.*

Definition 2.17. For a closed convex set $\mathscr{C}$ in X, we call the family of the open facets of $\mathscr{C}$ the *cellular structure* of $\mathscr{C}$.

At the end of this section, we prove the following two lemmas which are used later.

Lemma 2.18. *Let $\mathscr{C}$ be a closed convex set in X and D a convex set contained in $\partial\mathscr{C}$. Then D is contained in a closed facet of $\mathscr{C}$ contained in $\partial\mathscr{C}$.*

Proof. We may assume $D \neq \emptyset$. Then, by Lemma 2.10, we have $\text{int}(D) \neq \emptyset$. So, by Proposition 2.16, there is an open facet F of $\mathscr{C}$ such that $\text{int}(D) \cap F \neq \emptyset$. Since $\text{int}(D)$ is an open convex set contained in $\mathscr{C}$, $\text{int}(D)$ is contained in F by Lemma 2.14. This implies $D \subset \overline{F}$. Since $D \subset \partial\mathscr{C}$, we see that $\overline{F}$ is a closed facet of $\mathscr{C}$ contained in $\partial\mathscr{C}$. □

Lemma 2.19. *Let V be a closed subset of X and $V_0 = \{v_n \mid n \in \mathbb{N}\}$ a countable dense subset of V. Then the following hold.*

1. *$\mathscr{C}(V)$ is equal to the closure of $\bigcup_{n\in\mathbb{N}}\mathscr{C}(v_1,\ldots,v_n)$ in X.*

2. *$\text{int}(\mathscr{C}(V))$ is equal to $\bigcup_{n\in\mathbb{N}}\text{int}(\mathscr{C}(v_1,\ldots,v_n),\Pi(\mathscr{C}(V)))$.*

Proof.

(i) Since $\mathscr{C}(V)$ is a closed subset of X containing $\mathscr{C}(v_1,\ldots,v_n)$ $(n \in \mathbb{N})$, $\mathscr{C}(V)$ contains the closure of $\bigcup_n \mathscr{C}(v_1,\ldots,v_n)$. To prove the converse, note that $\{\mathscr{C}(v_1,\ldots,v_n)\}_{n\in\mathbb{N}}$ is an ascending sequence of convex sets. Thus $\bigcup_n \mathscr{C}(v_1,\ldots,v_n)$ is also a convex set and hence its closure is. Moreover, by the assumption that V_0 is dense in V, the closure of $\bigcup_n \mathscr{C}(v_1,\ldots,v_n)$ contains V. Thus the closure of $\bigcup_n \mathscr{C}(v_1,\ldots,v_n)$ is a closed convex set which contains V. Hence, $\mathscr{C}(V)$ is contained in the closure of $\bigcup_n \mathscr{C}(v_1,\ldots,v_n)$.

(ii) Since $\mathscr{C}(V)$ contains $\mathscr{C}(v_1,\ldots,v_n)$ for any $n \in \mathbb{N}$, $\mathrm{int}(\mathscr{C}(V))$ contains $\bigcup_n \mathrm{int}(\mathscr{C}(v_1,\ldots,v_n),\Pi(\mathscr{C}(V)))$. To prove the converse, let x be a point in $\mathscr{C}(V)$ which is not contained in $\mathrm{int}(\mathscr{C}(v_1,\ldots,v_n),\Pi(\mathscr{C}(V)))$ for any $n \in \mathbb{N}$. Since $\mathrm{int}(\mathscr{C}(v_1,\ldots,v_n),\Pi(\mathscr{C}(V)))$ is the intersection of the interiors of finitely many closed half spaces, there exists a half space H_n of X which contains $\mathscr{C}(v_1,\ldots,v_n)$ but does not contain x in the interior. Let l be the line segment $\mathscr{C}(x,v_1)$. Then ∂H_n $(n \geq 2)$ intersects the compact set l. Thus, by taking a subsequence, we may assume that $\{H_n\}$ converges to a half space H of X with respect to the Chabauty topology (cf. [BP92, Section E.1]). Then we can easily see $x \notin \mathrm{int}(H)$ and $\mathscr{C}(v_1,\ldots,v_n) \subset H$. By (1), $\mathscr{C}(V)$ is the closure of $\bigcup_n \mathscr{C}(v_1,\ldots,v_n)$. (Note that this holds even after taking a subsequence, because $\{\mathscr{C}(v_1,\ldots,v_n)\}_n$ is an ascending sequence.) Thus H contains $\mathscr{C}(V)$ but does not contain x in the interior. Hence x is not contained in $\mathrm{int}(\mathscr{C}(V))$ and hence $\mathrm{int}(\mathscr{C}(V)) \subset \bigcup_n \mathrm{int}(\mathscr{C}(v_1,\ldots,v_n),\Pi(\mathscr{C}(V)))$.

$\square$

3. Basic properties of convex cores

Let $M = M(\Gamma) = \mathbb{H}^d/\Gamma$ $(\Gamma < \mathrm{Isom}(\mathbb{H}^d))$ be a hyperbolic manifold such that Γ contains parabolic transformations. Recall that a *parabolic* transformation is an element of $\mathrm{Isom}^+(\mathbb{H}^d) < SO(1,d)$ which has a unique eigenvector (ray) in the positive light cone L^d and no eigenvector in the inside T^d of L^d.

Definition 3.1.

(i) $\Lambda = \Lambda(\Gamma)$ denotes the limit set of Γ.

(ii) $\Lambda_p = \{v_n \,|\, n \in \mathbb{N}\}$ denotes the set of the parabolic fixed points of Γ.

Then Λ_p is dense in Λ, because Λ is the minimal non-empty closed Γ-invariant set in $\partial\mathbb{H}^d$ (see [Thu02, Proposition 8.1.2]).

Since we study convex sets in two different ambient spaces, we introduce the following terminology, and use the symbol G instead of F to denote a facet of $\mathscr{C}(\Lambda)$. (Note that the terminology is slightly different from that in [EM87].)

Definition 3.2. By a *closed* (resp. *open*) *flat piece* of $\mathscr{C}(\Lambda)$, we mean a closed (resp. open) facet of $\mathscr{C}(\Lambda)$.

Lemma 3.3. *Let G be an open flat piece of $\mathscr{C}(\Lambda)$. Then G is either contained in $\mathbb{H}^d$ or a singleton consisting of a point in Λ.*

Proof. The assertion obviously holds when $\dim G = 0$. So, we may assume that $\dim G \geq 1$. Then $\Pi(G) \cap \mathbb{H}^d \neq \emptyset$, and hence

$$G = \operatorname{int}(\Pi(G) \cap \mathscr{C}(\Lambda), \Pi(G)) \subset \operatorname{int}(\Pi(G) \cap \overline{\mathbb{H}^d}, \Pi(G)) = \Pi(G) \cap \mathbb{H}^d \subset \mathbb{H}^d.$$

$\square$

Definition 3.4. Since the cellular structure of $\mathscr{C}(\Lambda)$ is Γ-equivariant, it projects to a decomposition of the convex core $M_0 = M_0(\Gamma)$ into

$$\{p(G) \,|\, G \text{ is an open flat piece of } \mathscr{C}(\Lambda) \text{ contained in } \mathbb{H}^d\}.$$

We call a member of this family an *open flat piece* of M_0 and call this family the *cellular structure* of M_0. By a *closed flat piece* of M_0, we mean a subset of $\mathscr{C}(\Lambda)$ of the form $p(\mathbb{H}^d \cap G)$ where G is a closed flat piece of $\mathscr{C}(\Lambda)$.

By Proposition 2.16 and Lemma 3.3, we have the following.

Proposition 3.5. *The convex core M_0 is the disjoint union of the open flat pieces of M_0.*

Since $\Lambda_p = \{v_n \,|\, n \in \mathbb{N}\}$ is a countable dense subset of Λ, Lemma 2.19 implies the following.

Lemma 3.6.

1. $\mathscr{C}(\Lambda)$ is equal to the closure of $\bigcup_{n \in \mathbb{N}} \mathscr{C}(v_1, \dots, v_n)$ in $\overline{\mathbb{H}^d}$.

2. $\operatorname{int}(\mathscr{C}(\Lambda))$ is equal to $\bigcup_{n \in \mathbb{N}} \operatorname{int}(\mathscr{C}(v_1, \dots, v_n), \Pi(\mathscr{C}(\Lambda)))$.

We recall a lemma from [EM87].

Lemma 3.7 ([EM87, Lemma 1.6.2]). *Let V be a closed non-empty subset of $\overline{\mathbb{H}^d}$, and let W be a support plane for $\mathscr{C}(V)$. Then $W \cap \mathscr{C}(V) = \mathscr{C}(W \cap V)$.*

By repeatedly using this lemma, we obtain the following lemma.

Lemma 3.8. *For every closed flat piece G of $\mathscr{C}(\Lambda)$, we have $G = \mathscr{C}(G \cap \Lambda)$.*

Lemma 3.9. *Let G be an open flat piece of $\mathscr{C}(\Lambda)$ and v a point in $\overline{G} \cap \Lambda$. Then G is a subset of*

$$\bigcup \{\mathscr{C}(v, w_1, \ldots, w_{d'}) - \mathscr{C}(w_1, \ldots, w_{d'}) \mid w_1, \ldots, w_{d'} \in \Pi(G) \cap \Lambda\},$$

where d' is the dimension of G.

Proof. By Lemma 3.8, $\overline{G}$ is equal to the closed convex set $\mathscr{C}(\overline{G} \cap \Lambda)$ in $\Pi(G)$. Choose a countable dense subset $V = \{u_n \mid n \in \mathbb{N}\}$ of $\overline{G} \cap \Lambda = \Pi(G) \cap \Lambda$. Then, by Lemma 2.19(2),

$$G = \bigcup_{n \in \mathbb{N}} \mathrm{int}(\mathscr{C}(v, u_1, \ldots, u_n), \Pi(G)).$$

Note that $\mathrm{int}(\mathscr{C}(v, u_1, \ldots, u_n), \Pi(G))$ is contained in

$$\bigcup \{\mathscr{C}(v, w_1, \ldots, w_{d'}) - \mathscr{C}(w_1, \ldots, w_{d'}) \mid w_1, \ldots, w_{d'} \in \{u_1, \ldots, u_n\}\}$$

Thus

$$G \subset \bigcup \{\mathscr{C}(v, w_1, \ldots, w_{d'}) - \mathscr{C}(w_1, \ldots, w_{d'}) \mid w_1, \ldots, w_{d'} \in V\}$$
$$\subset \bigcup \{\mathscr{C}(v, w_1, \ldots, w_{d'}) - \mathscr{C}(w_1, \ldots, w_{d'}) \mid w_1, \ldots, w_{d'} \in \Pi(G) \cap \Lambda\}.$$

$\square$

4. Generalization of Epstein–Penner decompositions

Convention 4.1. Throughout the remainder of this paper, $M = M(\Gamma) = \mathbb{H}^d / \Gamma$ with $\Gamma < \mathrm{Isom}(\mathbb{H}^d)$ denotes a hyperbolic manifold satisfying the following conditions.

(i) Γ has no proper invariant subspace of $\mathbb{H}^d$.

(ii) Γ contains parabolic transformations.

(iii) There exists a Γ-invariant family of mutually disjoint horoballs $\mathscr{H}$ in $\mathbb{H}^d$ such that Λ_p is equal to the set of the centers of horoballs in $\mathscr{H}$.

Recall that for any horoball H in $\mathbb{H}^d$, there exists a unique point $v \in L^d$ such that $H = \{x \in \mathbb{H}^d \mid \langle v, x \rangle \geq -1\}$ (see [EP88, Section 1]). The center of the horoball H corresponds to the ray thorough v, and as v moves away from the origin along the ray, the horoball contracts towards the center of the horoball.

Let $\mathscr{B}$ be the Γ-invariant set of the points in L^d corresponding to the horoballs in $\mathscr{H}$, which is prescribed to M by Convention 4.1. The following lemma follows from the arguments in the proof of [EP88, Theorem 2.4].

Lemma 4.2.

1. *For any positive real number h, there exist at most finite points in $\mathscr{B}$ whose heights, i.e., the 0th coordinates, are less than h.*

2. *The origin, O, is not contained in $\mathscr{C}(\mathscr{B})$.*

Definition 4.3.

(i) A *ray* from a point x is a closed half geodesic r with $\partial r = \{x\}$.

(ii) Let r be a ray and z a point in r. We denote the unbounded component of $r - \{z\}$ by $r_{>z}$, and denote the closure of $r_{>z}$ by $r_{\geq z}$.

Notation 4.4. T^d denotes the set of time-like vectors in the Minkowski space inside the positive light cone L^d, and $\pi: T^d \cup L^d \to \overline{\mathbb{H}^d}$ denotes the radial projection from O. For a subset Z of $\mathbb{E}^{1,d}$, we abbreviate $\pi(Z \cap (T^d \cup L^d))$ as $\pi(Z)$.

Lemma 4.5. $\mathrm{int}(\mathscr{C}(\mathscr{B})) \subset T^d$.

Proof. It suffices to prove that $\mathrm{int}(\mathscr{C}(\mathscr{B})) \cap L^d = \emptyset$, because $\mathscr{C}(\mathscr{B}) \subset T^d \cup L^d$ from the definition. Since Γ has no proper invariant subspace, we see that for any $x \in \mathrm{int}(\mathscr{C}(\mathscr{B}))$, there exists a $(d+1)$-dimensional ball D in $\mathscr{C}(\mathscr{B})$ centered at x. On the other hand, for any point $x \in L^d$ and any $(d+1)$-dimensional ball D centered at x, D contains a point in $\mathbb{E}^{1,d} - (T^d \cup L^d) \subset \mathbb{E}^{1,d} - \mathscr{C}(\mathscr{B})$. Thus $\mathrm{int}(\mathscr{C}(\mathscr{B})) \cap L^d = \emptyset$. $\square$

Lemma 4.6. $\mathscr{C}(\mathscr{B}) \cap L^d$ is the disjoint union of the rays $\{tb \mid t \geq 1\}$ ($b \in \mathscr{B}$). Moreover, every singleton in $\mathscr{B}$ is a 0-dimensional face of $\mathscr{C}(\mathscr{B})$.

Proof. This lemma is proved by the argument in the proof of [EP88, Lemma 3.3]. $\square$

The following two lemmas are proved simultaneously.

Lemma 4.7. $\pi(\mathrm{int}(\mathscr{C}(\mathscr{B}))) = \mathrm{int}(\mathscr{C}(\Lambda))$.

Lemma 4.8. *Let r be a ray in $\mathbb{E}^{1,d}$ from O which intersects $\mathscr{C}(\mathscr{B})$. Then the following hold.*

1. *If $\pi(r)$ is contained in $\mathrm{int}(\mathscr{C}(\Lambda))$, then r intersects $\partial\mathscr{C}(\mathscr{B})$ exactly in one point, say z, and $r_{>z} \subset \mathrm{int}(\mathscr{C}(\mathscr{B}))$.*

2. *If $\pi(r)$ is contained in $\partial \mathscr{C}(\Lambda)$, then there exists $z \in r$ such that $r \cap \mathscr{C}(\mathscr{B}) = r \cap \partial \mathscr{C}(\mathscr{B}) = r_{\geq z}$.*

Proof of Lemma 4.8(1). Suppose $\pi(r) \in \mathrm{int}(\mathscr{C}(\Lambda))$. Since $O \notin \mathscr{C}(\mathscr{B})$ by Lemma 4.2(2) and since $r \cap \mathscr{C}(\mathscr{B}) \neq \emptyset$ by the assumption, there exists a point, say z, in $r \cap \mathscr{C}(\mathscr{B})$ such that $r \cap \mathscr{C}(\mathscr{B}) \subset r_{\geq z}$. Then we have $z \in \partial \mathscr{C}(\mathscr{B})$.

In what follows, we prove that $r_{>z} \subset \mathrm{int}(\mathscr{C}(\mathscr{B}))$. By Lemma 3.6(2), $\pi(r)$ is contained in $\mathrm{int}(\mathscr{C}(v_1, \dots, v_n), \Pi(\mathscr{C}(\Lambda)))$ for some finite set $\{v_1, \dots, v_n\} \subset \Lambda_p$. Since any point in Λ_p is an accumulation point in Λ_p, there exist sequences $\{b_{i,j}\}_{i \in \mathbb{N}} \subset \mathscr{B}$ ($j \in \{1, \dots, n\}$) such that $\pi(b_{i,j}) \neq v_j$ and $\lim_{i \to \infty} \pi(b_{i,j}) = v_j$. Since $\lim_{i \to \infty} \pi(b_{i,j}) = v_j$ for each j, r intersects $\mathrm{int}(\mathscr{C}(\pi(b_{i,1}), \dots, \pi(b_{i,n})), \Pi(\mathscr{C}(\Lambda)))$ for sufficiently large i. Now, let x be a point in $r_{>z}$. Then, since the height of $b_{i,j}$ diverges as $i \to \infty$ by Lemma 4.2(1), the heights of $b_{i,1}, \dots, b_{i,n}$ are greater than that of x for sufficiently large i. Hence we see that x is contained in $\mathrm{int}(\mathscr{C}(z, b_{i,1}, \dots, b_{i,n}), \Pi(\mathscr{C}(\mathscr{B}))) \subset \mathrm{int}(\mathscr{C}(\mathscr{B}))$ for sufficiently large i. Thus we have proved $r_{>z} \subset \mathrm{int}(\mathscr{C}(\mathscr{B}))$, and hence we obtain Lemma 4.8(1). □

Proof of Lemma 4.7. By Lemma 4.8(1), we have $\mathrm{int}(\mathscr{C}(\Lambda)) \subset \pi(\mathrm{int}(\mathscr{C}(\mathscr{B})))$. To prove the converse, suppose that $\pi(\mathrm{int}(\mathscr{C}(\mathscr{B}))) - \mathrm{int}(\mathscr{C}(\Lambda))$ contains a point, say x. Then x belongs to $\mathbb{H}^d$ by Lemma 4.5. Thus there exists a closed half space K of $\overline{\mathbb{H}^d}$ such that K contains $\mathscr{C}(\Lambda)$ but does not contain x in the interior. Let H be the closed half space of $\mathbb{E}^{1,d}$ such that $H \supset K$ and $\partial H = \Pi(\pi^{-1}(\partial K))$. Then $\mathscr{C}(\mathscr{B}) \subset H$ and $\pi^{-1}(x)$ is contained in the closure of $\mathbb{E}^{1,d} - H$. Hence $\pi^{-1}(x) \cap \mathrm{int}(\mathscr{C}(\mathscr{B})) = \emptyset$, which implies $x \notin \pi(\mathrm{int}(\mathscr{C}(\mathscr{B})))$, a contradiction. This completes the proof of Lemma 4.7. □

Proof of Lemma 4.8(2). Suppose $\pi(r) \in \partial \mathscr{C}(\Lambda)$. We first show $r \cap \mathscr{C}(\mathscr{B}) = r \cap \partial \mathscr{C}(\mathscr{B})$. Suppose contrary that $r \cap \mathrm{int}(\mathscr{C}(\mathscr{B})) \neq \emptyset$. Then $\pi(r)$ (cf. Notation 4.4) is contained in $\pi(\mathrm{int}(\mathscr{C}(\mathscr{B})))$, which is equal to $\mathrm{int}(\mathscr{C}(\Lambda))$ by Lemma 4.7. This contradicts the assumption. So we have the desired equality.

As in the proof of Lemma 4.8(1), there is a point $z \in \partial \mathscr{C}(\mathscr{B})$ such that $r \cap \mathscr{C}(\mathscr{B}) \subset r_{\geq z}$. In what follows, we prove $r \cap \mathscr{C}(\mathscr{B}) = r_{\geq z}$. By Lemma 3.6(1), there exist a sequence of finite sided ideal polyhedra $\{\sigma_i\}$ in $\mathbb{H}^d$, each of which is spanned by a finite subset of Λ_p, and a sequence $\{y_i\}$ with $y_i \in \sigma_i$, such that $\lim y_i = \pi(r)$. Since y_i is contained in some (possibly degenerate) ideal d-simplex contained in σ_i, we may suppose that each σ_i is an ideal d-simplex spanned by parabolic fixed points. Then there exist $b_{i,0}, \dots, b_{i,d} \in \mathscr{B}$ such that $\sigma_i = \mathscr{C}(\pi(b_{i,0}), \dots, \pi(b_{i,d}))$. Since Λ is compact, by taking a subsequence, we may assume that each sequence $\{\pi(b_{i,j})\}_{i \in \mathbb{N}}$ converges to a point, say v_j, in Λ. Since any parabolic fixed point is an accumulation point of Λ_p, there exists, for each $i \in \mathbb{N}$ and $j \in \{1, \ldots, n\}$, a sequence $\{b_{i,j}^{(k)}\}_{k \in \mathbb{N}} \subset \mathscr{B}$ such that $b_{i,j}^{(k)} \neq b_{i,j}$ and $\lim_{k \to \infty} \pi(b_{i,j}^{(k)}) = \pi(b_{i,j})$.

For a while, we fix $i \in \mathbb{N}$ arbitrarily. Then there exists $K_1 \in \mathbb{N}$ such that for any $j \in \{0, \ldots, n\}$ and $k \geq K_1$, the height of $b_{i,j}^{(k)}$ is greater than i. Set $\sigma_i^{(k)} = \mathscr{C}(b_{i,0}^{(k)}, \ldots, b_{i,n}^{(k)})$. Then the height of any point in $\sigma_i^{(k)}$ is greater than i for any $k \geq K_1$. Note that $\pi(\sigma_i^{(k)})$ converges to σ_i as $k \to \infty$ with respect to the Chabauty topology. Thus there exists $K_2 \in \mathbb{N}$ such that for any $k \geq K_2$, there exists $x_i^{(k)} \in \sigma_i^{(k)}$ such that $d(\pi(x_i^{(k)}), y_i) \leq 1/i$. (Here $d(\cdot, \cdot)$ denotes the hyperbolic distance.) Put $K = \max\{K_1, K_2\}$ and $x_i = x_i^{(K)}$. Then $d(\pi(x_i), y_i) \leq 1/i$ and the height of x_i is greater than i. Moreover, x_i is a point in $\mathscr{C}(\mathscr{B})$ from the construction.

Let ε be an arbitrary positive number. Since $\lim y_i = \pi(r)$, there exists $N_1 \in \mathbb{N}$ such that $d(y_i, \pi(r)) < \varepsilon/2$ for any $i \geq N_1$. Put $N = \max\{N_1, 2/\varepsilon\}$. Suppose that $i \geq N$. Then

$$d(\pi(x_i), \pi(r)) \leq d(\pi(x_i), y_i) + d(y_i, \pi(r))$$
$$\leq 1/i + \varepsilon/2 \leq 1/N + \varepsilon/2 \leq \varepsilon/2 + \varepsilon/2 = \varepsilon.$$

Hence the sequence $\{\pi(x_i)\}$ converges to $\pi(r)$.

Since all z and x_i ($i \in \mathbb{N}$) are contained in $\mathscr{C}(\mathscr{B})$, the line segments $\mathscr{C}(z, x_i)$ are contained in $\mathscr{C}(\mathscr{B})$. Since $\{\pi(x_i)\}$ converges to $\pi(r)$ and the height of x_i tends to ∞ as $i \to \infty$, the sequence $\{\mathscr{C}(z, x_i)\}$ converges to $r_{\geq z}$ with respect to the Chabauty topology. Thus $r_{\geq z}$ is contained in $\mathscr{C}(\mathscr{B})$ because $\mathscr{C}(\mathscr{B})$ is closed. □

Definition 4.9.

(i) A facet F of $\mathscr{C}(\mathscr{B})$ is said to be *visible* if $\Pi(F)$ does not contain O.

(ii) A visible facet is said to be *elliptic* (resp. *parabolic*, *hyperbolic*) if $\Pi(F)$ is elliptic (resp. parabolic, hyperbolic), i.e., the restriction of the bilinear form $\langle \cdot, \cdot \rangle$ on $\mathbb{E}^{1,d}$ to $\Pi(F)$ is positive definite (resp. singular, of type $(1, d')$ for some $d' \geq 0$). A visible facet which is elliptic is also said to be *Euclidean*.

Lemma 4.10. *A facet of $\mathscr{C}(\mathscr{B})$ is mapped homeomorphically into $\overline{\mathbb{H}^d}$ by π if and only if it is a visible facet.*

Proof. Let F be a facet of $\mathscr{C}(\mathscr{B})$. First, suppose that F is visible. Then since $\Pi(F)$ does not contain O, the restriction of π to $F \cap (T^d \cup L^d)$ is proper and injective. Thus F is mapped homeomorphically into $\overline{\mathbb{H}^d}$ by π.

Next, suppose that F is not visible. Let x be a point in F and r the ray from O which contains x. Then $\Pi(F)$ contains r as it contains both x and O. By Lemma 4.8, $r_{\geq x}$ is contained in $\mathscr{C}(\mathscr{B})$ and hence it is contained in $\overline{F} = \Pi(F) \cap \mathscr{C}(\mathscr{B})$. If F is a closed facet, then it follows from this fact that π is not injective on F. Suppose F is an open facet. Then, since F is open in $\Pi(F)$, there is a neighborhood U of x in $\Pi(F)$ which is contained in F. In particular, $U \cap r$ is contained in F. Hence π is not injective on F. □

Lemma 4.11. *Let F be a visible open facet of $\mathscr{C}(\mathscr{B})$ with $\dim F \geq 1$. Then*

$$F = T^d \cap \operatorname{int}(\overline{F}, \Pi(F) \cap (T^d \cup L^d)).$$

In particular, F is an open subset of $\Pi(F) \cap T^d$ and $\pi(F) \subset \mathbb{H}^d$.

Remark 4.12. If $\dim F = 0$, then $F = \{b\}$ for some $b \in \mathscr{B}$ by Lemma 9.1.

Proof. For simplicity, set $Y = \operatorname{int}(\overline{F}, \Pi(F) \cap (T^d \cup L^d))$. Since $\overline{F} \subset \Pi(F) \cap (T^d \cup L^d) \subset \Pi(F)$, we have $F = \operatorname{int}(\overline{F}) \subset Y$. We also have $F \subset \Pi(F) \cap T^d$, because $\dim \Pi(F) \geq 1$ and $\Pi(F) \not\subset L^d$. Hence $F \subset T^d \cap Y$.

Next, we see the converse inclusion. It is clear that $T^d \cap Y \subset \overline{F}$. Since Y is open in $\Pi(F) \cap (T^d \cup L^d)$, there exists an open subset Y_1 of $\Pi(F)$ such that $Y_1 \cap (T^d \cup L^d) = Y$. Thus

$$T^d \cap Y = T^d \cap (Y_1 \cap (T^d \cup L^d)) = T^d \cap Y_1 = T^d \cap \Pi(F) \cap Y_1.$$

Hence $T^d \cap Y$ is open in $\Pi(F)$. Therefore $T^d \cap Y \subset \operatorname{int}(\overline{F}) = F$. □

Lemma 4.13. *Let r be a ray in $\mathbb{E}^{1,d}$ from O which intersects $\mathscr{C}(\mathscr{B})$ and z the point in r such that $r_{\geq z} = r \cap \mathscr{C}(\mathscr{B})$. Then the following hold.*

1. *The open facet of $\mathscr{C}(\mathscr{B})$ which contains z is visible.*

2. *Any facet of $\mathscr{C}(\mathscr{B})$ which contains a point in $r_{>z}$ is not visible.*

Proof.   (i) Let F be the open facet of $\mathscr{C}(\mathscr{B})$ which contains z. Suppose contrary that F is not visible. Then $r \subset \Pi(F)$. Since z is contained in the closure of $r - r_{\geq z} \subset \Pi(F) - F$, z cannot be contained in the open subset F of $\Pi(F)$, a contradiction.

(ii) Let F be a facet of $\mathscr{C}(\mathscr{B})$ containing a point, say y, of $r_{>z}$. Then, by Proposition 2.16, there is an open facet F' of $\mathscr{C}(\mathscr{B})$ such that $y \in F' \subset F$. Since $r_{>z}$ is an open convex set, $r_{>z}$ is contained in F' by Lemma 2.14. Hence $O \in r \subset \Pi(F') \subset \Pi(F)$, and therefore F is not visible. □

Lemma 4.14. *Let F be a visible closed facet of $\mathscr{C}(\mathscr{B})$ and v a point in $\pi(F) \cap \partial \mathbb{H}^d$. Then $v \in \Lambda_p$ and $F \cap \pi^{-1}(v) = b$, where $b \in \mathscr{B}$ such that $\pi(b) = v$.*

Proof. Let x be a point in $\pi^{-1}(v) \cap F$. Then $x = tb$ for some $b \in \mathscr{B}$ and $t \geq 1$ by Lemma 4.6. Since F is visible, we have $t = 1$ by Lemma 4.13. So, we have the conclusion. $\square$

Proposition 4.15.

1. $\pi(\mathscr{C}(\mathscr{B}))$ is the disjoint union of the images by π of the visible open facets of $\mathscr{C}(\mathscr{B})$.

2. For any visible closed facet F of $\mathscr{C}(\mathscr{B})$, $\pi(F)$ is the disjoint union of the images by π of the visible open facets contained in F.

Proof. (i) First, we prove that $\pi(\mathscr{C}(\mathscr{B}))$ is contained in the union of the images of visible open facets. Let x be a point in $\pi(\mathscr{C}(\mathscr{B}))$ and r a ray from O such that $\pi(r) = x$. Then, by Lemma 4.8, $r \cap \mathscr{C}(\mathscr{B}) = r_{\geq z}$ for some $z \in r$. By Proposition 2.16, there exists an open facet F of $\mathscr{C}(\mathscr{B})$ which contains z. Then, by Lemma 4.13, F is visible. Thus x is contained in the image of a visible open facet.

Next, we prove that for any distinct visible open facets F and F', $\pi(F)$ and $\pi(F')$ are disjoint. Suppose contrary that $\pi(F)$ and $\pi(F')$ contain a common point $x \in \pi(\mathscr{C}(\mathscr{B}))$. Let r be the ray from O such that $\pi(r) = x$. Then both F and F' intersect r. Let z be the point in r such that $r_{\geq z} = r \cap \mathscr{C}(\mathscr{B})$. Then, by Lemma 4.13, both F and F' contain z. Thus, by Proposition 2.16, we have $F = F'$.

(ii) Let F be a visible closed facet. By Proposition 2.16, F is the disjoint union of the open facets contained in F. Since F is visible, any (open) facet contained in F is also visible by the definition. Hence $\pi(F)$ is the union of images by π of the open visible facets contained in F. Thus we obtain the conclusion by (1).

$\square$

Definition 4.16 (EPH-decomposition (I)). We set

$$\widetilde{\Delta} = \widetilde{\Delta}(\Gamma) = \{\pi(\widehat{F}) \mid \widehat{F} \text{ is a visible open facet of } \mathscr{C}(\mathscr{B})\},$$

and call it the *EPH-decomposition* (with respect to $\mathscr{H}$) of the convex hull $\mathscr{C}(\Lambda)$. For a visible open (resp. closed) facet $\widehat{F}$ of $\mathscr{C}(\mathscr{B})$, $\widetilde{F} := \pi(\widehat{F})$ is called an *open facet* (resp. a *closed facet*) of $\widetilde{\Delta}$. $\widetilde{F}$ is said to be *elliptic* (or *Euclidean*), *parabolic*, or *hyperbolic* according as $\widehat{F}$ is so. The *dimension* of $\widetilde{F}$ is defined to be the dimension of $\widehat{F}$. The *support* of the EPH-decomposition $\widetilde{\Delta}$ is the union of the open facets of $\widetilde{\Delta}$, and denoted by $|\widetilde{\Delta}|$.

Lemma 4.17. *Let $\widetilde{F}$ be an open facet of $\widetilde{\Delta}$. Then $\widetilde{F}$ is either contained in $\mathbb{H}^d$ or a singleton consisting of a point in Λ_p.*

Proof. Suppose that $\dim \widetilde{F} = 0$ and $\widetilde{F} \not\subset \mathbb{H}^d$. Then, by Lemma 4.14, $\widetilde{F} = \{v\}$ for some $v \in \Lambda_p$. If $\dim \widetilde{F} \geq 1$, then $\widetilde{F} \subset \mathbb{H}^d$ by Lemma 4.11.   □

Definition 4.18 (EPH-decomposition (II)). We set

$$\Delta = \Delta(\Gamma) = \{p(\widetilde{F}) \,|\, \widetilde{F} \text{ is an open facet of } \widetilde{\Delta} \text{ contained in } \mathbb{H}^d\},$$

and call it the *EPH-decomposition* (with respect to $\mathscr{H}$) of the convex core $M_0 = M_0(\Gamma)$. For an open facet $\widetilde{F}$ of $\widetilde{\Delta}$, $F := p(\widetilde{F})$ is called an *open facet* of Δ. F is said to be *elliptic* (or *Euclidean*), *parabolic*, or *hyperbolic* according as $\widetilde{F}$ is so. The *dimension* of F is defined to be the dimension of $\widetilde{F}$. The *support* of the EPH-decomposition Δ is the union of the open facets of Δ, and denoted by $|\Delta|$. For a closed facet $\widetilde{F}$ of $\widetilde{\Delta}$, $F := p(\mathbb{H}^d \cap \widetilde{F})$ is called a *closed facet* of Δ.

Remark 4.19.

(i) Suppose M is of finite volume. Then the EPH-decomposition Δ of $M_0 = M$ is nothing other than the decomposition introduced by Epstein and Penner [EP88]. In this case, every facet of Δ is Euclidean (i.e., elliptic) and it is a finite sided ideal polyhedron.

(ii) As is the case of the Epstein-Penner decompositions, the EPH-decompositions depend on the choices of the Γ-invariant families of horoballs $\mathscr{H}$.

At the end of this section, we prove the following proposition which explains the reason why we call Δ (resp. $\widetilde{\Delta}$) a decomposition of the convex core M_0 (resp. the convex hull $\mathscr{C}(\Lambda)$).

Proposition 4.20. *The following hold.*

1. $\text{int}(\mathscr{C}(\Lambda)) \subset |\widetilde{\Delta}| \subset \mathscr{C}(\Lambda)$ *and* $|\widetilde{\Delta}| \cap \partial \mathbb{H}^d = \Lambda_p$.

2. $\text{Int}M_0 \subset |\Delta| \subset M_0$, *where* $\text{Int}M_0$ *denotes the interior of the manifold* M_0.

Proof.

(i) By using Lemma 4.7 and Proposition 4.15(1), we have

$$\mathrm{int}(\mathscr{C}(\Lambda)) = \pi(\mathrm{int}(\mathscr{C}(\mathscr{B}))) \subset \pi(\mathscr{C}(\mathscr{B})) = |\widetilde{\Delta}|.$$

Moreover we have

$$\begin{aligned}
|\widetilde{\Delta}| &= \pi(\mathscr{C}(\mathscr{B})) \quad &\text{(by Proposition 4.15(1))} \\
&= \pi(\mathrm{cl}(\mathrm{int}(\mathscr{C}(\mathscr{B})), \mathbb{E}^{1,d})) \quad &\text{(cf. Lemma 2.10)} \\
&\subset \mathrm{cl}(\pi(\mathrm{int}(\mathscr{C}(\mathscr{B}))), \overline{\mathbb{H}^d}) \quad &\text{(by the continuity of } \pi) \\
&= \mathrm{cl}(\mathrm{int}(\mathscr{C}(\Lambda)), \overline{\mathbb{H}^d}) \quad &\text{(by Lemma 4.7)} \\
&= \mathscr{C}(\Lambda) \quad &\text{(cf. Lemma 2.10).}
\end{aligned}$$

Thus we have the first assertion of (1). The second assertion follows from Lemma 4.6 and the fact that $|\widetilde{\Delta}| = \pi(\mathscr{C}(\mathscr{B}))$.

(ii) By (1), we have $\mathrm{int}(\mathscr{C}(\Lambda)) \subset |\widetilde{\Delta}| \cap \mathbb{H}^d \subset \mathscr{C}(\Lambda) \cap \mathbb{H}^d$. On the other hand, we can see $p(|\widetilde{\Delta}| \cap \mathbb{H}^d) = |\Delta|$ by Lemma 4.17. Hence $\mathrm{Int} M_0 \subset |\Delta| \subset M_0$.

$\square$

5. Statement of the main results

In this section, we state basic properties of the facets of Δ and then state the main theorem which describes the support $|\Delta|$ of the EPH-decomposition in terms of the cellular structure of M_0. The proofs are given in Sections 6–9.

Proposition 5.1. *Any open facet of $\widetilde{\Delta}$ is contained in an open flat piece of $\mathscr{C}(\Lambda)$.*

Proposition 5.2.

1. *The set of 0-dimensional closed (resp. open) facets of $\widetilde{\Delta}$ is equal to the set of singletons in Λ_p.*

2. *Let F be an open facet of $\widetilde{\Delta}$ of dimension ≥ 1. Then $F \subset \mathbb{H}^d$.*

3. *Let F be a closed facet of $\widetilde{\Delta}$ of dimension ≥ 1. Then $F \cap \partial \mathbb{H}^d$ is a non-empty subset of Λ_p.*

If the volume of M is finite, then every closed facet of $\widetilde{\Delta}$ is a finite sided ideal polyhedron, and hence it is a compact subset of $\overline{\mathbb{H}^d}$ and its stabilizer with respect to Γ

is trivial; in particular, every open facet of Δ is simply connected. However, this is not the case when the volume of M is infinite, that is, a closed facet of $\widetilde{\Delta}$ is not necessarily closed in $\overline{\mathbb{H}^d}$ and an open facet of Δ is not necessarily simply connected. In particular, we cannot expect a property for $\mathscr{C}(\mathscr{B})$ corresponding to Lemma 3.8. The facets of $\widetilde{\Delta}$ and Δ have the following properties.

Proposition 5.3. *Let F be a closed facet of $\widetilde{\Delta}$ of dimension $d' \geq 1$. Set $\delta F = \overline{F} - F$, where $\overline{F}$ is the closure of F in $\overline{\mathbb{H}^d}$. Then $\delta F \cap |\widetilde{\Delta}| = \emptyset$ and the following hold.*

1. *If F is elliptic, then F is compact, $\delta F = \emptyset$, and $\mathrm{Stab}_\Gamma(F)$, the stabilizer of G with respect to Γ, is trivial.*

2. *If F is parabolic, then δF consists of at most one point in $\Lambda - \Lambda_p$, and $\mathrm{Stab}_\Gamma(F)$ is trivial.*

3. *If F is hyperbolic, then δF is contained in a closed flat piece of $\mathscr{C}(\Lambda)$ of dimension $\leq (d' - 1)$, and $\mathrm{Stab}_\Gamma(F)$ is conjugate in $\mathrm{Isom}(\mathbb{H}^d)$ to a discrete subgroup of $\mathrm{Isom}(\mathbb{H}^{d'-1})$. Moreover F has at most one end.*

Corollary 5.4. *Let F be a d'-dimensional open facet of Δ. Then the following hold.*

1. *If F is elliptic or parabolic, then F is simply connected.*

2. *If F is hyperbolic, then $\pi_1(F) < \Gamma$ is conjugate in $\mathrm{Isom}(\mathbb{H}^d)$ to a discrete subgroup of $\mathrm{Isom}(\mathbb{H}^{d'-1})$.*

Remark 5.5. In Proposition 5.3(2) and Corollary 5.4 for parabolic facets, the assumption is essential that the set Λ_p of parabolic fixed points is precisely equal to the set of the centers of horoballs in $\mathscr{H}$ prescribed to M by Convention 4.1. Even if the set of the centers of horoballs in $\mathscr{H}$ is a proper subset of Λ_p, the EPH-decomposition of M_0 with respect to $\mathscr{H}$ is well-defined, as long as $\mathscr{H}$ is Γ-invariant. However, in this case, some parabolic open facet can be non-simply connected (see [ASWY, Conjecture 8.3 and Theorem 9.1]).

Recall that $\mathrm{Int}M_0 \subset |\Delta| \subset M_0$ by Proposition 4.20. The main Theorem 5.7 below describes explicitly the support $|\Delta|$ in terms of the cellular structure of M_0. To state the theorem, we need the following definition.

Definition 5.6. We say that a flat piece $G = p(\widetilde{G})$ of M_0 *intersects a cusp of M* if the closed flat piece of $\mathscr{C}(\Lambda)$ obtained as the closure of $\widetilde{G}$ in $\overline{\mathbb{H}^d}$ contains a point in Λ_p.

Theorem 5.7. *The support $|\Delta|$ of the EPH-decomposition Δ is the union of the open flat pieces of M_0 which intersect a cusp of M.*

The above theorem is obtained from the following theorem.

Theorem 5.8.

1. *Let G be a closed flat piece of $\mathscr{C}(\Lambda)$ such that $G \cap \Lambda_p = \emptyset$. Then $G \cap |\widetilde{\Delta}| = \emptyset$.*

2. *Let G be an open flat piece of $\mathscr{C}(\Lambda)$ such that $\overline{G} \cap \Lambda_p \neq \emptyset$. Then $G \subset |\widetilde{\Delta}|$.*

Corollary 5.9.

1. *If $\dim M = 2$, then $|\Delta|$ is equal to $\mathrm{Int} M_0$.*

2. *If $\dim M = 3$, then $|\Delta|$ is the union of $\mathrm{Int} M_0$ and the 2-dimensional open flat pieces of M_0 which intersect a cusp.*

6. Proof of Proposition 5.1

We begin by noting the following elementary observation.

Lemma 6.1. *Let A be a d'-dimensional affine subspace of $\mathbb{E}^{1,d}$ such that $A \cap T^d \neq \emptyset$ and $O \notin A$. Let V be the d'-dimensional vector subspace of $\mathbb{E}^{1,d}$ parallel to A and E the vector subspace of $\mathbb{E}^{1,d}$ spanned by A. Note that $(E, \langle \, , \, \rangle_E)$ is isomorphic to $\mathbb{E}^{1,d'}$ and determines a plane $\overline{\mathbb{H}^{d'}}$ of $\overline{\mathbb{H}^d}$.*

1. *Suppose A is elliptic. Then $\pi(A) = \overline{\mathbb{H}^{d'}}$ (see Notation 4.4).*

2. *Suppose A is parabolic. Then $\pi(A) = \overline{\mathbb{H}^{d'}} - \{v\}$, where v is the point of $\partial \mathbb{H}^{d'}$ determined by the ray $V \cap (T^d \cup L^d) = V \cap L^d$.*

3. *Suppose A is hyperbolic. Then $\pi(A)$ is an open half space of $\overline{\mathbb{H}^{d'}}$ bounded by the hyperplane determined by the d'-dimensional subspace $V \cap (T^d \cup L^d)$.*

In particular, $\pi(A) \cap \mathbb{H}^d$ is open in $\Pi(\pi(A))$.

Lemma 6.2. *Any open facet F of $\widetilde{\Delta}$ is an open subset of $\Pi(F)$.*

Proof. Since the assertion obviously holds when $\dim F = 0$, we may assume that $\dim F \geq 1$. Let $\widehat{F}$ be a visible open facet of $\mathscr{C}(\mathscr{B})$ such that $\pi(\widehat{F}) = F$, and set $A = \Pi(\widehat{F})$. Then, by Lemma 4.11, $\widehat{F}$ is an open subset of $A \cap T^d$. Since the map $\pi|_{A \cap T^d} : A \cap T^d \to \pi(A \cap T^d)$ is a homeomorphism, $F = \pi(\widehat{F})$ is open in $\pi(A \cap T^d) = \pi(A) \cap \mathbb{H}^d$. In particular, it follows $\Pi(F) = \Pi(\pi(A) \cap \mathbb{H}^d) = \Pi(\pi(A))$. On the other hand, $\pi(A) \cap \mathbb{H}^d$ is open in $\Pi(\pi(A))$ by Lemma 6.1. So, F is open in $\Pi(\pi(A)) = \Pi(F)$. $\qquad\square$

Proof of Proposition 5.1. Let F be an open facet of $\widetilde{\Delta}$. Then, by Propositions 4.20(1) and 2.16, there is an open flat piece, G, of $\mathscr{C}(\Lambda)$ such that $F \cap G \neq \emptyset$. Since F is an open convex set by Lemma 6.2, we have $F \subset G$ by Lemma 2.14. $\qquad\square$

7. Proof of Proposition 5.3

Lemma 7.1. *Let F be a closed facet of $\widetilde{\Delta}$ and $\widehat{F}$ a visible closed facet of $\mathscr{C}(\mathscr{B})$ such that $\pi(\widehat{F}) = F$, and set $A = \Pi(\widehat{F})$. Then*

$$\delta F \subset \overline{F} - \pi(A) \subset \mathrm{cl}(\pi(A), \overline{\mathbb{H}^d}) - \pi(A).$$

Proof. We first prove that $\overline{F} \cap \pi(A) \subset F$. Let x be a point in $\overline{F} \cap \pi(A)$. Then there is a sequence $\{x_n\}$ in F such that $\lim x_n = x$, and there is a unique point $\widehat{x}$ in $A \cap (T^d \cup L^d)$ such that $\pi(\widehat{x}) = x$. Let $\widehat{x}_n$ be the point in $\widehat{F} \subset A \cap (T^d \cup L^d)$ such that $\pi(\widehat{x}_n) = x_n$. Since the restriction of π to $A \cap (T^d \cup L^d)$ is a homeomorphism onto $\pi(A)$, we have $\widehat{x} = \lim \widehat{x}_n \in \widehat{F}$. Hence we have $x \in F$. Thus we have proved $\overline{F} \cap \pi(A) \subset F$. This implies that

$$\delta F = \overline{F} - F \subset \overline{F} - (\overline{F} \cap \pi(A)) \subset \overline{F} - \pi(A) \subset \mathrm{cl}(\pi(A), \overline{\mathbb{H}^d}) - \pi(A).$$

$\square$

The assertion of Proposition 5.3 that $\delta F \cap |\widetilde{\Delta}| = \emptyset$ is proved by the following lemma.

Lemma 7.2. *For any closed facet F of $\widetilde{\Delta}$, we have $\delta F \cap |\widetilde{\Delta}| = \emptyset$.*

Proof. Let $\widehat{F}$ be a visible closed facet of $\mathscr{C}(\mathscr{B})$ such that $\pi(\widehat{F}) = F$, and let $\{\widehat{F}_i\}_{i=0}^k$ be a face sequence for $\widehat{F}$. Set $A = \Pi(\widehat{F})$ and $E = \Pi(A \cup \{O\})$. Then A is a hyperplane of E because $\widehat{F}$ is a visible facet.

In what follows, we prove that A is a support plane for $V \cap \mathscr{C}(\mathscr{B})$ in E. Since $\Pi(\widehat{F}_0) = \mathbb{E}^{1,d}$, E is contained in $\Pi(\widehat{F}_0)$. On the other hand, E is not contained in $\Pi(\widehat{F}_k) = A$. Thus there exists $i_0 \in \{0, \ldots, k-1\}$ such that E is contained in $\Pi(\widehat{F}_{i_0})$ but not in $\Pi(\widehat{F}_{i_0+1})$. Let H be a supporting half space for $\widehat{F}_{i_0}$ in $\Pi(\widehat{F}_{i_0})$ such that $\partial H \cap \widehat{F}_{i_0} = \widehat{F}_{i_0+1}$. Then ∂H contains A because $\Pi(\widehat{F}_{i_0+1})$ contains $\Pi(\widehat{F}_k) = A$. Moreover, since $\Pi(\widehat{F}_{i_0}) \supset E$ and $\Pi(\widehat{F}_{i_0+1}) \not\supset E$, we have $\Pi(\widehat{F}_{i_0}) \cap E = E$ and $\partial H \cap V = A$. Then $H \cap E$ is a closed half space of E with boundary A, and $E \cap H \supset E \cap (\Pi(\widehat{F}_{i_0}) \cap \mathscr{C}(\mathscr{B})) = E \cap \mathscr{C}(\mathscr{B})$. Thus A is a support plane for $E \cap \mathscr{C}(\mathscr{B})$ in E.

By Lemma 4.8, $\mathscr{C}(\mathscr{B}) \cap V$ is contained in the closure, H, of the component of $V - A$ which does not contain O. Let x be a point in $\delta F = \overline{F} - F$. Then $x \notin \pi(A)$ by lemma 7.1. So the ray $\pi^{-1}(x)$ from O does not intersect A and hence is contained in $E - H$. Since $E \cap \mathscr{C}(\mathscr{B}) \subset H$, we have $x \notin \pi(\mathscr{C}(\mathscr{B})) = |\widetilde{\Delta}|$. $\square$

To prove the assertion of Proposition 5.3 on the stabilizers, we need the following lemma.

Lemma 7.3. *Let F be a visible open facet of $\mathscr{C}(\mathscr{B})$. Then*

$$\text{Stab}_\Gamma(\pi(F)) = \text{Stab}_\Gamma(F) = \text{Stab}_\Gamma(\overline{F}) = \text{Stab}_\Gamma(\pi(\overline{F})),$$

where $\text{Stab}_\Gamma(\cdot)$ denotes the stabilizer of $\cdot$ with respect to $\Gamma < \text{Isom}(\mathbb{H}^d) < O(1,d)$.

Proof. We first show $\text{Stab}_\Gamma(F) = \text{Stab}_\Gamma(\overline{F})$. Let γ be an element of $\text{Stab}_\Gamma(\overline{F})$. Then

$$\gamma(F) = \gamma(\text{int}(\overline{F}, \Pi(\overline{F}))) = \text{int}(\gamma(\overline{F}), \gamma(\Pi(\overline{F})))$$
$$= \text{int}(\gamma(\overline{F}), \Pi(\gamma(\overline{F}))) = \text{int}(\overline{F}, \Pi(\overline{F})) = F.$$

Hence $\text{Stab}_\Gamma(\overline{F}) \subset \text{Stab}_\Gamma(F)$. The converse inclusion can be proved by a similar argument.

Next, we prove $\text{Stab}_\Gamma(\pi(F)) = \text{Stab}_\Gamma(F)$. It is clear that $\text{Stab}_\Gamma(F) \subset \text{Stab}_\Gamma(\pi(F))$. To show the converse, pick an element $\gamma \in \text{Stab}_\Gamma(\pi(F))$. Then $\pi(\gamma(F)) = \gamma(\pi(F)) = \pi(F)$. On the other hand, since F is a visible open facet, $\gamma(F)$ is also a visible open facet. Hence, by Proposition 4.15(1), we have $F = \gamma(F)$ and hence $\gamma \in \text{Stab}_\Gamma(F)$.

Finally we prove $\text{Stab}_\Gamma(\overline{F}) = \text{Stab}_\Gamma(\pi(\overline{F}))$. It is clear that $\text{Stab}_\Gamma(\overline{F}) \subset \text{Stab}_\Gamma(\pi(\overline{F}))$. To show the converse, pick an element $\gamma \in \text{Stab}_\Gamma(\pi(\overline{F}))$. Set $F' = \gamma(F)$. Then $\overline{F'}$ is a visible closed facet of $\mathscr{C}(\mathscr{B})$, and $\pi(\overline{F'}) = \pi(\gamma(\overline{F})) = \gamma(\pi(\overline{F})) = \pi(\overline{F})$. By Lemma 2.10, $\pi(\overline{F})$ (resp. $\pi(\overline{F'})$) is a manifold, whose interior as a manifold is equal to $\pi(F)$ (resp. $\pi(F')$). Hence we have $\pi(F) = \pi(F')$, which is equal to $\gamma(\pi(F))$. This implies $\gamma \in \text{Stab}_\Gamma(\pi(F)) = \text{Stab}_\Gamma(F) = \text{Stab}_\Gamma(\overline{F})$. $\qquad\square$

Proof of Proposition 5.3. Let F be a closed facet of $\widetilde{\Delta}$ of dimension $d' \geq 1$, and let $\widehat{F}$ be the visible closed facet of $\mathscr{C}(\mathscr{B})$ such that $\pi(\widehat{F}) = F$. Set $A = \Pi(\widehat{F})$ and $E = \Pi(A \cup \{O\})$. Then E is a vector subspace of $\mathbb{E}^{1,d}$ which contains a point in T^d, because $d' \geq 1$. So, the space $(E, \langle \, , \, \rangle_E)$ is identified with $\mathbb{E}^{1,d'} = \mathbb{E}^{1,0} \oplus \mathbb{E}^{0,d'}$, where $\langle \, , \, \rangle_E$ is the restriction of the Minkowski metric $\langle \, , \, \rangle$ to E. Thus $E \cap \mathbb{H}^d$ (resp. $E \cap T^d$, $E \cap L^d$) is identified with $\mathbb{H}^{d'}$ (resp. $T^{d'}$, $L^{d'}$). Since A is a hyperplane of E which does not contain O, there exists a unique element $w \in E$ such that $A = \{x \in E \mid \langle w, x \rangle = -1\}$.

Case 1. F is Euclidean. Then the intersection of A and $T^{d'} \cup L^{d'}$ is compact. So $\widehat{F} = A \cap \mathscr{C}(\mathscr{B}) = (A \cap (T^{d'} \cup L^{d'})) \cap \mathscr{C}(\mathscr{B})$ is compact. Hence F is also compact and thus $\delta F = \emptyset$. Let γ be a non-trivial element of $\text{Stab}_\Gamma(F) = \text{Stab}_\Gamma(\widehat{F})$ (see Lemma 7.3). Then $\gamma(A) = A$ and hence $\gamma(E) = E$, because E is the smallest vector subspace of $\mathbb{E}^{1,d}$ containing A. Hence, $w = \gamma(w)$. This contradicts the assumption that Γ acts freely on $\mathbb{H}^d$, because w is contained in $T^{d'} \subset T^d$ and hence $\pi(w) \in \mathbb{H}^d$.

Case 2. F is parabolic. We define a sequence $\{w_n\}$ in E by $w_n = (w^- - 1/n, w^+)$, where $w = (w^-, w^+) \in \mathbb{E}^{1,0} \oplus \mathbb{E}^{0,d'} = E$. Let H_n be the subspace defined by $H_n = \{x \in E \mid \langle w_n, x \rangle \leq 0\}$, and set $\widehat{F}_n = \widehat{F} \cap H_n$.

Claim 7.4. $\{\widehat{F}_n\}$ *is a monotone increasing sequence of compact subsets of* $\widehat{F}$ *such that* $\cup \widehat{F}_n = \widehat{F}$.

Proof. Let $x = (x^-, x^+) \in \mathbb{E}^{1,0} \oplus \mathbb{E}^{0,d'} = E$ be any point in $\widehat{F}_n$. Since x is contained in both $\widehat{F} \subset A$ and H_n, we have the following:

$$\langle w, x \rangle = -1, \quad \langle w_n, x \rangle \leq 0,$$

which is equivalent to

$$-w^- x^- + \langle w^+, x^+ \rangle = -1, \quad -(w^- - 1/n)x^- + \langle w^+, x^+ \rangle \leq 0.$$

Thus we obtain $\widehat{F}_n = \widehat{F} \cap \{(x^-, x^+) \in E \mid x^- \leq n\}$. Since $\widehat{F}$ is a subset of $T^{d'} \cup L^{d'}$, $\widehat{F}_n$ is contained in a compact subset. Since both $\widehat{F}$ and H_n are closed, $\widehat{F}_n$ is also closed. So, $\widehat{F}_n$ is compact. Moreover, by the above observation, $\{\widehat{F}_n\}$ is an increasing sequence of compact subsets such that $\cup \widehat{F}_n = \widehat{F}$. □

Since H_n is a closed half space of E, its complement is an open half space and hence convex. Thus $\widehat{F} - \widehat{F}_n$ is convex and hence the number of the components of $\widehat{F} - \widehat{F}_n$ is at most one for any $n \in \mathbb{N}$. Therefore $\widehat{F}$ has at most one end. Since $\pi|_{\widehat{F}}$ is a homeomorphism onto its image by Lemma 4.10, $F = \pi(\widehat{F})$ also has at most one end. Moreover, by Lemmas 7.1 and 6.1, we have $\delta F \subset \mathrm{cl}(\pi(A), \overline{\mathbb{H}^d}) - \pi(A) \subset \{\pi(w)\}$. This implies $\delta F \subset \overline{F} \cap \partial \mathbb{H}^d$, and the latter set is contained in $\mathscr{C}(\Lambda) \cap \partial \mathbb{H}^d = \Lambda$ by Proposition 4.20. Hence δF is either empty or a singleton in Λ.

Next, we show $\pi(w) \notin \Lambda_p$. Suppose contrary that $\pi(w) \in \Lambda_p$. Then there exists $\lambda_0 > 0$ such that $\lambda_0 w \in \mathscr{B}$. Let H be a supporting half space for $\mathscr{C}(\mathscr{B}) \cap E$ such that $\partial H = A$. Then H is equal to $\{x \in E \mid \langle w, x \rangle \geq -1\}$, because $\lambda_0 w \in \mathscr{B} \cap E \subset \mathscr{C}(\mathscr{B}) \cap E$ and $\langle \lambda_0 w, w \rangle = \lambda_0 \langle w, w \rangle = 0$. Since $\widehat{F}$ is a visible facet of dimension ≥ 1, $\widehat{F}$ contains a point, say x_0, of $T^{d'}$. Then we have $\langle w, x_0 \rangle < 0$. Thus there is a real number $\lambda > 1$ such that $\langle w, \lambda x_0 \rangle < -1$ and hence $\lambda x_0 \notin H$. Therefore the point λx_0 in the ray $\{t x_0 \mid t \geq 0\}$ from O passing through $x_0 \in \widehat{F} \subset \mathscr{C}(\mathscr{B})$ is not contained in $\mathscr{C}(\mathscr{B})$. This contradicts Lemma 4.8. Thus we have proved that δF consists of at most one point in $\Lambda - \Lambda_p$.

Finally, we show that $\mathrm{Stab}_\Gamma(F) = \mathrm{Stab}_\Gamma(\widehat{F})$ is trivial. Note that $\mathrm{Stab}_\Gamma(\widehat{F})$ preserves A and E and hence $\mathbb{H}^{d'}$. Since Γ is discrete and torsion-free, $\mathrm{Stab}_\Gamma(\widehat{F})$ acts effectively on the horosphere $A \cap (E \cap \mathbb{H}^d) = A \cap \mathbb{H}^{d'}$ of $\mathbb{H}^{d'}$. Now, suppose contrary that $\mathrm{Stab}_\Gamma(\widehat{F})$ is nontrivial. Then $\mathrm{Stab}_\Gamma(\widehat{F}) \cap \mathrm{Isom}^+(\mathbb{H}^{d'})$ is a non-trivial parabolic subgroup of $\mathrm{Isom}^+(\mathbb{H}^{d'})$ (i.e., any nontrivial element is parabolic), because it preserves the horosphere $A \cap \mathbb{H}^{d'}$ and because $\mathrm{Stab}_\Gamma(\widehat{F})$ is discrete and torsion-free. This implies that $\mathrm{Stab}_\Gamma(\widehat{F}) \cap \mathrm{Isom}^+(\mathbb{H}^d)$ contains a non-trivial parabolic subgroup with parabolic fixed point $\pi(w)$. This contradicts the fact that $\pi(w) \notin \Lambda_p$. Hence $\mathrm{Stab}_\Gamma(F)$ is trivial.

Case 3. F is hyperbolic. We define a sequence $\{w_n\}$ in E by $w_n = (w^- - 1/n, w^+)$, where $w = (w^-, w^+) \in \mathbb{E}^{1,0} \oplus \mathbb{E}^{0,d'} = E$. Let H_n be the subspace defined by $H_n = \{x \in E \mid \langle w_n, x \rangle \leq 0\}$, and set $F_n = F \cap H_n$.

By the argument in the proof of Claim 7.4, we can show the following.

Claim 7.5. $\{\widehat{F_n}\}$ *is a monotone increasing sequence of compact subsets of F such that* $\cup \widehat{F_n} = \widehat{F}$.

By the same argument as in Case (2), we can see that F has at most one end. Moreover, by Lemma 7.1 and 6.1, we have $\delta F \subset \mathrm{cl}(\pi(A), \overline{\mathbb{H}^d}) - \pi(A) \subset A_0 \cap (E \cap \overline{\mathbb{H}^d}) = A_0 \cap \overline{\mathbb{H}^{d'}}$, where A_0 denotes the vector subspace of E which is parallel to A. Thus δF is equal to the intersection of the two closed convex sets $\overline{F}$ and a plane in $\overline{\mathbb{H}^d}$. So δF is a convex set. Since $\delta F \cap |\widetilde{\Delta}| = \emptyset$ (Lemma 7.2), we have $\delta F \subset \partial \mathscr{C}(\Lambda)$ by Proposition 4.20. Therefore, by Lemma 2.18, δF is contained in a closed flat piece of $\mathscr{C}(\Lambda)$ contained in $\partial \mathscr{C}(\Lambda)$.

Note that $\mathrm{Stab}_\Gamma(F) = \mathrm{Stab}_\Gamma(\widehat{F})$ preserves A, E, and $\mathbb{H}^{d'} = E \cap \mathbb{H}^d$. So it preserves $A \cap \mathbb{H}^{d'}$ and hence $A_0 \cap \mathbb{H}^{d'}$. Since Γ is discrete and torsion-free, $\mathrm{Stab}_\Gamma(\widehat{F})$ acts effectively on $A_0 \cap \mathbb{H}^{d'}$, which is identified with $\mathbb{H}^{d'-1}$. Hence we see that $\mathrm{Stab}_\Gamma(\widehat{F})$ is a discrete subgroup of $\mathrm{Isom}(\mathbb{H}^{d'-1})$. $\square$

8. Technical lemmas

Lemma 8.1. *Let x be a point in $T^d \cup L^d$. Let $\{b_n\}$ be a sequence in $\mathscr{B}$ and b a point in L^d such that for any $n \in \mathbb{N}$, $\pi(b_n) \neq \pi(b)$ and $\lim \pi(b_n) = \pi(b)$. Then for any $x \in \mathbb{E}^{1,d}$, the sequence of line segments $\{\mathscr{C}(x, b_n)\}$ converges to the ray $r := \{x + \lambda b \mid \lambda \geq 0\}$ with respect to the Chabauty topology. Moreover, if $x \in \mathscr{C}(\mathscr{B})$, then $r \subset \mathscr{C}(\mathscr{B})$.*

Proof. Since $\pi(b_n) \neq \pi(b)$ and $\lim \pi(b_n) = \pi(b)$, the height of b_n diverges as $n \to \infty$ by Lemma 4.2(1). By using this fact, we can see that the sequence $\{\mathscr{C}(x, b_n)\}$ converges to r. If $x \in \mathscr{C}(\mathscr{B})$, then each $\mathscr{C}(x, b_n)$ is contained in $\mathscr{C}(\mathscr{B})$. Since $\mathscr{C}(\mathscr{B})$ is closed, we have $r \subset \mathscr{C}(\mathscr{B})$. $\square$

Lemma 8.2. *Let $V_1 \subset \mathscr{B}$ be a finite subset and $V_2 = \{c_1, \ldots, c_n\} \subset L^d$. Suppose that for each $j \in \{1, \ldots, n\}$, there exists a sequence $\{b_{i,j}\}_{i \in \mathbb{N}} \subset \mathscr{B}$ such that $\pi(b_{i,j}) \neq \pi(c_j)$ $(i \in \mathbb{N})$ and that $\pi(b_{i,j}) \to \pi(c_j)$ as $i \to \infty$. Put $V_{i,2} = \{b_{i,1}, \ldots, b_{i,n}\}$ $(i \in \mathbb{N})$. Then the sequence of closed sets $\{\mathscr{C}(V_1 \cup V_{i,2})\}$ converges to $D := \{x + \lambda y \mid x \in \mathscr{C}(V_1), y \in \mathscr{C}(V_2), \lambda \geq 0\}$ with respect to the Chabauty topology.*

Proof. For any subsequence of $\{\mathscr{C}(V_1 \cup V_{i,2})\}$, there exists its subsequence which converges to a certain closed set C_∞ with respect to the Chabauty topology. For simplicity, we also denote the converging subsequence by $\{\mathscr{C}(V_1 \cup V_{i,2})\}$ and prove that $C_\infty = D$.

First, we prove that $D \subset C_\infty$. Since each $\mathscr{C}(V_1 \cup V_{i,2})$ is convex, the limit C_∞ is also closed and convex. Note that D is the union of the sets $\{x + \lambda y \mid y \in \mathscr{C}(V_2), \lambda \geq 0\}$ ($x \in \mathscr{C}(V_1)$), and the set $\{x + \lambda y \mid y \in \mathscr{C}(V_2), \lambda \geq 0\}$ is the closed convex hull of $\bigcup_{j=1}^n \{x + \lambda c_j \mid \lambda \geq 0\}$ ($j \in \{1,\dots,n\}$). By Lemma 8.1, the limit of the sequence of line segments $\{\mathscr{C}(x, b_{i,j})\}_{i \in \mathbb{N}}$ is equal to $\{x + \lambda c_j \mid \lambda \geq 0\}$ for any $x \in \mathscr{C}(V_1)$ and $j \in \{1,\dots,n\}$. Since each $\mathscr{C}(x, b_{i,j})$ ($i \in \mathbb{N}$) is contained in $\mathscr{C}(V_1 \cup V_{i,2})$, the limit $\{x + \lambda c_j \mid \lambda \geq 0\}$ is contained in C_∞, and hence

$$D = \bigcup_{x \in \mathscr{C}(V_1)} \mathscr{C}\left(\bigcup_{j=1}^n \{x + \lambda c_j \mid \lambda \geq 0\}\right) \subset C_\infty.$$

To prove $C_\infty \subset D$, let z be a point in C_∞. Then there exists $\{z_i\}_{i \in \mathbb{N}}$ which converges to z such that $z_i \in \mathscr{C}(V_1 \cup V_{i,2})$ ($i \in \mathbb{N}$). Since $\mathscr{C}(V_1 \cup V_{i,2})$ is equal to $\{(1-\mu)x + \mu y \mid x \in \mathscr{C}(V_1), y \in \mathscr{C}(V_{i,2}), 0 \leq \mu \leq 1\}$, there exist sequences $\{x_i\} \subset \mathscr{C}(V_1)$, $\{y_i\}$ with $y_i \in \mathscr{C}(V_{i,2})$ and $\{\mu_i\} \subset [0,1]$ such that $z_i = (1-\mu_i)x_i + \mu_i y_i$. Since $\mathscr{C}(V_1)$ and $[0,1]$ are compact, by taking a subsequence, we may assume that there exist $x \in \mathscr{C}(V_1)$ and $\mu \in [0,1]$ such that $x_i \to x$ and $\mu_i \to \mu$ as $i \to \infty$. Since the sequence $\{\pi(\mathscr{C}(V_{i,2}))\}$ converges to $\pi(\mathscr{C}(V_2))$ with respect to the Chabauty topology on the set of closed subsets of $\overline{\mathbb{H}^d}$ and since $\pi(\mathscr{C}(V_2))$ is compact, by taking a further subsequence, there exists $y \in \mathscr{C}(V_2)$ such that $\pi(y_i) \to \pi(y)$ as $i \to \infty$.

Put $r_i = |y_i|/|y|$. Then $z_i = (1-\mu_i)x_i + r_i\mu_i(|y|/|y_i|)y_i$. Since the sequence of the heights of y_i ($i \in \mathbb{N}$) diverges by Lemma 4.2(1), the sequence $\{r_i\}$ also diverges. On the other hand, since $x_i \to x$, $\mu_i \to \mu$ and $(|y|/|y_i|)y_i \to y$ as $i \to \infty$, the sequence $\{r_i\mu_i\}$ converges to some $\lambda \in \mathbb{R}_{\geq 0}$. Thus $\{\mu_i\}$ converges to 0 and hence $z = \lim_{i \to \infty} z_i = x + \lambda y \in D$. Therefore C_∞ is contained in D. $\square$

Definition 8.3. The symbol $\mathscr{C}_0(\mathscr{B})$ denotes the union of the finite sided polyhedra in $\mathbb{E}^{1,d}$ spanned by finite subsets of $\mathscr{B}$.

Lemma 8.4. *For any $z \in \mathscr{C}(\mathscr{B})$, there exists a ray r from a point in $\mathscr{C}_0(\mathscr{B})$ such that $z \in r$ and $r \subset \mathscr{C}(\mathscr{B})$.*

Proof. First, suppose that $z \in \mathscr{C}_0(\mathscr{B})$. Let r be the ray from O which contains z. Then, by Lemma 4.8, the ray $r_{\geq z}$ from z is contained in $\mathscr{C}(\mathscr{B})$. Thus $r_{\geq z}$ has the desired property.

Next, suppose that $z \notin \mathscr{C}_0(\mathscr{B})$. Note that $z \in \overline{\mathscr{C}_0(\mathscr{B})}$ by Lemma 2.19(1). Since each finite sided convex polyhedron whose vertices are contained in $\mathscr{B}$ is subdivided into (possibly degenerate) $(d+1)$-simplices whose vertices are contained in $\mathscr{B}$, there exist sequences $\{b_{i,j}\}_{i \in \mathbb{N}}$ ($j \in \{0,\dots,d+1\}$) and $\{z_i\}_{i \in \mathbb{N}}$ such that $b_{i,j} \in \mathscr{B}$, $z_i \in \mathscr{C}(b_{i,0},\dots,b_{i,d+1})$, and $\lim z_i = z$. By taking a subsequence, we may suppose that there exists a partition of $\{0,\dots,d+1\}$ into two subsets J_1 and J_2 such that

(i) $J_1 \neq \emptyset$, and for any $j \in J_1$ and $i \in \mathbb{N}$, $b_{i,j} = b_{1,j}$;

(ii) $J_2 \neq \emptyset$, and for any $j \in J_2$, there exists $c_j \in L^d$ such that $\pi(b_{i,j}) \neq \pi(c_j)$ and that $\pi(b_{i,j}) \to \pi(c_j)$ as $i \to \infty$.

We remark that the first condition can be attained by Lemma 4.2(1) and that the second condition can be attained because $z \notin \mathscr{C}_0(\mathscr{B})$. Then, by Lemma 8.2, $\{\mathscr{C}(b_{i,0}, \dots, b_{i,d+1})\}$ converges to $D = \{x + \lambda y \mid x \in \mathscr{C}(\{b_{1,j} \mid j \in J_1\}), y \in \mathscr{C}(\{c_j \mid j \in J_2\}), \lambda \geq 0\}$ with respect to the Chabauty topology. Since each $\mathscr{C}(b_{i,0}, \dots, b_{i,d+1})$ is contained in the closed set $\mathscr{C}(\mathscr{B})$, its limit D is also contained in $\mathscr{C}(\mathscr{B})$. Since $z \in D$, we have $z = x_0 + \lambda_0 y_0$ for some $x_0 \in \mathscr{C}(\{b_{1,j} \mid j \in J_1\}) \subset \mathscr{C}_0(\mathscr{B})$, $y_0 \in \mathscr{C}(\{c_j \mid j \in J_2\})$ and $\lambda_0 \geq 0$. Then the ray $r := \{x_0 + \lambda y_0 \mid \lambda \geq 0\}$ satisfies the desired property. $\square$

9. Proofs of Theorems 5.7, 5.8, Corollary 5.9 and Proposition 5.2

Lemma 9.1. *For any closed facet F of $\mathscr{C}(\mathscr{B})$, we have $F \cap \mathscr{B} \neq \emptyset$.*

Proof. We first prove $F \cap \mathscr{C}_0(\mathscr{B}) \neq \emptyset$. Pick a point $x \in \mathrm{int}(F)$. By Lemma 8.4, there is a ray r from a point $x_0 \in \mathscr{C}_0(\mathscr{B})$ passing through x such that $r \subset \mathscr{C}(\mathscr{B})$. If $x = x_0$, then $x_0 \in \mathrm{int}(F) \cap \mathscr{C}_0(\mathscr{B})$ and hence $F \cap \mathscr{C}_0(\mathscr{B}) \neq \emptyset$. Suppose that $x \neq x_0$. Then the open convex set $\mathrm{int}(r)$ intersects $\mathrm{int}(F)$, and hence $\mathrm{int}(r) \subset \mathrm{int}(F)$ by Lemma 2.14. So, $x_0 \in r \subset F$ and therefore $F \cap \mathscr{C}_0(\mathscr{B}) \neq \emptyset$.

By the above observation, there exists a finite subset V of $\mathscr{B}$ such that $F \cap \mathscr{C}(V) \neq \emptyset$. Let $\{F_i\}_{i=0}^k$ be a face sequence for F. We show by induction that $F_i \cap \mathscr{C}(V) = \mathscr{C}(V_i)$, where $V_i = F_i \cap V$. Once this is proved, we have $F \cap \mathscr{B} \neq \emptyset$, because the non-empty set $F \cap \mathscr{C}(V)$ is equal to $\mathscr{C}(V_k)$ and hence $F \cap \mathscr{B} \supset F \cap V = V_k \neq \emptyset$.

Since $F_0 = \mathscr{C}(\mathscr{B})$, we have $V_0 = V$ and hence $F_0 \cap \mathscr{C}(V) = \mathscr{C}(V_0)$. Suppose $F_i \cap \mathscr{C}(V) = \mathscr{C}(V_i)$ for some $i \in \{0, \dots, k-1\}$. Let H be a supporting half space for F_i in $\Pi(F_i)$ such that $\partial H \cap F_i = F_{i+1}$. Then $F_{i+1} \cap \mathscr{C}(V) = \partial H \cap F_i \cap \mathscr{C}(V) = \partial H \cap \mathscr{C}(V_i)$. Since H is a supporting half space for F_i, $\mathscr{C}(V_i)$ is contained in H. This implies $\partial H \cap \mathscr{C}(V_i) = \mathscr{C}(\partial H \cap V_i)$ because V_i is a finite set. Moreover $\mathscr{C}(\partial H \cap V_i) = \mathscr{C}(\partial H \cap F_i \cap V) = \mathscr{C}(F_{i+1} \cap V)$. Hence we have proved $F_i \cap \mathscr{C}(V) = \mathscr{C}(V_i)$ for any $i \in \{0, \dots, k\}$. This completes the proof of Lemma 9.1. $\square$

Proof of Theorem 5.8(1). Let G be a closed flat piece of $\mathscr{C}(\Lambda)$ such that $G \cap \Lambda_p = \emptyset$. Suppose contrary that $G \cap |\widetilde{\Delta}| \neq \emptyset$. Then, by Proposition 4.15(1), there exists a visible open facet F of $\mathscr{C}(\mathscr{B})$ such that $\pi(F) \cap G \neq \emptyset$. Since $\pi(F)$ is an open convex set by Lemma 6.2, $\pi(F)$ is contained in G by Lemma 2.14. Since $\overline{F} \cap \mathscr{B} \neq \emptyset$ by Lemma 9.1, we have $\emptyset \neq \pi(\overline{F}) \cap \Lambda_p \subset \overline{\pi(F)} \cap \Lambda_p \subset G \cap \Lambda_p$, a contradiction. $\square$

To prove Theorem 5.8(2), we prepare two lemmas.

Lemma 9.2. *For any $v \in \Lambda_p$ and $w \in \Lambda$, $\mathscr{C}(v,w) - \{w\}$ is contained in $\pi(\mathscr{C}(\mathscr{B}))$.*

Proof. We may suppose that $v \neq w$. Since v is a point in $\Lambda_p = \pi(\mathscr{B})$, there exists $b \in \mathscr{B}$ such that $\pi(b) = v$.

Suppose first that w is also contained in $\pi(\mathscr{B})$. Then there exists $b' \in \mathscr{B}$ such that $\pi(b') = w$. Since $\mathscr{C}(b,b') \subset \mathscr{C}(\mathscr{B})$, we have $\mathscr{C}(v,w) = \pi(\mathscr{C}(b,b')) \subset \pi(\mathscr{C}(\mathscr{B}))$.

Next, suppose that $w \notin \pi(\mathscr{B})$. Since $\pi(\mathscr{B})$ is dense in Λ, there exists a sequence $\{b_n\} \subset \mathscr{B}$ such that $\{\pi(b_n)\}$ converges to w. By the assumption that $w \notin \pi(\mathscr{B})$, all $\pi(b_n)$ are distinct from w. Let c be a point in L^d such that $\pi(c) = w$. Then, by Lemma 8.1, the ray $\{b + \lambda c \mid \lambda \geq 0\}$ is contained in $\mathscr{C}(\mathscr{B})$. Since $\pi(\{b + \lambda c \mid \lambda \geq 0\}) = \mathscr{C}(v,w) - \{w\}$, we have $\mathscr{C}(v,w) - \{w\} \subset \pi(\mathscr{C}(\mathscr{B}))$. □

Lemma 9.3. *For any $v \in \Lambda_p$ and $w_1, \ldots, w_{d'} \in \Lambda$ $(d' \in \mathbb{N})$, $\mathscr{C}(v,w_1,\ldots,w_{d'}) - \mathscr{C}(w_1, \ldots, w_{d'})$ is contained in $\pi(\mathscr{C}(\mathscr{B}))$.*

Proof. By Lemma 9.2, $\mathscr{C}(v,w_j) - \{w_j\}$ is contained in $\pi(\mathscr{C}(\mathscr{B}))$ for each $j \in \{1,\ldots, d'\}$. Since $\mathscr{C}(v,w_1,\ldots,w_{d'}) - \mathscr{C}(w_1,\ldots,w_{d'})$ is contained in the convex hull of $\bigcup_{j=1}^{d'}(\mathscr{C}(v,w_j) - \{w_j\})$, it is contained in $\pi(\mathscr{C}(\mathscr{B}))$. □

Proof of Theorem 5.8(2). Let G be an open flat piece of $\mathscr{C}(\Lambda)$ such that $\overline{G} \cap \Lambda_p \neq \emptyset$. Let v be a parabolic fixed point contained in $\overline{G}$. Then, by Lemma 3.9,

$$G \subset \bigcup \{\mathscr{C}(v,w_1,\ldots,w_{d'}) - \mathscr{C}(w_1,\ldots,w_{d'}) \mid w_1,\ldots,w_{d'} \in \Pi(G) \cap \Lambda\},$$

where $d' = \dim G$. By Lemma 9.3, the set in the right hand side is contained in $\pi(\mathscr{C}(\mathscr{B})) = |\widetilde{\Delta}|$. Hence we have $G \subset |\widetilde{\Delta}|$. □

At the end of this section, we prove Theorem 5.7, Corollary 5.9, and Proposition 5.2.

Proof of Theorem 5.7. This directly follows from Proposition 3.5 and Theorem 5.8. □

Proof of Corollary 5.9. This follows from the following facts.

(i) Any flat piece of M_0 has dimension greater than 0 by Lemma 3.8.

(ii) If $\dim M = 2$, then every 1-dimensional flat piece of M_0 does not intersect a cusp of M.

(iii) If $\dim M = 3$, then every 1-dimensional flat piece does not intersect a cusp of M. This is observed by [KS93, p.725] (cf. [Ser85]).

□

Proof of Proposition 5.2.

(i) By Definition 4.16, any 0-dimensional facet of $\widetilde{\Delta}$ is equal to $\pi(\widehat{F})$ for some 0-dimensional visible facet $\widehat{F}$ of $\mathscr{C}(\mathscr{B})$. Since $\widehat{F}$ is 0-dimensional, $\widehat{F} = \Pi(\widehat{F}) \cap \mathscr{C}(\mathscr{B})$ is a singleton. By Lemma 9.1, $\Pi(\widehat{F}) \cap \mathscr{B} \neq \emptyset$. Thus $\widehat{F}$ consists of a point in $\mathscr{B}$, which projects to a point in $\Lambda_p \subset \partial\mathbb{H}^d$ by π. Thus every 0-dimensional facet of $\widetilde{\Delta}$ is a singleton in Λ_p. Conversely, let v be a point in Λ_p. Then $v = \pi(b)$ for some $b \in \mathscr{B}$ and $\{b\}$ is a 0-dimensional face of $\mathscr{C}(\mathscr{B})$ by Lemma 4.6. By Lemmas 4.6 and 4.13(1), the facet $\{b\}$ is visible. Hence $\{v\}$ is a 0-dimensional facet of $\widetilde{\Delta}$.

(ii) This follows from Lemma 4.11.

(iii) Let F be a closed facet of $\widetilde{\Delta}$ of dimension ≥ 1, and let $\widehat{F}$ be a visible closed facet of $\mathscr{C}(\mathscr{B})$ such that $\pi(\widehat{F}) = F$. Then $\widehat{F} \cap \mathscr{B} \neq \emptyset$ by Lemma 9.1, and hence $F \cap \partial\mathbb{H}^d \neq \emptyset$. Moreover, by Lemma 4.6, we have $F \cap \partial\mathbb{H}^d \subset \pi(\mathscr{C}(\mathscr{B})) \cap \partial\mathbb{H}^d = \Lambda_p$.

□

10. Relation with the Ford domain

In this section, we give a brief description of a generalization of the duality between the Epstein-Penner decomposition and the Ford domain of a cusped hyperbolic manifold of finite volume due to Epstein and Penner [EM87, Section 4] (cf. [ASWY, Section 2]).

Let M and $\mathscr{H}$ be as in Convention 4.1. Let $\widetilde{\mathscr{F}}$ be the subspace of $\mathbb{H}^d$ consisting of the points x such that there are at least two shortest geodesic segments from x to the horoballs in $\mathscr{H}$. For each point $x \in \widetilde{\mathscr{F}}$, let $\mathscr{H}_x$ be the subset of $\mathscr{H}$ consisting of the horoballs H such that $d(x, H) = d(x, \cup\mathscr{H})$. Then, by Lemma 4.2, we see that $\mathscr{H}_x$ is a finite set (cf. [EM87, Section 4]). By using $\mathscr{H}_x$ we obtain a natural partition of $\widetilde{\mathscr{F}}$ as follows. Consider the equivalence relation $\sim$ on $\widetilde{\mathscr{F}}$ such that $x \sim y$ if and only if $\mathscr{H}_x = \mathscr{H}_y$. This equivalence relation gives a Γ-invariant partition of $\widetilde{\mathscr{F}}$ into the equivalence classes. Moreover, we can see that this partition is locally finite (see [EM87, Section 3]).

The *Ford complex*, Ford(Γ), of M (with respect to $\mathscr{H}$) is defined to be the image $p(\widetilde{\mathscr{F}})$ in M together with the locally finite partition induced by that of $\widetilde{\mathscr{F}}$. (It is sometimes more convenient to define Ford(Γ) to be the closure of $p(\widetilde{\mathscr{F}})$ in $\overline{M} := (\mathbb{H}^d \cup \Omega)/\Gamma$, and we employ this alternative definition in [ASWY, Section 2].)

We now define a subcomplex, $\Delta_{\mathbb{E}}$, of Δ which is a geometric dual to Ford(Γ). Let $\widehat{\Delta}_{\mathbb{E}}$ be the family of the visible open facets $\widehat{F}$ of $\mathscr{C}(\mathscr{B})$ such that $\widehat{F} = \text{int}(W \cap \mathscr{C}(\mathscr{B})) \subset \mathbb{H}^d$ for some Euclidean support plane W for $\mathscr{C}(\mathscr{B})$. Set

$$\widetilde{\Delta}_{\mathbb{E}} = \{\pi(\widehat{F}) \,|\, \widehat{F} \in \widehat{\Delta}_{\mathbb{E}}\}, \quad \Delta_{\mathbb{E}} = \{p(\widetilde{F}) \,|\, \widetilde{F} \in \widetilde{\Delta}_{\mathbb{E}}\}.$$

Then we can see, by the arguments in [EP88, Section 3], that the closure in $\overline{\mathbb{H}^d}$ of each member $\widetilde{F}$ of $\widetilde{\Delta}_{\mathbb{E}}$ is a finite sided ideal polyhedron spanned by a finite subset of Λ_p and that $\widetilde{\Delta}_{\mathbb{E}}$ and $\Delta_{\mathbb{E}}$ are locally finite. Moreover, the arguments in [EP88, Section 4] imply that $\widetilde{\Delta}_{\mathbb{E}}$ is dual to $\widetilde{\mathscr{F}}$ and $\Delta_{\mathbb{E}}$ is dual to Ford(Γ) in the following sense. Let P be a piece of the partition of $\widetilde{\mathscr{F}}$ determined by a finite subset $\{H_0, H_1, \ldots, H_k\}$ of $\mathscr{H}$, i.e.,

$$P = \left\{x \in \mathbb{H}^d \,|\, \mathscr{H}_x = \{H_0, H_1, \ldots, H_k\}\right\}.$$

Then the interior of the finite sided ideal polyhedron spanned by the centers of the horoballs $H_0, H_1, \ldots, H_k$ is the closure of some open facet F of $\widetilde{\Delta}_{\mathbb{E}}$. The correspondence $P \mapsto F$ gives a one-to-one correspondence between the pieces of the partition of $\widetilde{\mathscr{F}}$ and the open facets of $\widetilde{\Delta}_{\mathbb{E}}$. Moreover, this induces a one-to-one correspondence between the pieces of the partition of Ford(Γ) and the open facets of $\Delta_{\mathbb{E}}$.

11. Punctured torus groups

Let Γ be a *quasifuchsian punctured torus group*, i.e., a Kleinian group satisfying the following conditions.

(i) Γ is freely generated by two isometries whose commutator is parabolic.

(ii) The domain of discontinuity Ω consists of exactly two simply connected components $\Omega^{\pm}$, whose quotients $\Omega^{\pm}/\Gamma$ are each homeomorphic to a punctured torus T.

Then the quotient triple $(\mathbb{H}^3 \cup \Omega, \Omega^-, \Omega^+)/\Gamma$ is identified with $(T \times [-1, 1], T \times \{-1\}, T \times \{1\})$. The boundary of the convex core M_0 of $M = \mathbb{H}^3/\Gamma$ has two components; the component facing to $\Omega^{\varepsilon}/\Gamma$ ($\varepsilon \in \{-, +\}$) is denoted by $\partial^{\varepsilon} M_0$. Then $\partial^{\varepsilon} M_0$ with the path metric has a structure of a hyperbolic punctured torus bent along a geodesic measured lamination, $pl^\varepsilon = pl^\varepsilon(\Gamma)$, called the *bending measured lamination* of $\partial^\varepsilon M_0$. The underlying geodesic lamination $|pl^\varepsilon|$ is called the *bending lamination* of $\partial^\varepsilon M_0$ (see [Thu02], [EM87], [KS93], [KS97], [KS98]). It is observed by [KS93, p.725] (cf. [Ser85]) that $|pl^\varepsilon|$ is compactly supported (i.e., disjoint from a cusp neighborhood), and we have the following proposition, which gives the cellular structure of the convex core (see [Thu02], [EM87], [KS93], [KS97], [KS98]).

Proposition 11.1. *For any quasifuchsian punctured torus group Γ, the following hold.*

1. *Suppose that $|pl^\varepsilon|$ is rational, i.e., a simple closed geodesic. Then the cellular structure of $\partial^\varepsilon M_0$ consists of a single 2-dimensional open flat piece, A, and a single 1-dimensional open flat piece, α, satisfying the following conditions.*

    a. *The flat piece α is the simple closed geodesic $|pl^\varepsilon|$.*

    b. *The flat piece A with the path metric is isometric to the interior of the convex core of a once-punctured annulus.*

    Moreover, the lifts of both A and α to $\partial\mathscr{C}(\Lambda)$ are faces of $\mathscr{C}(\Lambda)$ (cf. Definition 2.11(1) and Remark 2.12(2)).

2. *Suppose that $|pl^\varepsilon|$ is irrational. Then the cellular structure of $\partial^\varepsilon M_0$ consists of a single 2-dimensional open flat piece, B, and uncountably many 1-dimensional open flat pieces, $\{\beta_\iota\}_\iota$, satisfying the following conditions.*

    a. *The union $\cup\beta_\iota$ forms the geodesic lamination $|pl^\varepsilon|$.*

    b. *The flat piece B with the path metric is isometric to the interior of a punctured bigon.*

    Moreover, the lifts to $\partial\mathscr{C}(\Lambda)$ of all but the two 1-dimensional open flat pieces, forming the two boundary leaves of $|pl^\varepsilon|$, are faces of $\mathscr{C}(\Lambda)$.

Note that each open facet of the EPH-decomposition Δ of M_0 is either disjoint from $\partial^\varepsilon M_0$ or contained in an open flat piece of M_0 contained in $\partial^\varepsilon M_0$ (Proposition 5.1). By the *restriction* of Δ to $\partial^\varepsilon M_0$ we mean the collection of the open facets of Δ contained in $\partial^\varepsilon M_0$, and denote it by the symbol $\partial^\varepsilon \Delta$.

In this section, we prove the following theorem, which clarifies the structure of $\partial^\varepsilon \Delta$ for a quasifuchsian punctured torus group.

Theorem 11.2. *For any quasifuchsian punctured torus group Γ, the following hold.*

1. *Suppose that $|pl^\varepsilon|$ is rational. Then $|\partial^\varepsilon \Delta|$ is equal to the 2-dimensional open flat piece A. (See Proposition 11.1(1).) Moreover, the restriction $\partial^\varepsilon \Delta$ consists of the unique geodesic α' joining the puncture to itself and the two components of $A - \alpha'$. (See Figure 1.)*

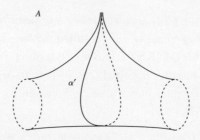

Figure 1: $\partial^{\varepsilon}\Delta$ for the case where $|pl^{\varepsilon}|$ is rational

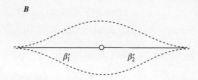

Figure 2: $\partial^{\varepsilon}\Delta$ for the case where $|pl^{\varepsilon}|$ is irrational

2. *Suppose that $|pl^{\varepsilon}|$ is irrational. Then $|\partial^{\varepsilon}\Delta|$ is equal to the 2-dimensional open flat piece B. (See Proposition 11.1(2).) Moreover, the restriction $\partial^{\varepsilon}\Delta$ consists of the two geodesics β_1' and β_2', each joining the puncture with a vertex of the bigon B, and the two components of $B - (\beta_1' \cup \beta_2')$. (See Figure 2.)*

To prove the above theorem, we need the following two lemmas.

Lemma 11.3. *For any 1-dimensional visible open facet $\widehat{e}$ of $\mathscr{C}(\mathscr{B})$, the following hold.*

1. *If $\widehat{e}$ is elliptic, then there exist $b_0, b_1 \in \mathscr{B}$ such that $\widehat{e} = \mathrm{int}(\mathscr{C}(b_0, b_1))$.*

2. *If $\widehat{e}$ is parabolic, then there exist $b_0 \in \mathscr{B}$ and $x_0 \in \Lambda - \Lambda_p$ such that $\widehat{e} = b_0 + \pi^{-1}(x_0)$.*

3. *If $\widehat{e}$ is hyperbolic, then there exist $b_0 \in \mathscr{B}$ and $x_0 \in \mathbb{H}^d$ such that $\widehat{e} = b_0 + \pi^{-1}(x_0)$.*

Proof. Since $\dim \widehat{e} = 1$, $\partial \widehat{e}$ consists of at most two points. On the other hand, by Lemma 9.1, the closure of $\widehat{e}$ intersects $\mathscr{B}$ and hence $\partial \widehat{e}$ contains at least one point in $\mathscr{B}$. Note that $\partial \widehat{e}$ is the union of 0-dimensional visible facets of $\mathscr{C}(\mathscr{B})$ (see Proposition 4.15(2)). Since every 0-dimensional visible facet consists of a point in $\mathscr{B}$ by Lemma 9.1, $\partial \widehat{e}$ is either $\{b_0, b_1\} \subset \mathscr{B}$ or $\{b_0\} \subset \mathscr{B}$.

If $\partial \widehat{e} = \{b_0, b_1\}$, then $\widehat{e}$ is equal to $\mathrm{int}(\mathscr{C}(b_0, b_1))$ and hence elliptic.

Suppose that $\partial \widehat{e} = \{b_0\}$. Then $\widehat{e}$ is equal to $\{b_0 + t\widehat{x_0} \mid t > 0\}$ for some $\widehat{x_0} \in \mathbb{E}^{1,d}$. Since $\widehat{e} \subset T^d \cup L^d$, we can see that $\widehat{x_0} \in T^d \cup L^d$ and hence $\widehat{e} = b_0 + \pi^{-1}(x_0)$ where

$x_0 = \pi(\widehat{x}_0)$. Moreover $\widehat{e}$ is parabolic or hyperbolic according as $x_0 \in \partial \mathbb{H}^d$ or $x_0 \in \mathbb{H}^d$. This completes the proof of Lemma 11.3. □

Lemma 11.4. *Let $\widehat{e}$ be a 1-dimensional visible open facet of $\mathscr{C}(\mathscr{B})$ such that $\partial \widehat{e}$ contains a point $b_0 \in \mathscr{B}$, and $\widehat{W}$ be a 2-dimensional hyperbolic affine subspace of $\mathbb{E}^{1,d}$ which contains b_0 and does not contain O. Set $\widetilde{W} = \pi(\widehat{W})$ and $\delta \widetilde{W} = \mathrm{cl}(\widetilde{W}, \overline{\mathbb{H}^d}) - \widetilde{W}$. Suppose that the point of $\partial \pi(\widehat{e})$ distinct from $\pi(b_0)$ belongs to $\delta \widetilde{W} - \Lambda_p$. Then $\widehat{e}$ is contained in $\widehat{W}$.*

Proof. Put $\widehat{e}' = \widehat{W} \cap \pi^{-1}(\pi(\widehat{e}))$. Then $\widehat{e}'$ is the interior of a ray from b_0 such that $\pi(\widehat{e}') = \pi(\widehat{e})$, because $\partial \pi(\widehat{e}) - \{\pi(b_0)\} \in \delta \widetilde{W}$. Since $\partial \pi(\widehat{e}) - \{\pi(b_0)\} \not\subseteq \Lambda_p$, $\widehat{e}$ is the interior of a ray from b_0 by Lemma 11.3. Thus both $\widehat{e}$ and $\widehat{e}'$ are the interiors of rays in $T^d \cup L^d$ from b_0 and $\pi(\widehat{e}') = \pi(\widehat{e})$. This implies $\widehat{e} = \widehat{e}'$. □

Proof of Theorem 11.2. Let $\widetilde{W}$ be a hyperplane of $\overline{\mathbb{H}^3}$ such that the open flat piece $\mathrm{int}(\widetilde{W} \cap \mathscr{C}(\Lambda))$ projects to the unique 2-dimensional open flat piece of $\partial^{\varepsilon} M_0$ (see Proposition 11.1). Set $\widehat{W} = \Pi(\pi^{-1}(\widetilde{W})) \subset \mathbb{E}^{1,3}$. Since $\widetilde{W}$ is a support plane for $\mathscr{C}(\Lambda)$ and $\widehat{W}$ contains O, one of the closed half spaces of $\mathbb{E}^{1,3}$ bounded by $\widehat{W}$ contains $\mathscr{C}(\mathscr{B})$. Since $\widetilde{W}$ contains a parabolic fixed point by Proposition 11.1, we have $\widehat{W} \cap \mathscr{B} \neq \emptyset$. So $\widehat{W}$ is a support plane for $\mathscr{C}(\mathscr{B})$. Hence, by Lemma 2.13 and the definition of Δ, the restriction $\partial^{\varepsilon} \Delta$ consists of the images by $p \circ \pi$ of the visible open facets of $\mathscr{C}(\mathscr{B})$ contained in $\widehat{W} \cap \mathscr{C}(\mathscr{B})$.

(1) By Corollary 5.9(2) and Proposition 11.1(1), we have $|\partial^{\varepsilon} \Delta| = A$. Since A contains the main cusp, $\pi_1(A) < \Gamma$ contains a parabolic transformation. Since the fundamental group of a 2-dimensional open facet of Δ is conjugate to a discrete subgroup of $\mathrm{Isom}(\mathbb{H}^1)$ by Corollary 5.4, A cannot be an open facet of Δ.

We can see that $\partial^{\varepsilon} \Delta$ contains an *edge*, i.e., a 1-dimensional open facet, as follows: Suppose this is not the case, then, by Proposition 5.2(1), A is the disjoint union of at least two 2-dimensional open facets. Let F be a 2-dimensional open facet of $\partial^{\varepsilon} \Delta$. Set $\widetilde{A} = \mathrm{int}(\widetilde{W} \cap \mathscr{C}(\Lambda))$ and let $\widetilde{F}$ be an open facet of $\widetilde{\Delta}$ such that $\widetilde{F} \subset \widetilde{A}$ and $F = p(\widetilde{F})$. Then $\widetilde{F}$ is open in $\Pi(\widetilde{F}) = \widetilde{W}$ by Lemma 6.2, and hence it is open in $\widetilde{A}$. Since $p|_{\widetilde{A}} : \widetilde{A} \to A$ is a covering projection, $F = p(\widetilde{F})$ is open in A. Thus A is a disjoint union of at least two open subsets. This contradicts the connectedness of A.

Let e be an edge of $\partial^{\varepsilon} \Delta$, $\widetilde{e}$ a 1-dimensional open facet of $\widetilde{\Delta}$ contained in $\widetilde{W}$ such that $p(\widetilde{e}) = e$, and $\widehat{e}$ the 1-dimensional visible open facet of $\mathscr{C}(\mathscr{B})$ such that $\pi(\widehat{e}) = \widetilde{e}$. In what follows, we show that $\partial \widehat{e}$ consists of two points in $\mathscr{B}$. Suppose contrary that this is not the case. Then, by Lemma 11.3, $\widehat{e} = b_0 + \pi^{-1}(x_0)$ for some $b_0 \in \mathscr{B}$ and $x_0 \in \mathbb{H}^d \cup (\Lambda - \Lambda_p)$. Then $\partial \widetilde{e} = \{v_0, x_0\}$, where $v_0 = \pi(b_0)$. Since $\widetilde{e} \subset \widetilde{A} := \mathrm{int}(\widetilde{W} \cap \mathscr{C}(\Lambda))$, we have $x_0 \in \partial \widetilde{A}$ by Proposition 5.3. Since $x_0 \notin \Lambda_p$, there is a 1-dimensional plane l_1 in $\widetilde{W} = \overline{\mathbb{H}^2}$ such that $x_0 \in l_1 \subset \partial \widetilde{A}$ (see Figure 3).

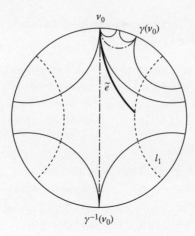

Figure 3: The action of $\pi_1(A)$ on $\widetilde{W} = \overline{\mathbb{H}^2}$: The hexagonal region is a fundamental domain for the action of $\pi_1(A)$ on $\widetilde{W}$.

Let γ be a primitive element of $\pi_1(A)$ which has l_1 as the axis. Since γ stabilizes $\widetilde{W}$, it also stabilizes $\widehat{W}$. Thus γ is a hyperbolic transformation of the isometry group of the $(2+1)$-dimensional Minkowski space $\widehat{W}$. Let $\widehat{V}$ be the 2-dimensional affine subspace of $\widehat{W}$ such that $b_0 \in \widehat{V}$ and that $\widehat{V}$ is parallel to the vector subspace $\Pi(\pi^{-1}(l_1))$. Then $\widehat{V}$ is γ-invariant, because γ is a hyperbolic transformation with axis l_1. Set $\widetilde{V} = \pi(\widehat{V})$. Then $\widetilde{V}$ is the open half space of $\widetilde{W} \cong \overline{\mathbb{H}^2}$ bounded by l_1 containing v_0. In particular, $\delta\widetilde{V} := \mathrm{cl}(\widetilde{V}, \overline{\mathbb{H}^3}) - \widetilde{V}$ is equal to l_1 (cf. Lemma 6.1). Since both $\widehat{V}$ and $\widehat{e}$ contain $b_0 \in \mathscr{B}$ and since $\widehat{e}$ is a visible open facet of $\mathscr{C}(\mathscr{B})$ such that $\partial\pi(\widehat{e}) - \{v_0\} \in l_1$, $\widehat{V}$ contains $\widehat{e}$ by Lemma 11.4.

Claim 11.5. *The points $\gamma^{\pm 1}(b_0)$ in the 2-dimensional plane $\widehat{V}$ are separated by the 1-dimensional plane $\Pi(\widehat{e})$.*

Proof. This follows from the fact that the points $\gamma^{\pm 1}(v_0) = \pi(\gamma^{\pm 1}(b_0))$ in the open half space $\widetilde{V}$ are separated by $\widetilde{e} = \pi(\widehat{e})$ (see Figure 3). $\qquad\square$

Let $\{\widehat{F}_i\}_{i=0}^k$ be a face sequence for $\widehat{e}$ extending the sequence $\{\widehat{F}_0, \widehat{F}_1\} = \{\mathscr{C}(\mathscr{B}), \widehat{W} \cap \mathscr{C}(\mathscr{B})\}$ (see Lemma 2.13).

Claim 11.6. $\Pi(\widehat{F}_2) = \widehat{V}$.

Proof. We can see that $\widehat{W} \cap \mathscr{B} \subset \widehat{W} \cap \mathscr{C}(\mathscr{B})$ cannot be contained in a 2-dimensional affine subspace of $\widehat{W}$ because $\pi_1(A)$ is a non-elementary fuchsian group. Thus $\Pi(\widehat{F}_1)$ is equal to $\widehat{W}$. Let $\widehat{V}'$ be a support plane for $\widehat{F}_1$ in $\widehat{W}$ such that $\widehat{V}' \cap \widehat{F}_1 = \widehat{F}_2$. Since both $\widehat{V}'$ and $\widehat{V}$ are 2-dimensional affine subspaces of the 3-dimensional vector space $\widehat{W}$ which

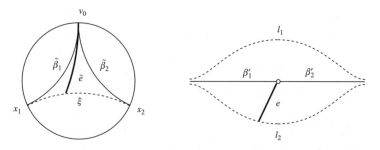

Figure 4: Part of $\widetilde{W} \cap \mathscr{C}(\Lambda)$ and B

contain the common line segment $\widehat{e}$, $\widehat{V}'$ is obtained from $\widehat{V}$ by rotating around the 1-dimensional plane $\Pi(\widehat{e})$. Since $\gamma^{\pm 1}(b_0) \in \widehat{W} \cap \mathscr{C}(\mathscr{B})$, both $\gamma^{\pm 1}(b_0)$ are contained in the same side of $\widehat{V}'$ in $\widehat{W}$. This together with Claim 11.5 implies $\widehat{V}' = \widehat{V}$. Finally, since $\widehat{F}_2 = \widehat{V}' \cap \widehat{F}_1$ contains $\mathscr{C}(\widehat{e} \cup \{\gamma^{-1}(b_0), \gamma(b_0)\})$, we have $\Pi(\widehat{F}_2) = \widehat{V}' = \widehat{V}$. $\qquad\square$

By Claim 11.5, we have $\widehat{e} \subset \mathrm{int}(\mathscr{C}(\widehat{e} \cup \{\gamma^{-1}(b_0), \gamma(b_0)\}))$. Since the 2-dimensional plane $\Pi(\widehat{F}_2) = \widehat{V}$ contains $\mathscr{C}(\widehat{e} \cup \{\gamma^{-1}(b_0), \gamma(b_0)\})$, there exists no half space of $\Pi(\widehat{F}_2)$ which contains $\widehat{e}$ in the boundary. This contradicts the assumption that $\{\widehat{F}_i\}_{i=0}^k$ is a face sequence for $\widehat{e}$. Thus we have proved that $\partial\widehat{e}$ consists of two points in $\mathscr{B}$. Hence e is equal to the geodesic α' in A joining the puncture to itself. This completes the proof of Theorem 11.2(1).

(2) By Corollary 5.9(2) and Proposition 11.1(2), we have $|\partial^\varepsilon \Delta| = B$. Then, by the argument in the proof of the assertion (1), we can see that B contains at least two open facets of $\partial^\varepsilon \Delta$, and at least one of them is 1-dimensional. Let l_1 and l_2 be the boundary leaves of $|pl^\varepsilon|$ (see Figure 4).

Claim 11.7. *Both β_1' and β_2' are edges of $\partial^\varepsilon \Delta$.*

Proof. Let $\mathscr{G}$ be the space of geodesics in B emanating from the puncture of B. Then two distinct members of $\mathscr{G}$ are disjoint, and $\mathscr{G}$ is identified with S^1. By Propositions 5.2(3) and 5.3, any edge of $\partial^\varepsilon \Delta$ belongs to $\mathscr{G}$.

Suppose that β_j' is not an edge of $\partial^\varepsilon \Delta$ for some $j \in \{1, 2\}$. Then there is a 2-dimensional open facet F of $\partial^\varepsilon \Delta$ containing β_j', because any edge of $\partial^\varepsilon \Delta$ is disjoint from β_j'. As in the proof of the assertion (1), we see that F is open in B. Thus F contains all geodesics belonging to some neighborhood of β_j' in $\mathscr{G}$. Hence the end of F "intersects" both l_1 and l_2. To be precise, the following holds. Let $\widetilde{F}$ be a closed facet of $\widetilde{\Delta}$ such that $F = p(\mathrm{int}(\widetilde{F}))$. Then $\delta\widetilde{F}$ intersects a lift of l_i for each $i \in \{1, 2\}$. This contradicts Proposition 5.3. $\qquad\square$

In what follows, we prove that $\partial^{\varepsilon}\Delta$ has no edges except for β_1' and β_2'. Suppose that there exists an edge e of $\partial^{\varepsilon}\Delta$ different from β_1' and β_2'. Since B is the unique flat piece of M_0 contained in $|\partial^{\varepsilon}\Delta|$, and since $\pi_1(B)$ is the infinite cyclic group generated by a parabolic transformation, $\widetilde{W} \cap \Lambda_p$ consists of a single point, v_0 (see Figure 4). Then there are lifts $\widetilde{e}$ and $\widetilde{\beta}_j$ ($j \in \{1,2\}$) of e and β_j' respectively such that

(i) $\widetilde{\beta}_1$ and $\widetilde{\beta}_2$ span an ideal triangle with vertex v_0 whose interior projects by p to the component of $B - (\beta_1' \cup \beta_2')$ containing e;

(ii) $\partial \widetilde{e}$ consists of v_0 and a point in the edge ξ of the ideal triangle spanned by $\widetilde{\beta}_1$ and $\widetilde{\beta}_2$ opposite to v_0.

Let $\widehat{e}$ and $\widehat{\beta}_j$ ($j \in \{1,2\}$) be the visible open facets of $\mathscr{C}(\mathscr{B})$ such that $\pi(\widehat{e}) = \widetilde{e}$ and $\pi(\widehat{\beta}_j) = \widetilde{\beta}_j$. Then all of $\widehat{e}$ and $\widehat{\beta}_j$ ($j \in \{1,2\}$) have b_0 as a vertex, where b_0 is the element of $\mathscr{B}$ such that $\pi(b_0) = v_0$ (see Lemma 11.3).

Let $\widehat{V}$ be the 2-dimensional affine subspace of $\widehat{W}$ spanned by $\widehat{\beta}_1 \cup \widehat{\beta}_2$.

Claim 11.8. $\widehat{V}$ *contains* $\widehat{e}$.

Proof. By the assumption on $\widehat{\beta}_j$ ($j \in \{1,2\}$) and Lemma 11.3, $\widehat{\beta}_j = b_0 + \pi^{-1}(x_j)$, where x_j is the point in $\partial \mathbb{H}^d$ such that $\partial \widetilde{\beta}_j = \{v_0, x_j\}$. Thus $\widehat{V}$ is a 2-dimensional hyperbolic affine subspace of $\widehat{W}$ and $\mathrm{cl}(\widetilde{V}, \overline{\mathbb{H}^3}) - \widetilde{V}$ is equal to ξ, where $\widetilde{V} = \pi(\widehat{V})$. Since $\mathrm{cl}(\widetilde{V}, \overline{\mathbb{H}^3}) - \widetilde{V}$ projects by p onto one of l_1 and l_2, it does not intersect Λ_p. Thus, by Lemma 11.4, $\widehat{V}$ contains $\widehat{e}$. □

Let $\{\widehat{F}_i\}_{i=0}^{k}$ be a face sequence for $\widehat{e}$ extending the sequence $\{\widehat{F}_0, \widehat{F}_1\} = \{\mathscr{C}(\mathscr{B}), \widehat{W} \cap \mathscr{C}(\mathscr{B})\}$ (see Lemma 2.13).

Claim 11.9. $\Pi(\widehat{F}_2) = \widehat{V}$.

Proof. Since $\pi_1(B) < \mathrm{Isom}^+(\widetilde{W}) < SO(1,2)$ is the infinite cyclic group generated by a parabolic transformation having b_0 as an eigenvector of eigenvalue 1, $\widehat{\beta}_1$ and $\widehat{\beta}_2$ cannot be contained in a $\pi_1(B)$-invariant 2-dimensional affine subspace of $\widehat{W} \cong \mathbb{E}^{1,2}$. Since $\widehat{F}_1 = \widehat{W} \cap \mathscr{C}(\mathscr{B}) \supset \widehat{\beta}_1 \cup \widehat{\beta}_2$ and since $\widehat{F}_1$ is $\pi_1(B)$-invariant, we have $\dim \Pi(\widehat{F}_1) = 3$ and hence $\Pi(\widehat{F}_1) = \widehat{W}$. Let $\widehat{V}'$ be a support plane for $\widehat{F}_1$ in $\widehat{W}$ such that $\widehat{V}' \cap \widehat{F}_1 = \widehat{F}_2$. Since both $\widehat{V}'$ and $\widehat{V}$ are 2-dimensional affine subspaces of the 3-dimensional vector space $\widehat{W}$ containing the common line segment $\widehat{e}$, $\widehat{V}'$ is obtained from $\widehat{V}$ by rotating around the 1-dimensional plane $\Pi(\widehat{e})$. Note that $\widehat{\beta}_1$ and $\widehat{\beta}_2$ lie in the different components of $\widehat{V} - \Pi(\widehat{e})$. Since $\widehat{\beta}_1 \cup \widehat{\beta}_2 \subset \widehat{W} \cap \mathscr{C}(\mathscr{B})$, both $\widehat{\beta}_1$ and $\widehat{\beta}_2$ are contained in the same side of $\widehat{V}'$ in $\widehat{W}$. Thus $\widehat{V}'$ must be equal to $\widehat{V}$. Finally, since $\widehat{F}_2 = \widehat{V}' \cap \widehat{F}_1$ contains $\mathscr{C}(\widehat{e} \cup \widehat{\beta}_1 \cup \widehat{\beta}_2)$, we have $\Pi(\widehat{F}_2) = \widehat{V}$. □

Since the 2-dimensional plane $\Pi(\widehat{F}_2) = \widehat{V}$ contains $\mathscr{C}(\widehat{e} \cup \widehat{\beta}_1 \cup \widehat{\beta}_2)$, which contains $\widehat{e}$ in the interior, there exists no half space of $\Pi(\widehat{F}_2)$ which contains $\widehat{e}$ in the boundary. This contradicts the assumption that $\{\widehat{F}_i\}_{i=0}^k$ is a face sequence for $\widehat{e}$. Hence we have proved that $\partial^\varepsilon \Delta$ has no edges except for β_1' and β_2'. This completes the proof of Theorem 11.2(2). $\square$

References

[ASWY] H. Akiyoshi, M. Sakuma, M. Wada and Y. Yamashita. Jørgensen's picture of punctured torus groups and its refinement, *this volume*.

[Ber87] M. Berger (1987). *Geometry II*, Universitext, Springer-Verlag.

[BP92] R. Benedetti and C. Petronio (1992). *Lectures on hyperbolic geometry*, Universitext, Springer-Verlag.

[EM87] D.B.A. Epstein and A. Marden (1987). Convex hulls in hyperbolic space, a theorem of Sullivan, and measured pleated surfaces. In *Analytical and Geometric Aspects of Hyperbolic Space (LMS Lecture Notes* **111***)*, Cambridge University Press, 113–253.

[EP88] D.B.A. Epstein and R.C. Penner (1988). Euclidean decompositions of noncompact hyperbolic manifolds, *J. Diff. Geom.* **27**, 67–80.

[FP] R. Frigerio and C. Petronio. Construction and recognition of hyperbolic 3-manifolds with geodesic boundary, *preprint*.

[Koj90] S. Kojima (1990). Polyhedral decomposition of hyperbolic manifolds with boundary, *Proc. Work. Pure Math.* **10**, 37–57.

[KS93] L. Keen and C. Series (1993). Pleating coordinates for the Maskit embedding of the Teichmüller space of punctured tori, *Topology* **32**, 719–749.

[KS97] L. Keen and C. Series (1997). How to bend pairs of punctured tori. In *Lipa's Legacy, Proceedings of the Bers Colloquium 1995, edited by J. Dodziuk and L. Keen, Contemp. Math.* **211**, 359–388.

[KS98] L. Keen and C. Series (1998). Pleating invariants for punctured torus groups, *to appear Topology*.

[Ser85] C. Series (1985). The geometry of Markoff numbers, *Math. Intelligencer* **7**, 20–29.

246   H. Akiyoshi & M. Sakuma

[Thu02] W.P. Thurston (2002). The geometry and topology of three-manifolds, *Electronic version.*
http://www.msri.org/publications/books/gt3m/

[Wee93] J. Weeks (1993). Convex hulls and isometries of cusped hyperbolic 3-manifolds, *Topology Appl.* **52**, 127–149.

Hirotaka Akiyoshi

Dept of Mathematics
Graduate School of Science
Osaka University
Machikaneyama-cho 1-16
Toyonaka
Osaka, 560-0043
Japan

akiyoshi@gaia.math.wani.
osaka-u.ac.jp

Makoto Sakuma

Dept of Mathematics
Graduate School of Science
Osaka University
Machikaneyama-cho 1-16
Toyonaka
Osaka, 560-0043
Japan

sakuma@math.wani.
osaka-u.ac.jp

AMS Classification: 57M50, 57N15, 57N16

Keywords: convex hull, convex core, Ford domain, Euclidean decomposition, bending lamination, punctured torus

Kleinian Groups and Hyperbolic 3-Manifolds
Lond. Math. Soc. Lec. Notes **299**, 247–273

Y. Komori, V. Markovic & C. Series (Eds.)
Cambridge Univ. Press, 2003

Jørgensen's picture of punctured torus groups and its refinement

Hirotaka Akiyoshi, Makoto Sakuma,
Masaaki Wada and Yasushi Yamashita

Dedicated to Professor Shin'ichi Suzuki on the occasion of his sixtieth birthday

Abstract

This is a sequel of the paper [AS]. We give a description of Jørgensen's theorem on the Ford domains of punctured torus groups from the 3-dimensional
view point, and propose conjectures which refine his theorem and relate it to the
bending laminations of the convex core boundaries of the quotient hyperbolic
manifolds. We also present partial results and experimental results supporting the
conjectures.

1. Introduction

Let $M(\Gamma) = \mathbb{H}^d / \Gamma$ be a cusped hyperbolic manifold. In [AS] we defined a decomposition $\Delta(\Gamma)$ of (a certain subspace of) the convex core $M_0(\Gamma) = \mathscr{C}(\Lambda(\Gamma))/\Gamma$ of $M(\Gamma)$
using the convex hull construction of Epstein and Penner [8], and studied its relation
with the structure of the boundary of the convex core. In particular we described the
relation between $\Delta(\Gamma)$ and the bending lamination $|pl(\Gamma)|$ in the 3-dimensional case,
and showed that $\Delta(\Gamma)$ determines $|pl(\Gamma)|$ if Γ is a quasifuchsian punctured torus group,
i.e., a Kleinian group satisfying the following conditions.

(i) Γ is freely generated by two isometries whose commutator is parabolic.

(ii) The domain of discontinuity Ω consists of exactly two simply connected components $\Omega^{\pm}$, whose quotients $\Omega^{\pm}/\Gamma$ are each homeomorphic to a punctured
torus T.

In this paper, we study the following problem.

Problem 1.1. Does $|pl(\Gamma)|$ determine the combinatorial structure of $\Delta(\Gamma)$ for a punc-
tured torus group Γ?

Jørgensen's work [Jør] which clarifies the beautiful structure of the Ford domains of quasifuchsian punctured torus groups gives us the starting point, because the Ford domain of a Kleinian group Γ with parabolic transformations can be regarded as a geometric dual to a certain subcomplex $\Delta_{\mathbb{E}}(\Gamma)$ of $\Delta(\Gamma)$ (see Section 2 and [AS, Section 10]). The works of Keen and Series [KS93],[KS97],[KS] which give detailed studies of $pl(\Gamma)$ for quasifuchsian punctured torus groups Γ also hold the key to the study of the above problem. The problem may be regarded as a generalization of the problem to compare the works of Jørgensen with those of Keen and Series.

This paper is organized as follows. In Section 2, we recall the definition of the Ford domain, $Ph(\Gamma)$, and describe the duality between $Ph(\Gamma)$ and the ideal polyhedral complex $\Delta_{\mathbb{E}}(\Gamma)$ which arises from the Epstein-Penner convex hull construction in the Minkowski space. We also recall the definition of the EPH-decomposition $\Delta(\Gamma)$ as a natural extension of $\Delta_{\mathbb{E}}(\Gamma)$. In Sections 3-6, we explain Jørgensen's results and related results from our view point. After giving a quick review to the works of Keen and Series on bending laminations of quasifuchsian punctured torus groups in Section 7, we propose a few conjectures expecting a positive answer to the problem in Section 8. In the last Section 9, we present partial results and experimental results supporting the conjectures.

Though this paper is a sequel of [AS], it is self-contained and can be read independently. In particular, it is our pleasure if this article is helpful for those readers who want to understand the beautiful work of Jørgensen [Jør], which is included in this proceedings. Since the proofs of the results in this paper are based on the arguments developed in the work announced in [ASWY00], they will be included in the forthcoming paper [ASWY] which gives the proofs to the announced work.

Acknowledgment. The authors would like thank Troels Jørgensen for his kind explanation of his results and for the mathematics he had produced. They would also like to thank David Epstein and Caroline Series for their interest on this work and for their warm hospitality in Warwick. Their thanks also go to Yohei Komori and Hideki Miyachi for stimulating conversations. Finally, we would like to thank the referee for his/her careful reading of the manuscript and helpful suggestions.

2. Punctured torus groups, Ford domains and EPH-decompositions

Let T be the topological (once) punctured torus. A *marked punctured torus group* is the image of a discrete faithful representation $\rho : \pi_1(T) \to \mathrm{PSL}(2, \mathbb{C})$ satisfying the following condition:

- If ω is represented by a loop around the puncture, then $\rho(\omega)$ is parabolic.

Two marked punctured torus groups $\Gamma = \rho(\pi_1(T))$ and $\Gamma' = \rho'(\pi_1(T))$ are *equivalent* if ρ is conjugate to ρ' by an element of $\mathrm{PSL}(2,\mathbb{C})$.

A *marked quasifuchsian punctured torus group* is a marked punctured torus group Γ such that the domain of discontinuity $\Omega(\Gamma)$ consists of exactly two simply connected components $\Omega^{\pm}(\Gamma)$, whose quotients $\Omega^{\pm}(\Gamma)/\Gamma$ are each homeomorphic to T. We employ a sign convention so that there is an orientation-preserving homeomorphism f from $T \times [-1, 1]$ to the quotient manifold $\bar{M}(\Gamma) = (\mathbb{H}^3 \cup \Omega(\Gamma))/\Gamma$ such that $f(T \times \{\pm 1\}) = \Omega^{\pm}(\Gamma)/\Gamma$ and that the isomorphism $f_* : \pi_1(T \times [-1, 1]) = \pi_1(T) \to \pi_1(M(\Gamma)) = G < \mathrm{PSL}(2,\mathbb{C})$ is identified with ρ. Since such a homeomorphism is unique up to isotopy, we can identify the topological triple $(\bar{M}(\Gamma), \Omega^-(\Gamma)/\Gamma, \Omega^+(\Gamma)/\Gamma)$ with $(T \times [-1, 1], T \times \{-1\}, T \times \{1\})$. The *quasifuchsian punctured torus space* $\mathcal{QF}$ is the space of the equivalence classes of marked quasifuchsian punctured torus groups. As a consequence of Minsky's celebrated theorem [Min99], the space of marked punctured torus groups is equal to the closure $\overline{\mathcal{QF}}$ of $\mathcal{QF}$ in the $\mathrm{PSL}(2,\mathbb{C})$-representation space of $\pi_1(T)$.

We now recall the definition of the Ford domain of a punctured torus group $\Gamma = \rho(\pi_1(T))$, and introduce a notion of the *Ford complex*. To this end, we normalize the group Γ by a conjugation in $\mathrm{PSL}(2,\mathbb{C})$ so that the stabilizer Γ_∞ in Γ of the point ∞ of the upper-half space model is the infinite cyclic group generated by a parabolic transformation $\rho(\omega)$, where ω is represented by a puncture. Then for each element $A = \begin{bmatrix} a & b \\ c & d \end{bmatrix}$ of $\Gamma - \Gamma_\infty$, we have $A(\infty) \neq \infty$, or equivalently, $c \neq 0$, and the *isometric circle $I(A)$* of A is defined by

$$I(A) = \{z \in \mathbb{C} \mid |A'(z)| = 1\} = \{z \in \mathbb{C} \mid |cz + d| = 1\}.$$

$I(A)$ is the circle in the complex plane whose center is $-d/c = A^{-1}(\infty)$ and has radius $1/|c|$. The *isometric hemisphere $Ih(A)$* is the hyperplane of the upper half-space $\mathbb{H}^3$ bounded by $I(A)$. We denote by $P(\Gamma)$ (resp. $Ph(\Gamma)$) the subset of the complex plane (resp. the upper half-space) which consists of all points lying exterior to each of the isometric circles (resp. isometric hemispheres) defined by Γ. These symbols are slightly different from those in [Jør], where the same sets are denoted by $\widetilde{P}(\Gamma)$ and $\widetilde{Ph}(\Gamma)$ respectively.

Definition 2.1.

(i) We call $Ph(\Gamma)$ the *Ford domain* of Γ.

(ii) The *Ford complex*, $Ford(\Gamma)$, of Γ is the subcomplex of the quotient manifold $\bar{M}(\Gamma)$ obtained as the closure of the image of $\partial Ph(\Gamma)$.

In order to describe a geometric meaning of the Ford domain, we pick a small horoball, H_∞, centered at ∞ which projects to a horospherical neighborhood of the (main) cusp in the quotient hyperbolic manifold $M(\Gamma)$. Then H_∞ is precisely invariant by (Γ, Γ_∞), that is, for any element $A \in \Gamma$, $A(H_\infty) \cap H_\infty \neq \emptyset$ if and only if $A \in \Gamma_\infty$. Then for each element $A \in \Gamma - \Gamma_\infty$, the isometric hemisphere $Ih(A)$ is equal to the set of points in $\mathbb{H}^3$ which are equidistant from H_∞ and $A^{-1}(H_\infty)$. This implies that $Ph(\Gamma)$ can be regarded as the "Dirichlet domain of Γ centered at ∞", because

$$Ph(\Gamma) = \{x \in \mathbb{H}^3 \mid d(x, H_\infty) \leq d(x, AH_\infty) \text{ for any } A \in \Gamma\}.$$

Thus $Ph(\Gamma)$ is a "fundamental domain of Γ modulo Γ_∞" in the sense that the intersection of $Ph(\Gamma)$ with a fundamental domain of Γ_∞ is a fundamental domain of Γ. As is noted in [EP88, Section 4], it is more natural to work with the quotient $Ph(\Gamma)/\Gamma_\infty$ in $\mathbb{H}^3/\Gamma_\infty$. In fact the hyperbolic manifold $M(\Gamma)$ is obtained from $Ph(\Gamma)/\Gamma_\infty$ by identifying pairs of faces by isometries.

The above geometric description of the Ford domain implies that the Ford complex is equal to the cut locus of a horospherical neighborhood of the cusp of $M(\Gamma)$, that is, $Ford(\Gamma) \cap M(\Gamma)$ consists of the points in $M(\Gamma)$ which have more than two shortest geodesics to a fixed horospherical neighborhood. Here is an intuitive description: Take a horospherical neighborhood of the cusp of $M(\Gamma)$ and let it expand. We regard the horospherical neighborhood as a balloon which is gently expanded, coming to rest where it meets itself. Then the collision locus is equal to $Ford(\Gamma) \cap M(\Gamma)$.

The above description of the Ford complex $Ford(\Gamma)$ yields an ideal polyhedral complex, $\Delta_\mathbb{E}(\Gamma)$, dual to $Ford(\Gamma)$ as follows. Let $\widetilde{Ford}(\Gamma)$ be the 2-dimensional complex in the hyperbolic space obtained as the inverse image of $Ford(\Gamma) \cap M(\Gamma)$. Let p be a vertex of $\widetilde{Ford}(\Gamma)$. Then p is the intersection of at least three isometric hemispheres, and hence it is equidistant from at least four horoballs in the orbit ΓH_∞. We regard the ideal polyhedron spanned by the centers of these horoballs as the geometric *dual* to the vertex p. Similarly, for each edge (resp. face) of $\widetilde{Ford}(\Gamma)$, we can associate an ideal polygon (an ideal edge) as its geometric dual. The family of these ideal polyhedra, ideal polygons and ideal edges dual to the cells of $\widetilde{Ford}(\Gamma)$ compose a Γ-invariant ideal polyhedral complex embedded in $\mathbb{H}^3$. This ideal polyhedral complex descends to an ideal polyhedral complex embedded in $M(\Gamma)$; this is the desired $\Delta_\mathbb{E}(\Gamma)$.

Following the argument of Epstein and Penner [EP88, Section 10], we explain that $\Delta_\mathbb{E}(\Gamma)$ arises from the Epstein-Penner convex hull construction in the Minkowski space. Let $\mathbb{E}^{1,3}$ be the 4-dimensional Minkowski space with the Minkowski product

$$\langle x,y \rangle = -x_0y_0 + x_1y_1 + x_2y_2 + x_3y_3.$$

Then

$$\mathbb{H}^3 = \{x \in \mathbb{E}^{1,3} \mid \langle x,x \rangle = -1, \ x_0 > 0\}$$

together with the restriction of the Minkowski product to the tangent space gives a hyperboloid model of the 3-dimensional hyperbolic space. Any horoball H in this model is represented by a vector, v, in the positive light cone (i.e., $\langle v,v \rangle = 0$ and $v_0 > 0$) as

$$H = \{x \in \mathbb{H}^3 \mid \langle v,x \rangle \geq -1\}.$$

The center of the horoball H corresponds to the ray thorough v, and as v moves away from the origin along the ray, the horoball contracts towards the center of the horoball. Let v_∞ denote the light-like vector representing the horoball H_∞. Then its orbit Γv_∞ is the set of light-like vectors corresponding to the horoballs in ΓH_∞. Let $\mathscr{C}$ be the closed convex hull of Γv_∞ in $\mathbb{E}^{1,3}$. Now consider the ideal polyhedron $\langle x_0, x_1, \cdots, x_k \rangle$ in $\mathbb{H}^3$ which is dual to a vertex p of $\widetilde{Ford}(\Gamma)$. Then there are horoballs $H_{x_0}, H_{x_1}, \cdots, H_{x_k}$ in the orbit ΓH_∞, such that:

(i) H_{x_i} is centered at x_i.

(ii) $d(p, H_{x_0}) = d(p, H_{x_1}) = \cdots = d(p, H_{x_k}) = d(p, \Gamma H_\infty)$.

(iii) $d(p, H') > d(p, \Gamma H_\infty)$ for any horoball H' in $\Gamma H_\infty - \{H_{x_0}, H_{x_1}, \cdots, H_{x_k}\}$.

Let v_{x_i} be the light-like vector representing the horoball H_{x_i}. After coordinate change, we may assume the vertex p corresponds to the vector $(1,0,0,0)$ in the hyperboloid model. Then the points $v_{x_0}, v_{x_1}, \cdots, v_{x_k}$ lie in a horizontal hyperplane $W : x_0 = $ constant, because of the second condition. Moreover, by the third condition, all points in $\Gamma v_\infty - \{v_{x_0}, v_{x_1}, \cdots, v_{x_k}\}$ lie above the hyperplane W. Hence we see that the hyperplane W is a support plane of the convex hull $\mathscr{C}$ (i.e., $\mathscr{C}$ is contained in one of the two closed half-space bounded by W), and $W \cap \partial \mathscr{C}$ in the polyhedron $\langle v_{x_0}, v_{x_1}, \cdots, v_{x_k} \rangle$. In other words, $\langle v_{x_0}, v_{x_1}, \cdots, v_{x_k} \rangle$ is a (top-dimensional) face of $\partial \mathscr{C}$. Moreover, it is *Euclidean* in the sense that the restriction of the Minkowski product to the hyperplane W is positive-definite. The ideal polyhedron $\langle x_0, x_1, \cdots, x_k \rangle$ dual to p is equal to the image of the Euclidean face $\langle v_{x_0}, v_{x_1}, \cdots, v_{x_k} \rangle$ of $\partial \mathscr{C}$ by the radial projection form the origin to $\mathbb{H}^3$. In conclusion, the ideal polyhedral complex $\Delta_{\mathbb{E}}(\Gamma)$ is obtained as follows. Consider the collection of faces of $\partial \mathscr{C}$ which has a Euclidean support plane, that is, the collection of the subset of $\mathbb{E}^{1,3}$ which is of the form $W \cap \mathscr{C}$ for some Euclidean support plane W of $\mathscr{C}$. Then their images by the radial projection compose a Γ-invariant ideal polyhedral complex embedded in $\mathbb{H}^3$, and $\Delta_\mathbb{E}(\Gamma)$ is equal to its image in $M(\Gamma)$ (see [AS, Section 10]).

Though $\Delta_\mathbb{E}(\Gamma)$ has the nice geometric meaning that it is dual to the Ford complex, its underlying space $|\Delta_\mathbb{E}(\Gamma)|$ looks far from nice. In general, it is not convex and is strictly smaller than the convex core $M_0(\Gamma)$. This is because we take only Euclidean faces of $\partial\mathscr{C}$ into account in the construction of $\Delta_\mathbb{E}(\Gamma)$. If the group Γ were a finitely generated Kleinian group of *cofinite volume* with parabolic transformations, then as is proved by Epstein and Penner [EP88], the above construction gives a finite ideal polyhedral decomposition of the whole quotient hyperbolic manifold. Moreover, every face of $\partial\mathscr{C}$ is Euclidean and hence each piece of the decomposition admits a natural Euclidean structure. Thus it is called the *Euclidean decomposition*. However, in general case, $\partial\mathscr{C}$ can have non-Euclidean faces. So, it is natural to try to construct an ideal polyhedral complex by taking all faces of $\partial\mathscr{C}$ into account. This was made explicit in [AS], and we call it the EPH-decomposition and denote it by $\Delta(\Gamma)$. Here the letters E, P and H, respectively, stand for Euclidean (or elliptic), parabolic and hyperbolic.

Finally, we give a brief description of $\Delta(\Gamma)$. (Those who are interested only in Jørgensen's work can skip this part.) To this end, we point out the following troublesome phenomena which we must be careful about.

(i) The cellular structure of $\partial\mathscr{C}$ is not necessarily locally finite, and we must be careful about the definition of a face (see Remark [AS, Remark 2.12]). This forces us to introduce the notion of a *facet* by refining the notion of a face (see [AS, Definition 2.11]).

(ii) Some part of $\partial\mathscr{C}$ may be "invisible" from the origin, that is, there may be a point of $\partial\mathscr{C}$ such that the line segment between the origin and the point contains some other points of $\partial\mathscr{C}$. So, we need to consider only the *visible* facets of $\partial\mathscr{C}$, i.e., those facets whose affine hulls do not contain the origin (see [AS, Definition 4.9 and Lemma 4.10]).

The EPH-decomposition $\Delta(\Gamma)$ is defined to be the image of visible open facets of $\partial\mathscr{C}$ in $M(\Gamma)$ (see [AS, Definition 4.18]). We call a member of $\Delta(\Gamma)$ an *open facet* of $\Delta(\Gamma)$. The support $|\Delta(\Gamma)|$ is the union of the open facet of $\Delta(\Gamma)$. Then the following are proved in [AS, Proposition 4.15 and Corollary 1.1].

(i) $|\Delta(\Gamma)|$ is the disjoint union of the open facets of $\Delta(\Gamma)$.

(ii) $|\Delta(\Gamma)|$ is equal to the convex core $M_0(\Gamma)$ minus the support of the bending lamination (if Γ is a quasifuchsian punctured torus group).

At the end of this section, we note that $Ford(\Gamma) \cap M(\Gamma)$ is a *spine* of $M(\Gamma)$, that is, it is a strong deformation retract of $M(\Gamma)$. Moreover, this spine is *canonical* in the sense that it is uniquely determined from the cusped hyperbolic manifold $M(\Gamma)$. We can apply the same construction to every cusped hyperbolic manifold, and in the special case when the manifold is of finite volume and has only one cusp (e.g. the complement of a hyperbolic knot), the combinatorial structure of the Ford complex is a complete invariant of the underlying topological 3-manifold by virtue of the Mostow rigidity theorem. So, the study of the Ford complex has an important meaning for the 3-manifold theory and the knot theory, too.

3. Jørgensen's theorem for quasifuchsian punctured torus groups, I

In this section, we explain Jørgensen's theorem which describes the combinatorial structures of the Ford domains of quasifuchsian punctured torus groups [Jør]. Before presenting the precise statement of Jørgensen's theorem, we give a brief intuitive description of the idea. Let Γ be a quasifuchsian punctured torus group. Then the quotient manifold $\bar{M}(\Gamma)$ is identified with $T \times [-1, 1]$. Since $P(\Gamma) \subset \mathbb{C}$ is a fundamental domain of the action of Γ on $\Omega(\Gamma) = \Omega^-(\Gamma) \cup \Omega^+(\Gamma)$, modulo Γ_∞, $P(\Gamma)$ is a disjoint union of $P^-(\Gamma)$ and $P^+(\Gamma)$ where $P^\pm(\Gamma) = P(\Gamma) \cap \Omega^\pm(\Gamma)$. The the first assertion of Jørgensen's theorem is that each $P^\pm(\Gamma)$ is simply connected. This implies that the image of $\partial P^\pm(\Gamma)$ in $T \times \{\pm 1\}$ is a spine of the punctured torus T. If Γ is fuchsian, then these two spines are identical, and the Ford complex $Ford(\Gamma)$ is equal to the product of the spine with the interval $[-1, 1]$. In general, these two spines are not isotopic to each other. However, they are related by a canonical sequence of Whitehead moves. (This is the main point why the punctured torus is so special.) The "trace" of the canonical sequence of Whitehead moves form a spine of $T \times [-1, 1]$. The main assertion of Jørgensen's theorem is that this spine is isotopic to the Ford complex $Ford(\Gamma)$. Thus we can say that $Ford(\Gamma)$ records the history of how the two boundary spines evolved.

Now let's give the precise statement. We begin by recalling basic topological facts on the punctured torus T. To this end, we identify T with the quotient space $(\mathbb{R}^2 - \mathbb{Z}^2)/\mathbb{Z}^2$. A simple loop in T is said to be *essential*, if it bounds neither a disk nor a once-punctured disk. Similarly, a simple arc in T having the puncture as end points is said to be *essential*, if it does not cut off a disk with a point on the boundary removed. Then the isotopy classes of essential simple loops (resp. essential simple arcs) in T are in one-to-one correspondence with $\hat{\mathbb{Q}} := \mathbb{Q} \cup \{1/0\}$: A representative of the isotopy class corresponding to $r \in \hat{\mathbb{Q}}$ is the projection of a line in $\mathbb{R}^2$ (the line being disjoint from $\mathbb{Z}^2$ for the loop case, and intersecting $\mathbb{Z}^2$ for the arc case). The element $r \in \hat{\mathbb{Q}}$

associated to a circle or an arc is called its *slope*. The representative of the isotopy class of an essential arc of slope r is denoted by β_r.

Consider the ideal triangle in the hyperbolic plane $\mathbb{H}^2 = \{z \in \mathbb{C} \mid \mathrm{Im}(z) > 0\}$ spanned by the ideal vertices $\{0/1, 1/1, 1/0\}$. Then the translates of this ideal triangle by the action of $SL(2, \mathbb{Z})$ form a tessellation of $\mathbb{H}^2$. This is called the *modular diagram* or the *Farey tessellation* and is denoted by $\mathscr{D}$. The abstract simplicial complex having the combinatorial structure of $\mathscr{D}$ is also denoted by the same symbol. The set of (ideal) vertices of $\mathscr{D}$ is equal to $\hat{\mathbb{Q}}$, and a typical (ideal) triangle σ of $\mathscr{D}$ is spanned by $\{\frac{p_1}{q_1}, \frac{p_1+p_2}{q_1+q_2}, \frac{p_2}{q_2}\}$ where $\begin{pmatrix} p_1 & p_2 \\ q_1 & q_2 \end{pmatrix} \in SL(2, \mathbb{Z})$.

Let $\sigma = \langle r_0, r_1, r_2 \rangle$ be a triangle of $\mathscr{D}$. Then the essential arcs β_{r_0}, β_{r_1}, β_{r_2} are mutually disjoint, and their union determines a *topological ideal triangulation* $\mathrm{trg}(\sigma)$ of T, in the sense that T cut open along $\beta_{r_0} \cup \beta_{r_1} \cup \beta_{r_2}$ is the disjoint union of two 2-simplices with all vertices deleted. Let $\mathrm{spine}(\sigma)$ be a 1-dimensional cell complex embedded in T which is dual to the 1-skeleton of $\mathrm{trg}(\sigma)$. Then $\mathrm{spine}(\sigma)$ consists of two vertices and three edges γ_i ($i = 0, 1, 2$), such that γ_i intersects the 1-skeleton of $\mathrm{trg}(\sigma)$ transversely precisely at a point of $\mathrm{int}\beta_{r_i}$. Note that $\mathrm{spine}(\sigma)$ is a deformation retract of T and hence is a *spine* of T. We define the *slope* of an edge γ_i of $\mathrm{spine}(\sigma)$ to be $r_i \in \hat{\mathbb{Q}}$, the slope of the ideal edge β_{r_i} of $\mathrm{trg}(\sigma)$ dual to γ_i.

Let $\tau = \langle r_0, r_1 \rangle$ be an edge of $\mathscr{D}$. Then the union $\beta_{r_0} \cup \beta_{r_1}$ determines a *topological ideal polygonal decomposition* of T, in the sense that T cut open along it is homeomorphic to a quadrilateral with all vertices deleted. Let $\mathrm{spine}(\tau)$ be a 1-dimensional cell complex embedded in T which is dual to the 1-skeleton of $\mathrm{trg}(\sigma)$. Then $\mathrm{spine}(\tau)$ consist of a single vertex and two edges, and it is also a spine of T. The slope of an edge of $\mathrm{spine}(\tau)$ is also defined as explained in the preceding paragraph.

Let $\mathscr{D}^{(i)}$ denote the set of i-simplices of $\mathscr{D}$. Then we have the following well-known fact.

Lemma 3.1. *For any spine C of T, there is a unique element δ of $\mathscr{D}^{(1)} \cup \mathscr{D}^{(2)}$ such that C is isotopic to $\mathrm{trg}(\delta)$.*

If $\tau = \langle r_0, r_1 \rangle$ is an edge of a triangle $\sigma = \langle r_0, r_1, r_2 \rangle$ of $\mathscr{D}$, then $\mathrm{spine}(\tau)$ is obtained from $\mathrm{spine}(\sigma)$ by collapsing the edge γ_2 of $\mathrm{spine}(\sigma)$ of slope r_2 to a point (see Figure 1). By an *elementary transformation*, we mean this transformation or its converse.

Let (δ^-, δ^+) be a pair of elements of $\mathscr{D}^{(1)} \cup \mathscr{D}^{(2)}$. Then, since the 1-skeleton of the dual to $\mathscr{D}$ is a tree, there is a unique sequence $\delta^- = \delta_0, \delta_1, \delta_2, \cdots, \delta_m = \delta^+$ in $\mathscr{D}^{(1)} \cup \mathscr{D}^{(2)}$ satisfying the following conditions.

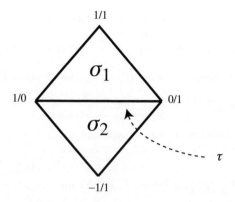

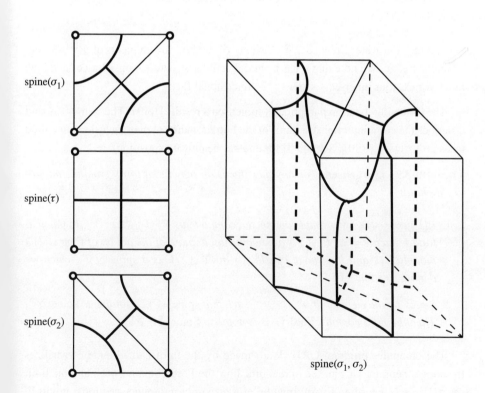

Figure 1

(i) For each $i \in \{0, 1, \cdots, m-1\}$, either δ_i is an edge of δ_{i+1} or δ_{i+1} is an edge of δ_i.

(ii) $\delta_i \neq \delta_j$ whenever $i \neq j$.

Thus we obtain a canonical sequence of elementary transformations

$$\text{spine}(\delta^-) = \text{spine}(\delta_0) \mapsto \text{spine}(\delta_1) \mapsto \cdots \mapsto \text{spine}(\delta_m) = \text{spine}(\delta^+).$$

Regard the sequence as a continuous family $\{C_t\}_{t \in [-1,1]}$ of spines of T, and set

$$\text{Spine}(\delta^-, \delta^+) = \cup_{t \in [-1,1]} C_t \subset T \times [-1,1].$$

Then $\text{Spine}(\delta^-, \delta^+)$ is a 2-dimensional subcomplex of $T \times [-1,1]$ satisfying the following conditions.

(i) $\text{Spine}(\delta^-, \delta^+) \cap (T \times \{\varepsilon 1\}) = \text{spine}(\delta^\varepsilon) \times \{\varepsilon 1\}$ for each $\varepsilon = \pm$.

(ii) There is a level-preserving deformation retraction from $T \times [-1,1]$ to $\text{Spine}(\delta^-, \delta^+)$.

Figure 1 illustrates $\text{Spine}(\delta^-, \delta^+)$, where δ^- and δ^+ are elements of $\mathscr{D}^{(2)}$ sharing a common edge. We note that it has a natural cellular structure, consisting of a unique inner-vertex, four inner-edges and six 2-dimensional faces.

The following theorem paraphrases Jørgensen's results [Jør97, Theorems 1-3], and describes the combinatorial structures of the Ford domains of quasifuchsian punctured torus groups (see [PS01, Section 3] for another beautiful exposition).

Theorem 3.2. (Jørgensen) *For any quasifuchsian punctured torus group Γ, the following hold:*

1. $P(\Gamma)$ *consists of two simply connected components* $P^\pm(\Gamma) \subset \Omega^\pm(\Gamma)$. *In particular, for each* $\varepsilon = \pm$, $P^\varepsilon(\Gamma)$ *is a fundamental domain of the action of* Γ *on* $\Omega^\varepsilon(\Gamma)$ *modulo* Γ_∞, *and the image of* $\partial P^\varepsilon(\Gamma)$ *in* $\Omega^\varepsilon(\Gamma)/\Gamma$ *is a spine of* T, *which we denote by* $\text{spine}^\varepsilon(\Gamma)$.

2. *Let* δ^ε *be the element of* $\mathscr{D}^{(1)} \cup \mathscr{D}^{(2)}$ *such that* $\text{spine}^\varepsilon(\Gamma)$ *is isotopic to* $\text{spine}(\delta^\varepsilon)$. *Then the Ford complex* $\text{Ford}(\Gamma)$ *is isotopic to* $\text{Spine}(\delta^-, \delta^+)$.

The computer program OPTi [Wad] made by the third named author visualizes the above theorem: we can see in real time how the Ford domain $Ph(\Gamma)$ and the limit set $\Lambda(\Gamma)$ vary according to deformation of a quasifuchsian punctured torus group Γ. Figure 2(a), which was drawn by using OPTi, illustrates a typical example of the Ford domain of a quasifuchsian punctured torus group Γ. We can observe the following (see [Jør],[JM79],[ASWY00]).

(i) Each face F of the Ford domain $Ph(\Gamma)$ is preserved by an elliptic transformation, P_F, of order 2.

(ii) The transformations $\{P_F\}$ where F runs over the faces of the $Ph(\Gamma)$ generate a Kleinian group $\tilde{\Gamma}$ which contains Γ as a normal subgroup of index 2. In fact Γ is identified with the orbifold fundamental group of the 2-dimensional orbifold which is the quotient of T by an involution with three fixed points.

(iii) There is a parabolic transformation, K, of $\tilde{\Gamma}$ such that $K(\infty) = \infty$ and K^2 is the element of Γ corresponding to a peripheral loop of T.

(iv) If F is a face of $Ph(\Gamma)$, then $F' = K(F)$ is also a face of $Ph(\Gamma)$ and the transformation $K \circ P_F$ is the element of Γ which sends F to F'.

(a)

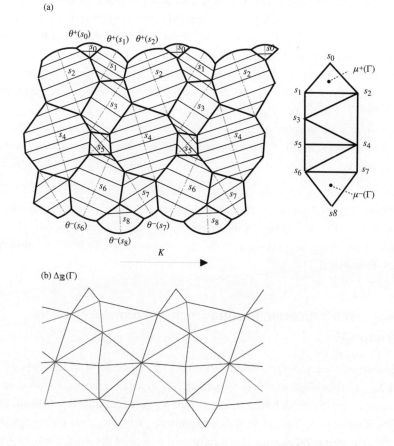

(b) $\Delta_{\mathbb{E}}(\Gamma)$

Figure 2

(v) There is a continuous family of bi-infinite periodic broken lines $\{L_t\}_{t \in (-1,1)}$ contained in the projection of $Ph(\Gamma)$ in $\mathbb{C}$ each of which is orthogonal to the projection of $F \cap \text{Axis}(P_F)$ whenever they intersect. Let S_t be the intersection of the broken vertical planes above L_t with $Ph(\Gamma)$. Then S_t projects to a torus, T_t, in $M(\Gamma) = T \times (-1, 1)$ isotopic to a fiber, and the image, C_t, of ∂S_t forms a spine of T_t. So $\{C_t\}$ gives a continuous family of spines of T. Moreover, C_t is generic or non-generic according as L_t is disjoint from the projections of the vertices of $Ph(\Gamma)$. This family $\{C_t\}$ (with certain modification near $t = \pm 1$) realizes the canonical sequence of elementary moves transforming $\text{spine}^-(\Gamma)$ to $\text{spine}^+(\Gamma)$.

Figure 2(b) illustrates the cross section of $\tilde{\Delta}_{\mathbb{E}}(\Gamma)$ along a small horosphere H_∞ centered at ∞, where $\tilde{\Delta}_{\mathbb{E}}(\Gamma)$ is the Γ-invariant ideal polyhedral complex in $\mathbb{H}^3$ obtained as the inverse image of $\Delta_{\mathbb{E}}(\Gamma)$. It is also regarded as the projection of $\tilde{\Delta}_{\mathbb{E}}(\Gamma) \cap \partial H_\infty$ to the complex plane $\mathbb{C}$. Then the vertices are identified with the centers of the isometric hemispheres supporting faces of the Ford domain $Ph(\Gamma)$. To be more explicit, if v is a projection of a vertex of $\tilde{\Delta}_{\mathbb{E}}(\Gamma) \cap \partial H_\infty$, then the vertical geodesic $\overline{v\infty}$ joining v to ∞ is an edge of $\tilde{\Delta}_{\mathbb{E}}(\Gamma)$, and v is the center of the isometric hemisphere supporting a face of $Ph(\Gamma)$ dual to the edge $\overline{v\infty}$ of $\tilde{\Delta}_{\mathbb{E}}(\Gamma)$. Similarly, each triangle $\langle x_0, x_1, x_2 \rangle$ of $\tilde{\Delta}_{\mathbb{E}}(\Gamma) \cap \partial H_\infty$ is a horospherical cross section of an ideal tetrahedron in $\tilde{\Delta}_{\mathbb{E}}(\Gamma)$ dual to the vertex of $Ph(\Gamma)$ obtained as the intersection of the isometric hemispheres centered at x_0, x_1 and x_2.

Let $\tilde{\Delta}(\Gamma)$ be the inverse image in $\mathbb{H}^3$ of the EPH-decomposition of $\Delta(\Gamma)$. Then $\tilde{\Delta}(\Gamma) \cap \partial H_\infty$ gives a (not necessarily locally finite) decomposition of the infinite strip which arises as the intersection of the convex hull of the limit set with ∂H_∞, because the underlying space $|\Delta(\Gamma)|$ is equal to the convex core minus the bending laminations by [AS, Corollary 1.1].

4. Jørgensen's theorem for quasifuchsian punctured torus groups, II

In this section, we explain Jørgensen's theorem [Jør, Theorem 4] which refines Theorem 3.2. A *weighted spine* of T is a spine of T with an assignment of a positive real number to each edge, which we call the *weight* on the edge, such that the sum of the weights is equal to 1. We call such an assignment a *weight system* on the spine. By regarding the weight on an edge as a weight on the slope of the edge, a weight system on a spine is regarded as a barycentric coordinate of a point in $|\mathscr{D}| - |\mathscr{D}^{(0)}|$. By fixing a $\text{PSL}(2, \mathbb{Z})$-equivariant injective continuous map from the underlying space $|\mathscr{D}|$ (of the abstract simplicial complex $\mathscr{D}$) onto $\mathbb{H}^2 \cup \hat{\mathbb{Q}} \subset \bar{\mathbb{H}}^2$, we regard a point in $|\mathscr{D}| - |\mathscr{D}^{(0)}|$ as a point in $\mathbb{H}^2$. Thus each weighted spine corresponds to a unique point of $\mathbb{H}^2$. For each $\nu \in \mathbb{H}^2$, we denote by $\text{spine}(\nu)$ the weighted spine of T corresponding to ν.

Let Γ be a marked quasifuchsian punctured torus group. For each $\varepsilon = \pm$ and for each edge e of spine$^\varepsilon(\Gamma)$, let $t^\varepsilon(e)$ be $1/\pi$ times the angle, $\theta^\varepsilon(e)$, of a circular arc component of the inverse image of e in $\partial P^\varepsilon(\Gamma)$ (see Figure 2). Then we have the following (see [Jør, Section 4]):

Lemma 4.1. *The sum of $t^\varepsilon(e)$ where e runs over the edges of* spine$^\varepsilon(\Gamma)$ *is equal to* 1.

Thus spine$^\varepsilon(\Gamma)$ has the structure of a weighted spine of T such that the weight of an edge e is $t^\varepsilon(e)$. Let $v^\varepsilon(\Gamma)$ be the point of $\mathcal{D}$ corresponding to the weighted spine spine$^\varepsilon(\Gamma)$, and put $v(\Gamma) = (v^-(\Gamma), v^+(\Gamma))$. We call it the *side parameter* of Γ following [Jør]. The following theorem is a refinement of Theorem 3.2(1) (see [Jør, Theorem 4]):

Theorem 4.2. (Jørgensen) *The map* $v : \mathcal{QF} \to \mathbb{H}^2 \times \mathbb{H}^2$ *is continuous and onto.*

Remark 4.3.

(i) Though [Jør, Theorem 1] asserts that v is also injective, we have not been able to confirm the assertion.

(ii) We do not know any relation between the side-parameter $v(\Gamma)$ and the usual end invariant of Γ (see [Min99]), which records the conformal structure $(\Omega^-(\Gamma)/\Gamma, \Omega^+(\Gamma)/\Gamma) \in Teich(T) \times Teich(T) = \mathbb{H}^2 \times \mathbb{H}^2$.

To combine Theorems 3.2 and 4.2, we introduce the following concept:

Definition 4.4. (1) A *weighted relative spine* of $T \times [-1, 1]$ is a 2-dimensional sub-complex C of $T \times [-1, 1]$ satisfying the following conditions.

(i) There is a level-preserving deformation retraction from $T \times [-1, 1]$ to Spine(δ^-, δ^+).

(ii) A weight system is specified on each of $\partial^\pm C := C \cap T \times \{\pm 1\}$.

Two weighted relative spines are *equivalent*, if the underlying relative spines are iso-topic and the weight systems coincide (after the isotopy).

(2) For $v = (v^-, v^+) \in \mathbb{H}^2 \times \mathbb{H}^2$, Spine$(v)$ denotes the weighted relative spine, such that the underlying relative spine is Spine$(\delta^-(v), \delta^+(v))$, where $\delta^\varepsilon(v)$ denotes the triangle or edge of $\mathcal{D}$ whose interior contains v^ε, and the weight system on the boundary $\partial^\pm$Spine(v) is given by $v^\pm$.

Then we can summarize Theorems 3.2 and 4.2 as follows:

Theorem 4.5. (Jørgensen) *For each* $\Gamma \in \mathscr{QF}$, *the Ford complex* $\mathrm{Ford}(\Gamma)$ *has a natural structure of weighted relative spine of* $T \times [-1,1]$. *Moreover, there is a continuous onto map* $v : \mathscr{QF} \to \mathbb{H}^2 \times \mathbb{H}^2$, *such that the weighted spine* $\mathrm{Ford}(\Gamma)$ *is equivalent to* $\mathrm{Spine}(v(\Gamma))$ *for any* $\Gamma \in \mathscr{QF}$.

5. The topological ideal polyhedral complex $\mathrm{Trg}(v)$ dual to $\mathrm{Spine}(v)$

As explained in Section 2, the Ford complex $\mathrm{Ford}(\Gamma)$ is a dual to the subcomplex $\Delta_{\mathbb{E}}(\Gamma)$ of $\Delta(\Gamma)$ consisting of the Euclidean (or elliptic) facets. In this section, we describe the structure of $\Delta_{\mathbb{E}}(\Gamma)$ following the exposition by Floyd-Hatcher [FH82] of Jørgensen's ideal triangulation of punctured torus bundles over S^1.

For each element $v = (v^-, v^+)$ of $(\mathscr{D} - \mathscr{D}^{(0)}) \times (\mathscr{D} - \mathscr{D}^{(0)})$, we construct a topological ideal triangulation $\mathrm{Trg}(v)$ as follows. Let $\Sigma(v) = \{\sigma_1, \sigma_2, \cdots, \sigma_n\}$ $(n \geq 0)$ be the triangles of $\mathscr{D}$ whose interiors intersect the oriented geodesic segment joining v^- with v^+ in this order. Note that $n = 0$ if and only if $v^\pm$ is contained in a single edge τ of $\mathscr{D}$. In this case we redefine $\Sigma(v) = \{\tau\}$.

Case 1. $n \geq 2$. Let $\widetilde{\mathrm{trg}}(\sigma_i)$ be the ideal triangulation of $\mathbb{R}^2 - \mathbb{Z}^2$ obtained as the lift of $\mathrm{trg}(\sigma_i)$. By superimposing $\widetilde{\mathrm{trg}}(\sigma_{i+1})$ upon $\widetilde{\mathrm{trg}}(\sigma_i)$, we obtain an array of ideal tetrahedra whose bottom faces compose $\widetilde{\mathrm{trg}}(\sigma_i)$ and whose top faces compose $\widetilde{\mathrm{trg}}(\sigma_{i+1})$. We denote this array of ideal tetrahedra by $\widetilde{\mathrm{Trg}}(\sigma_i, \sigma_{i+1})$. By stacking $\widetilde{\mathrm{Trg}}(\sigma_1, \sigma_2), \cdots,$ $\widetilde{\mathrm{Trg}}(\sigma_{n-1}, \sigma_n)$ up in order, we obtain a set of layers whose bottom faces form $\widetilde{\mathrm{trg}}(\sigma_1)$ and whose top faces form $\widetilde{\mathrm{trg}}(\sigma_n)$. The covering transformation group $\mathbb{Z}^2$ of the covering $\mathbb{R}^2 - \mathbb{Z}^2 \to T$ naturally acts on the above topological ideal simplicial complex, and we define $\mathrm{Trg}(v)$ to be the quotient topological ideal simplicial complex.

Case 2. $n = 1$. Then $\mathrm{Trg}(v)$ is defined to be the 2-dimensional topological ideal triangulation $\mathrm{trg}(\sigma_1)$.

Case 3. $n = 0$. Then $\delta^-(v) = \delta^+(v)$ is an edge of $\mathscr{D}$. We define $\mathrm{Trg}(v)$ to be the 2-dimensional topological ideal triangulation $\mathrm{Trg}(\delta^\pm(v))$.

Note that the underlying space $|\mathrm{Trg}(v)|$ is homeomorphic to the the quotient space of $T \times [-1,1]$ by an equivalence relation $\sim$ such that $(x,s) \sim (y,t)$ only if $x = y$. In particular, $|\mathrm{Trg}(v)|$ is homotopy equivalent to T and has a natural embedding into $T \times (-1,1)$. Then we have the following theorem by Theorem 3.2.

Theorem 5.1. *For any* $\Gamma \in \mathscr{QF}$, $\Delta_{\mathbb{E}}(\Gamma)$ *is isotopic to* $\mathrm{Trg}(v(\Gamma))$ *in the convex core* $M_0(\Gamma)$ *of* $M(\Gamma)$.

Figure 2(b) illustrates the cross section of $\Delta_{\mathbb{E}}(\Gamma)$ along a horosphere centered at ∞ for the quasifuchsian punctured torus group Γ in Figure 2(a).

6. Generalization of Jørgensen's theorem to the groups in $\overline{\mathscr{QF}}$

We first generalize the constructions of $\text{Spine}(v)$ and $\text{Trg}(v)$ in the previous sections. Let $v = (v^-, v^+)$ be an element of $v \in \bar{\mathbb{H}}^2 \times \bar{\mathbb{H}}^2 - \text{diag}(\partial\mathbb{H}^2)$. Let $\Sigma(v) = \{\cdots, \sigma_i, \sigma_{i+1}, \cdots\}$ be the possibly (bi-)infinite sequence of triangles of the modular diagram $\mathscr{D}$ whose interior intersect the oriented geodesic joining v^- with v^+ in this order. Then we construct $\text{Spine}(v)$ and $\text{Trg}(v)$ as in Sections 3 and 4 by using $\Sigma(v)$, where we introduce the following modification in the construction of $\text{Spine}(v)$:

- Suppose v^ε is equal to a rational point $s_0^\varepsilon \in \hat{\mathbb{Q}} \subset \partial\mathbb{H}^2$. Let $\sigma^\varepsilon = \langle s_0^\varepsilon, s_1^\varepsilon, s_2^\varepsilon \rangle$ be the triangle in $\Sigma(v)$ having s_0^ε as a vertex. Consider the loop α in $\text{spine}(\sigma^\varepsilon)$ obtained as the union of the edges of slopes s_1^ε and s_2^ε. (Note that (i) the slope of α in T is s_0^ε and that (ii) the assumption $v^\varepsilon = s_0^\varepsilon$ means that the element of Γ corresponding to α is parabolic.) Then we shrink α to a point and remove it. (See [Jør, Figure 6] for the reason of this modification.)

Then the following gives a generalization of Theorems 3.2, 4.2, and 4.5.

Theorem 6.1. *The side parameter map* $v : \mathscr{QF} \to \mathbb{H}^2 \times \mathbb{H}^2$ *is extended to a map* $v : \overline{\mathscr{QF}} \to \bar{\mathbb{H}}^2 \times \bar{\mathbb{H}}^2 - \text{diag}(\partial\mathbb{H}^2)$ *which satisfies the following conditions.*

1. *For any* $\Gamma \in \overline{\mathscr{QF}}$, $\text{Ford}(\Gamma)$ *and* $\Delta_{\mathbb{E}}(\Gamma)$ *are isotopic to* $\text{Spine}(v(\Gamma))$ *and* $\text{Trg}(v(\Gamma))$, *respectively.*

2. $v^\varepsilon(\Gamma) \in \partial\mathbb{H}^2$ *if and only if the end invariant of the* ε-*end lies in* $\partial\mathbb{H}^2$. *In this case, the end invariant is equal to* $v^\varepsilon(\Gamma)$.

We continue to call the extended v the *side parameter* map. The second assertion of the above theorem implies the following:

(i) $\Omega^\varepsilon(\Gamma)/\Gamma$ is a triply punctured sphere if and only if $v^\varepsilon(\Gamma)$ is a rational point of $\partial\mathbb{H}^2$. In this case, the simple loop $v^\varepsilon(\Gamma)$ corresponds to the accidental parabolic transformation.

(ii) $\Omega^\varepsilon(\Gamma) = \emptyset$, i.e., the ε-end is degenerate, if and only if $v^\varepsilon(\Gamma)$ is an irrational point of $\partial\mathbb{H}^2$. In this case, $v^\varepsilon(\Gamma)$ is equal to the ending lamination of the ε-end.

For geometrically finite boundary groups, the above theorem had been obtained by Jørgensen (cf. [Jør, Theorem 5]). In [Jør], he also studied the Ford domains of geometrically infinite punctured torus groups. Moreover, Jørgensen-Marden [JM79] gave an explicit construction of the Ford domains for the fiber groups of the two simplest punctured torus bundles over S^1. Their construction was generalized by Helling [Hel] to an explicit construction of the Ford domains of a certain infinite family of punctured torus bundles over S^1. For the general hyperbolic punctured torus bundles over S^1, Parker [Par] gave a geometric description of the canonical decompositions, and Lackenby [Lac] gave a purely topological proof to the description of their canonical decompositions due to Jørgensen (cf. [FH82]). For a proof of Theorem 6.1, please see the first author's announcement [Aki99] and his forthcoming paper.

7. Pleating invariants for punctured torus groups

In this section, we give a quick review of the works of Keen and Series [KS93], [KS97], [KS] on bending laminations of quasifuchsian punctured torus groups. Let Γ be a marked quasifuchsian punctured torus group. Then the boundary of the convex core $M_0(\Gamma) = \mathscr{C}(\Lambda(\Gamma))/\Gamma$ of $M(\Gamma) = \mathbb{H}^3/\Gamma$ has two components, the component $\partial^+ M_0(\Gamma)$ facing $\Omega^+(\Gamma)/\Gamma$ and the component $\partial^- M_0(\Gamma)$ facing $\Omega^-(\Gamma)/\Gamma$. Then $\partial^\pm M_0(\Gamma)$ has a structure of a complete hyperbolic punctured torus bent along a measured geodesic lamination, $pl^\pm(\Gamma)$, called the *bending measured lamination* (see Thurston [Thu02] and Epstein-Marden [EM87]). Set $pl(\Gamma) = (pl^-(\Gamma), pl^+(\Gamma))$, and let $[pl(\Gamma)]$ be the pair of projective measured laminations $([pl^-(\Gamma)], [pl^+(\Gamma)])$. For a pair $\mu = (\mu^-, \mu^+)$ of distinct projective measured laminations on T, set

$$\mathscr{P}(\mu) = \{\Gamma \in \mathscr{QF} - \mathscr{F} \mid [pl(\Gamma)] = \mu\}.$$

Then the following results have been proved by Keen-Series [KS] (cf. [KS93], [KS97]).

(i) Let $\lambda_{\mu\pm} : \mathscr{QF} \to \mathbb{C}$ be the complex length function (see [KS93, Section 6.2]). Then $\lambda_{\mu\pm}$ is real valued on $\mathscr{P}(\mu)$ and $\lambda_{\mu^-} \times \lambda_{\mu^+} : \mathscr{P}(\mu) \to \mathbb{R}_+ \times \mathbb{R}_+$ is a diffeomorphism onto the region bounded by the two positive axes in $\mathbb{R}_+ \times \mathbb{R}_+$ and the graph of a continuous function $f_\mu : \mathbb{R}_+ \to \mathbb{R}_+$. The function f_μ is monotone decreasing, $\lim_{t \to +0} f_\mu(t) = +\infty$ and $\lim_{t \to +\infty} f_\mu(t) = 0$.

(ii) The three components of the boundary of the image of $\mathscr{P}(\mu)$ in $\mathbb{R}_+ \times \mathbb{R}_+$ correspond to three distinct parts of the closure $\bar{\mathscr{P}}(\mu)$ in $\overline{\mathscr{QF}}$. The component corresponding to the graph of f_μ represents groups on the Kerckhoff's line of minima in the Fuchsian space $\mathscr{F}$. For the groups corresponding to the axis $\lambda_{\mu^\pm} = 0$, the component $\Omega^\pm$ is degenerated and the support $|\mu^\pm|$ of $\mu^\pm$ is an ending lamination. The boundary point $(0,0)$ represents a doubly degenerate group, unique by [Min99], with the two real bending lamination μ^- and μ^+.

(iii) The complement of the image of $\mathscr{P}(\mu)$ in $\mathbb{R}_+ \times \mathbb{R}_+$ corresponds to Fuchsian groups. (We thank Y. Komori and H. Miyachi for informing us of this fact.)

8. Does $pl(\Gamma)$ determine $\Delta(\Gamma)$?

In this section, we propose the following conjecture, and explain its refinements and related conjectures.

Conjecture 8.1. The combinatorial structure of the EPH-decomposition of a quasifuchsian punctured torus group Γ is determined by the pair $\mu = (\mu^-, \mu^+)$ of its projective bending measured laminations.

In the following, we explain the conjectural picture of the EPH-decomposition $\Delta(\Gamma)$ of a group Γ in the pleating variety $\mathscr{P}(\mu)$, which we denote by $\mathrm{Trg}(\mu, \emptyset)$. (The symbol $\emptyset$ in the above notation represents the fact that both ends of $M(\Gamma)$ are geometrically finite and have no accidental parabolics.) Note that $\mu \in \partial \mathbb{H}^2 \times \partial \mathbb{H}^2 - \mathrm{diag}(\partial \mathbb{H}^2)$ and hence we have the topological ideal simplicial complex $\mathrm{Trg}(\mu)$, which we introduced in Section 6. The desired complex $\mathrm{Trg}(\mu, \emptyset)$ is obtained from $\mathrm{Trg}(\mu)$ by attaching a new piece, $Q(\mu^+)$ and $Q(\mu^-)$, to the $(+)$-side and $(-)$-side, respectively, as explained below. For simplicity, we explain only the (generic) case where $\Sigma(\mu)$ contains a triangle (see the second paragraph in Section 5).

Case 1. μ^ε is rational. Let $\sigma^\varepsilon = \langle s_0^\varepsilon, s_1^\varepsilon, s_2^\varepsilon \rangle$ be the triangle in $\Sigma(\mu)$ such that s_0^ε is the slope of μ^ε. Consider a triangular prism, one of the quadrangular face, A, is triangulated into two triangles, and let $Q(\mu^\varepsilon)$ be the space obtained from it by applying the following operations (see Figure 3 (a)).

(i) Remove the vertices and the ridge line opposite to the triangulated quadrangular face A.

(ii) Identify the two triangular faces of the triangular prism through a translation. Note that $A - \{\text{vertices}\}$ projects to an annulus, A', with one point removed from each boundary component, and the triangulation of A induces a topological ideal triangulation of A'.

(iii) Identify the two ideal edges of A' so as to obtain a punctured torus. Identify the induced topological ideal triangulation on the punctured torus with $\mathrm{trg}(\sigma^\varepsilon)$ so that the edge of slope s_0^ε corresponds to the image of the two ideal edges of A'.

The cellular structure of the triangular prism induces the "cellular structure" on $Q(\mu^\varepsilon)$ with

(i) one 3-dimensional (non-simply connected) facet,

(ii) four 2-dimensional facets, two of which are topological ideal triangles of A and the remainders are homeomorphic to $S^1 \times (0, 1]$ with one point in $S^1 \times 1$ removed (note that we do not regard the images of the two triangular faces of the triangular prism as facets), and

(iii) three ideal edges.

Note that $\mathrm{trg}(\sigma^\varepsilon)$ is the boundary component of $\mathrm{Trg}(\mu)$ on the ε-side. We attach $Q(\mu^\varepsilon)$ to $\mathrm{Trg}(\mu)$ along $\mathrm{trg}(\sigma^\varepsilon)$. We note that the removed ridge line of the triangular prism corresponds the axis of the purely hyperbolic transformation representing the rational bending locus μ^ε.

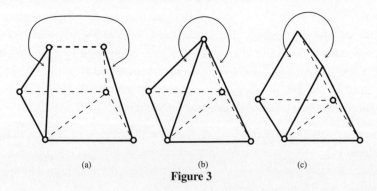

(a)   (b)   (c)

Figure 3

Case 2. μ^ε is irrational. Fix a complete hyperbolic structure on a punctured torus, and continue to denote the hyperbolic punctured torus by T. Then $Q(\mu^\varepsilon) = T - |\mu^\varepsilon|$, that is, the complement of the underlying geodesic lamination $|\mu^\varepsilon|$, and its cellular structure consists of

(i) two edges each joining the puncture and one of the two ends of the open once-punctured bigon $T - |\mu^\varepsilon|$ with path metric, and

(ii) two 2-dimensional facets obtained as the connected components of the complements of the above two edges.

In the following, we explain the way the 2-dimensional piece $Q(\mu^\varepsilon)$ is attached to the ε-end of $\mathrm{Trg}(\mu)$. For simplicity, we assume $\varepsilon = +$. Pick a triangle σ_0 in $\Sigma(\mu)$ and consider the triangles $\sigma_0, \sigma_1, \sigma_2, \cdots$ of $\Sigma(\mu)$ starting from σ_0. Let f be a surjective continuous map from $T \times [0, \infty)$ to the subcomplex $\cup_{i \geq 0}\mathrm{Trg}(\sigma_i, \sigma_{i+1})$ of $\mathrm{Trg}(\mu)$ which satisfy the following conditions.

(i) $f(T \times [i, i+1]) = \mathrm{Trg}(\sigma_i, \sigma_{i+1})$.

(ii) The restriction of f to $T \times i$ is a homeomorphism onto $\mathrm{trg}(\sigma_i) = \mathrm{Trg}(\sigma_{i-1}, \sigma_i)$ $\cap \mathrm{Trg}(\sigma_i, \sigma_{i+1})$ such that the inverse image of each edge of $\mathrm{trg}(\sigma_i)$ is the geodesic of the hyperbolic punctured torus $T \times i$ of the same slope.

Let g be the homeomorphism $g : T \times [0, 1) \to T \times [0, \infty)$ defined by $g(x, t) = (x, t/(1 - t))$, and set $h = f \circ g : T \times [0, 1) \to \cup_{i \geq 0}\mathrm{Trg}(\sigma_i, \sigma_{i+1})$. Let $\sim$ be the equivalence relation on $T \times [0, 1]$ such that $x \sim y$ if and only if (i) $x = y$ or (ii) $x, y \in T \times [0, 1)$ and $h(x) = h(y)$. Then we can identify $\cup_{i \geq 0}\mathrm{Trg}(\sigma_i, \sigma_{i+1})$ with $T \times [0, 1)/\sim$, which is embedded in $(T \times [0, 1] - (T - |\mu^+|) \times 1)/\sim$. This gives the way to attach $Q(\mu^\varepsilon) = T - |\mu^+|$ to $\mathrm{Trg}(\mu)$ along the $(+)$-end of $\mathrm{Trg}(\mu)$.

The following conjecture is a refinement of Conjecture 8.1.

Conjecture 8.2. Let Γ be a quasifuchsian punctured torus group in the pleating variety $\mathscr{P}(\mu)$. Then the EPH-decomposition $\Delta(\Gamma)$ is isotopic to $\mathrm{Trg}(\mu, \emptyset)$.

To extend the above conjecture to that for all punctured torus groups, recall that the map $\Gamma \mapsto \mu = (\mu^-, \mu^+)$ from $\mathscr{QF}$ to $\partial\mathbb{H}^2 \times \partial\mathbb{H}^2 - \mathrm{diag}(\partial\mathbb{H}^2)$ has a natural extension to a map from the closure $\overline{\mathscr{QF}}$. Namely, if $\Omega^\varepsilon(\Gamma)/\Gamma$ is not a punctured torus, then we define $\mu^\varepsilon(\Gamma) = \mu^\varepsilon \in \partial\mathbb{H}^2$ to be the end invariant of the ε-end (see [Min99]), which is equal to the side parameter $\nu^\varepsilon(\Gamma)$ by Theorem 6.1. For each group $\Gamma \in \overline{\mathscr{QF}}$, let $\iota(\Gamma)$ be the subset of $\{-, +\}$ defined by

$$\iota(\Gamma) = \{\varepsilon \,|\, \mu^\varepsilon(\Gamma) = \nu^\varepsilon(\Gamma) \in \partial\mathbb{H}^2\}$$
$$= \{\varepsilon \,|\, \partial^\varepsilon M(\Gamma) \text{ is a triply punctured sphere or the empty set}\}.$$

For each subset $\iota \subset \{-, +\}$, we construct yet another topological ideal polyhedral complex $\mathrm{Trg}(\mu, \iota)$ from $\mathrm{Trg}(\mu)$, by generalizing the construction of $\mathrm{Trg}(\mu, \emptyset)$, as follows. Let $\varepsilon \in \{-, +\}$.

Case 1. $\varepsilon \notin \iota$. Then we attach the piece $Q(\mu^\varepsilon)$ to $\mathrm{Trg}(\mu)$ as in the construction of $\mathrm{Trg}(\mu, \emptyset)$.

Case 2. $\varepsilon \in \iota$ and μ^ε is rational. Then we also attach the 3-dimensional piece $Q(\mu^\varepsilon)$ to $\mathrm{Trg}(\mu)$ as in the construction of $\mathrm{Trg}(\mu, \emptyset)$. However, we should regard that the construction starts with a pyramid which is obtained from the triangular prism in Figure 3(a) by shrinking the removed ridge line to a point (and remove it) (see Figure 3(b)). The removed point corresponds to the parabolic fixed point of the accidental parabolic transformation representing the rational lamination μ^ε.

Case 3. $\varepsilon \in \iota$ and μ^ε is irrational. In this case $Q^\varepsilon(\mu, \iota)$ is the empty set, i.e., we leave the ε-end of $\mathrm{Trg}(\mu)$ as it is.

The following conjecture is a generalization of Conjecture 8.1.

Conjecture 8.3. Let Γ be a punctured torus group which lies in the closure $\overline{\mathscr{P}}(\mu)$ of the pleating variety $\mathscr{P}(\mu)$. Then the EPH-decomposition $\Delta(\Gamma)$ is isotopic to the complex $\mathrm{Trg}(\mu, \iota(\Gamma))$.

Recall that, for each pair $\mu = (\mu^-, \mu^+)$, the distinct projective laminations on the punctured torus T, $\Sigma(\mu)$ denotes the set $\{\cdots, \sigma_i, \sigma_{i+1}, \cdots\}$ of possibly (bi-)infinite sequence of triangles of the modular diagram $\mathscr{D}$ whose interior intersect the oriented geodesic joining μ^- with μ^+ in this order. Let $|\Sigma(\mu)|$ be the union of triangles in $\Sigma(\mu)$. Since the Ford domain is dual to the subcomplex $\Delta_{\mathbb{E}}(\Gamma)$ of $\Delta(\Gamma)$, the above conjecture implies the following conjecture on the Ford domain of Γ, which relates the works of Jørgensen with those of Keen-Series.

Conjecture 8.4. Let Γ be a punctured torus group which lies in the closure $\overline{\mathscr{P}}(\mu)$. Then $v^\pm(\Gamma) \in |\Sigma(\mu)|$.

We obtain a partition of $\overline{\mathscr{P}}(\mu)$ according to the combinatorial structures of $\mathrm{Ford}(\Gamma)$. The above conjecture is refined to a conjecture on the structure of this partition. For simplicity, we state the conjecture only for the rational case.

Conjecture 8.5. Let $\mu = (\mu^-, \mu^+)$ be a pair of distinct rational projective measured laminations on T, and let $\sigma_1, \cdots, \sigma_n$ be the members of $\Sigma(\mu)$. For each pair (σ_i, σ_j) $(1 \le i \le j \le n)$, set

$$\overline{\mathscr{P}}(\mu; \sigma_i, \sigma_j) = \{\Gamma \in \overline{\mathscr{P}}(\mu) \mid v^-(\Gamma) \in \sigma_i, \ v^+(\Gamma) \in \sigma_j\}.$$

Then $\{\overline{\mathscr{P}}(\mu; \sigma_i, \sigma_j) \mid 1 \le i \le j \le n\}$ gives a partition of $\overline{\mathscr{P}}(\mu)$ as illustrated in Figure 4. Here the region labeled (i, j) corresponds to $\overline{\mathscr{P}}(\mu; \sigma_i, \sigma_j)$.

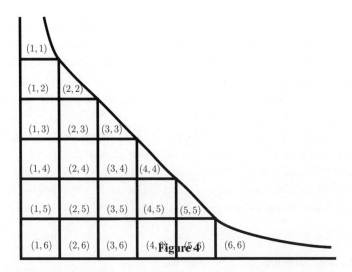

Figure 4

Let us explain the meaning of the above conjecture. If Γ corresponds to a point on the diagonal edge, then Γ is fuchsian, and the side parameter satisfies $v^-(\Gamma) = v^+(\Gamma)$. So the label for this group is (i,i) for some i (provided that Conjecture 8.4 is true). We expect that the number i changes monotonically as Γ moves on the diagonal edge. If Γ corresponds to the origin, then Γ is a double cusp group, where the simple loops μ^- and μ^+ are pinched. So, by Theorem 6.1, we have $v^\pm(\Gamma) = \mu^\pm$ and hence the label for Γ must be $(1,n)$. If Γ corresponds to a point on the y-axis then Γ lies in the Maskit slice (see [KS93]) where μ^- is pinched. So, by Theorem 6.1, the label for Γ should be $(1,i)$ for some i (provided that Conjecture 8.4 is true). We conjecture that the number i changes monotonically as Γ moves on the y-axis. Likewise, the label for a group Γ corresponding to a point on the x-axis is (i,n) for some i, and we expect that the number i changes monotonically. In general, Conjecture 8.5 predicts the following: As a group $\Gamma \in \overline{\mathscr{P}}(\mu)$ evolves from a fuchsian group (on the diagonal edge) to the most complicated double cusp group (at the origin), the Ford domain becomes complicated "monotonically". All possible shapes of the Ford domains are determined as long as Γ evolves in the pleating variety $\overline{\mathscr{P}}(\mu)$. Even if Γ is just an infant, i.e., very near to a fuchsian group, one should be able to see the possible shapes of the Ford domains in its future. Because Conjecture 8.3 predicts that the EPH-decomposition $\Delta(\Gamma)$ is uniquely determined by μ and because the Ford complex is dual to the subcomplex $\Delta_{\mathbb{E}}(\Gamma)$ of $\Delta(\Gamma)$. Even a very young Γ should have the same EPH-decomposition as the ultimate (double cusp) group.

Finally, we present yet another conjecture. In the example of a Ford domain illustrated in Figure 2, we see that the axes of the faces incline to the left or right
"alternately". To be precise, the axes of the faces in Figure 2(a) corresponding to the
vertices on the left side of Figure 2(b) (i.e., s_1, s_3, s_5, s_6) are inclined one way and those
corresponding to the vertices on the right (i.e., s_2, s_4, s_7) are inclined the other way.
Here by an axis of a face F, we mean (the projection to $\mathbb{C}$) of $F \cap \mathrm{Axis}(P_F)$, where P_F
is the order 2 elliptic transformation which maps F to itself (see the paragraph after
Theorem [Jør]). Recall the parabolic transformation K such that $K(\infty) = \infty$ and K^2
corresponds to the puncture of T (see Section 3). Set $A_F = KP_F$. Then $A_F(F) = K(F)$
is also a face of the Ford domain $Ph(\Gamma)$, and hence A_F is a face pairing for $Ph(\Gamma)$.
Then we can easily see the following:

$$\arg(\text{the projection of } \mathrm{Axis}(P_F) \text{ to } \mathbb{C}) \equiv \pi - \arg(\mathrm{tr} A_F) \quad (\mathrm{mod} \ \pi).$$

So the following conditions are equivalent.

(i) A_F is a right screw motion, i.e., $\arg(\mathrm{tr} A_F) \in (0, \pi/2) \cup (\pi, 3\pi/2)$.

(ii) The axis of F inclines to the left, i.e., $\arg(\text{the projection of } \mathrm{Axis}(P_F)) \in (\pi/2, \pi)$
$\cup(3\pi/2, 2\pi)$.

The above observation was brought to us by Jørgensen through his lecture in Osaka
[Jør97], where he suggested that it would be an interesting and challenging problem
to study this phenomenon. The following conjecture formulates his proposal in con-
junction with the bending laminations.

Conjecture 8.6. Let Γ be an element of $\bar{\mathscr{P}}(\mu)$. Let $\Sigma(\mu)_L^{(0)}$ (resp. $\Sigma(\mu)_R^{(0)}$) be the
subset of the vertex set $\Sigma(\mu)^{(0)}$ of $\Sigma(\mu)$ which lies on the left (resp. right) of the
oriented geodesic $\ell(\mu)$ joining μ^- to μ^+. Then for any element s of $\Sigma(\mu)_L^{(0)}$ (resp.
$\Sigma(\mu)_R^{(0)}$), A_s is a right (resp. left) screw motion. Here A_s denotes an element of Γ
corresponding to a simple loop on T of slope s.

9. Partial positive answers and experimental results

Theorem 6.1 implies that Conjectures 8.3 and 8.4 are valid for doubly degenerate
groups. Moreover, [AS, Theorem 11.2] shows that Conjecture 8.3 holds for the restriction of $\Delta(\Gamma)$ to the boundary of the convex core $M_0(\Gamma)$ for a quasifuchsian punctured
torus group Γ. In addition to these partial results, we have the following partial result.

Theorem 9.1. *Suppose μ is rational, and let Γ_0 be the double cusp group in $\bar{\mathscr{P}}(\mu)$.*
Then there is a neighborhood U of Γ_0 in $\bar{\mathscr{P}}(\mu)$, such that Conjectures 8.1–8.4 are
valid for any group in U.

We note that an analogy of Conjecture 8.3 for the groups on the outside of the quasifuchsian space has already been established by [2] as follows. If μ is rational, the $\mathscr{P}(\mu)$ has a natural extension to the outside of the quasifuchsian space and each group in the extension can be regarded as the holonomy group of a hyperbolic cone-manifold with a cusp. Moreover, it has been proved that the "Ford complex" of the cusped hyperbolic cone-manifold is dual to the complex constructed as in Cases 1 and 2 in Section 8 with the following minor modification:

(i) We use the triangular prism which is obtained from the pyramid in Case 2 by expanding the peak vertex (that was obtained by shrinking the ridge line in Case 1 to a point) to an edge in a direction perpendicular to the original ridge line (see Figure 3 (c)). This new ridge line corresponds to the axis of an elliptic transformation $A_{s^{\varepsilon}}$, and hence it is a component of the cone axis.

(ii) We do not delete the new ridge line.

For the full description of this result, please see the announcement [ASWY00]. For the proof of this result and the results announced in this paper, please see the forthcoming paper [ASWY].

For Conjecture 8.6, we have the following partial result.

Proposition 9.2. *Suppose μ^{ε} is rational for some $\varepsilon = \pm$, and let s^{ε} be the rational number corresponding to μ^{ε}. Then for any $\Gamma \in \bar{\mathscr{P}}(\mu)$, the following holds. Let s be a rational number such that s and s^{ε} span an edge of $\mathscr{D}$. Then A_s is a right or left screw motion according as s lies on the left or right of the oriented geodesic, $\ell(\mu)$, joining μ^{-} to μ^{+}. In particular, Conjecture 8.6 is valid for the elements s of $\Sigma(\mu)^{(0)}$ such that s and s^{ε} span an edge of $\mathscr{D}$.*

Finally, we explain some experimental results towards Conjectures 8.5 and 8.6. Suppose that μ is rational. Then the diffeomorphism $L\mu = \lambda_{\mu^{-}} \times \lambda_{\mu^{+}}$ in Section 7 is defined by

$$\lambda_{\mu^{\pm}}(\Gamma) = \text{the translation length of } A^{\pm},$$

where $A^{\pm}$ denote the elements of Γ represented by the rational lamination $|\mu^{\pm}|$. Since the translation length is determined by the trace, we may replace $L\mu$ with the map $T\mu = \text{tr}_{\mu^{-}} \times \text{tr}_{\mu^{+}} : \mathscr{P}(\mu) \to (2, \infty) \times (2, \infty)$ defined by $\text{tr}_{\mu^{\pm}}(\Gamma) = \text{tr}(\tilde{A}^{\pm})$, where $\tilde{A}^{\pm}$ are the element of $SL(2, \mathbb{C})$ with positive real trace projecting to the elements $A^{\pm}$ of Γ. (Recall that $A^{\pm}$ are purely hyperbolic by [KS93, Lemma 4.6].) $T\mu$ extends to a map $\overline{\mathscr{P}}(\mu) \to [2, \infty) \times [2, \infty)$, and for every point $(x, y) \in [2, \infty) \times [2, \infty)$ there is a punctured torus group Γ in $\overline{\mathscr{P}}(\mu) \cup \mathscr{F}$ such that $T_\mu(\Gamma) = (x,y)$: moreover, such a group is unique if $(x,y) \in T_\mu(\overline{\mathscr{P}}(\mu))$. Since there is an effective method to determine the Ford complex of Γ (cf. [Jør],[ASWY00],[KSWY]), we can do an experimental study on Conjecture 8.5. The computer experiments made by the last named author support the following results:

- Conjectures 8.5 and 8.6 are valid for $\mu = (1/0, \mu^+)$ with

$$\mu^+ \in \{0/1, 1/2, 1/3, 1/4, 2/5, 3/7, 3/8\}.$$

We note that Conjectures 8.5 and 8.6 for $\mu = (1/0, \mu^+)$ with $\mu^+ = 0/1, 1/2$ can be easily proved. Figure 5 (a) and (b), respectively, are the output for $\mu = (1/0, 1/4)$ and $(1/0, 2/5)$.

The idea to compare the two convex hull constructions for punctured torus groups lead us to a refinement of McShane's identity [AMS], which seems to have some relation with Conjecture 8.6.

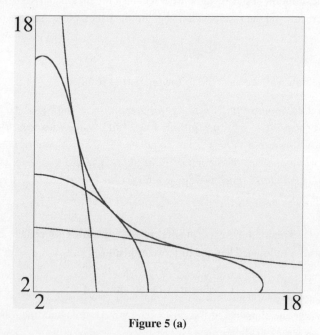

Figure 5 (a)

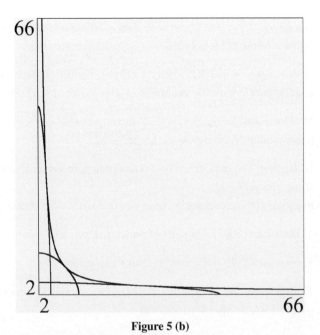

66

2

2   66

Figure 5 (b)

References

[Aki99]   H. Akiyoshi (1999). On the Ford domains of once-punctured torus groups. In *Hyperbolic spaces and related topics, RIMS, Kyoto, Kokyuroku* **1104**, 109–121.

[AMS]   H. Akiyoshi, H. Miyachi and M. Sakuma. A refinement of McShane's identity for quasifuchsian punctured torus groups, *to appear Proceedings of the 2002 Ahlfors-Bers Colloquium.*

[AS]   H. Akiyoshi and M. Sakuma. Comparing two convex hull constructions for cusped hyperbolic manifolds, *this volume.*

[ASWY00]   H. Akiyoshi, M. Sakuma, M. Wada and Y. Yamashita (2000). Ford domains of punctured torus groups and two-bridge knot groups. In *Knot Theory, Proceedings of the workshop held in Toronto*, 14–71.

[ASWY]   H. Akiyoshi, M. Sakuma, M. Wada and Y. Yamashita. *in preparation.*

[EM87]   D.B.A. Epstein and A. Marden (1987). Convex hulls in hyperbolic space, a theorem of Sullivan, and measured pleated surfaces. In *Analytical and*

272   H. Akiyoshi, M. Sakuma, M. Wada & Y. Yamashita

Geometric Aspects of Hyperbolic Space, edited by D.B.A. Epstein (LMS Lecture Notes 111), 112–253.

[EP88]   D.B.A. Epstein and R.C. Penner (1988). Euclidean decompositions of noncompact hyperbolic manifolds, *J. Diff. Geom.* **27**, 67–80.

[FH82]   W. Floyd and A. Hatcher (1982). Incompressible surfaces in punctured torus bundles, *Topology Appl.* **13**, 263–282.

[Hel]   H. Helling. The trace fields of a series of hyperbolic manifolds, *preprint, Univ. Bielefeld.*
http://www.mathematik.uni-bielefeld.de/sfb343/

[Jør]   T. Jørgensen (2002). On pairs of punctured tori, *this volume.*

[Jør97]   T. Jørgensen (1997). Lecture at Osaka University.

[JM79]   T. Jørgensen and A Marden (1979). Two doubly degenerate groups, *Quart. J. Math.* **30**, 143–156.

[KMS93]   L. Keen, B. Maskit and C. Series (1993). Geometric finiteness and uniqueness for Kleinian groups with circle packing limit sets, *J. reine angew. Math.* **436**, 209–219.

[KS93]   L. Keen and C. Series (1993). Pleating coordinates for the Maskit embedding of the Teichmüller space of punctured tori, *Topology* **32**, 719–749.

[KS97]   L. Keen and C. Series (1997). How to bend pairs of punctured tori. In *Lipa's Legacy, Proceedings of the Bers Colloquium 1995, edited by J. Dodziuk and L. Keen, Contemp. Math.* **211**, 359–388.

[KS]   L. Keen and C. Series. Pleating invariants for punctured torus groups, *to appear Topology.*

[KSWY]   Y. Komori, T. Sugawa, M. Wada and Y. Yamashita. Drawing Bers embeddings of the Teichmüller space of once punctured torus, *in preparation.*

[Lac]   M. Lackenby. The canonical decomposition of once-punctured torus bundles, *preprint.* http://www.maths.ox.ac.uk/ lackenby/

[Min99]   Y. Minsky (1999). The classification of punctured-torus groups, *Ann. Math.* **149**, 559–626.

[Par]   J.R. Parker. Tetrahedral decomposition of punctured torus bundles, *this volume.*

[PS01]   J.R. Parker and B.O. Stratmann (2001). Kleinian groups with two singly cusped parabolic fixed points, *Kodai Mathematical Journal* **24**, 169–206.

[Thu02]   W.P. Thurston (2002). The geometry and topology of three-manifolds, *Electronic version.* http://www.msri.org/publications/books/gt3m/

[Wad]   M. Wada. Opti. http://vivaldi.ics.nara-wu.ac.jp/ wada/OPTi/

Hirotaka Akiyoshi

Dept of Mathematics
Graduate School of Science
Osaka University
Machikaneyama-cho 1-16
Toyonaka
Osaka, 560-0043
Japan

akiyoshi@gaia.math.wani.
          osaka-u.ac.jp

Masaaki Wada

Department of Information and
Computer Sciences
Faculty of Science
Nara Women's University
Kita-Uoya Nishimachi Nara,
630-8506
Japan

wada@ics.nara-wu.ac.jp

Makoto Sakuma

Dept of Mathematics
Graduate School of Science
Osaka University
Machikaneyama-cho 1-16
Toyonaka
Osaka, 560-0043
Japan

sakuma@math.wani.osaka-u.ac.jp

Yasusi Yamasita

Department of Information and
Computer Sciences
Faculty of Science
Nara Women's University
Kita-Uoya Nishimachi Nara,
630-8506
Japan

yamasita@ics.nara-wu.ac.jp

AMS Classification: 57M50, 57N15, 57N16

Keywords: convex hull, convex core, Ford domain, Euclidean decomposition, bending lamination, punctured torus

Tetrahedral decomposition of punctured torus bundles

J. R. Parker

Abstract

We consider hyperbolic manifolds which fibre over the circle with fibre the once punctured torus. Normalising so that ∞ is a parabolic fixed point, we analyse the Ford domain of such a manifold. Using the cutting surfaces associated to this domain, we give a canonical decomposition of the manifold into ideal tetrahedra

1. Introduction

A well known result of Thurston [Thu86b] says that if Σ is a surface of negative Euler characteristic and ϕ is a pseudo-Anosov diffeomorphism of Σ to itself then the mapping torus M of ϕ carries a finite volume hyperbolic structure (see also McMullen [McM96], Otal [Ota01]). In the case where Σ is a once punctured torus its fundamental group is a free group of rank 2 and the mapping class group of Σ is the classical modular group $\Gamma = \mathrm{PSL}(2, \mathbb{Z})$. An automorphism ϕ of Σ is *pseudo-Anosov* if and only if, as an element of Γ, it is hyperbolic.

We consider manifolds M which are mapping tori of pseudo-Anosov diffeomorphisms of the once punctured torus. We can associate a combinatorial ideal triangulation of M by decomposing the automorphism of $\pi_1(\Sigma)$ induced by ϕ into elementary Nielsen moves (see for example page 328 of [Bow97]). The main problem addressed in this paper is to show that when we give M its unique hyperbolic structure then this combinatorial triangulation is realised as an ideal triangulation of M by ideal hyperbolic tetrahedra. The point is that we must show that the orientation of these hyperbolic tetrahedra are consistent so that they fit together without overlap. Our method is to use cutting surfaces, which were developed by Parker and Stratmann in [PS01] to solve a different problem. The techniques developed in [PS01] give the tetrahedral decomposition directly from the combinatorics of the Ford domain, which we find using work of Jørgensen [Jør03].

Some concrete examples of these tetrahedral decompositions are given by Helling in [Hel99]. He also gives an example of such a "triangulation" where the tetrahedra do not fit together consistently. However, this example is a for a non-discrete representation of M which is constructed by taking the wrong root of the polynomial $f_m(z)$ used to find the trace field of G.

I would like to thank the referee for making constructive suggestions. These have greatly improved the paper.

2. Background

2.1. Punctured torus groups

The material in this section is completely standard, see Section 2 of Minsky [Min99] for example. A Kleinian group $G < \mathrm{PSL}(2, \mathbb{C})$ is called a *punctured torus group* if and only if it is freely generated by two maps whose commutator is parabolic. Therefore G is a discrete, faithful representation of the fundamental group of a punctured torus, the commutator corresponding to a loop around the puncture. Any element g of G that corresponds to a non-trivial, non-peripheral simple closed curve on the punctured torus will be called a *generator*. Two generators corresponding to a pair of simple closed curves with intersection number 1 will be called *neighbours*. (The intersection number of two curves is the minimal number of intersections of any curves in their homotopy class.) Any pair of neighbours g and h generate G and their commutator corresponds to a loop around the puncture, and so is parabolic. When necessary we fix a pair of neighbours g and h and write $G = \langle g, h \rangle$.

If two generators are conjugate within G then they correspond to the same simple closed curve. A generator and its inverse also correspond to the same curve with opposite orientation. Similarly, a pair of neighbours (g, h) is only defined up to conjugation of both elements, interchanging the order and taking inverses of either element.

The complex plane punctured at every Gaussian integer, $\mathbb{C} - \mathbb{Z}[i]$, is a covering surface of the punctured torus. The lift of a generator is homotopic to a line of rational slope p/q. Two generators are neighbours if and only if the corresponding rational numbers p/q and r/s satisfy $ps - qr = \pm 1$. We may take our fixed pair of generators to have slope $0 = 0/1$ and $\infty = 1/0$.

We can move between pairs of neighbours using *Nielsen moves*. The elementary Nielsen moves send the pair of neighbours (g, h) to one of (g, gh), $(g, g^{-1}h)$, (gh, h) or (gh^{-1}, h). Topologically these Nielsen moves correspond to performing a Dehn twist (either clockwise or anticlockwise) about the curve corresponding to either g or h. A classical result of Nielsen [Nie18] says that any automorphism of G may be written as a product of Nielsen moves.

If g and h are neighbours, then gh is also a generator and, moreover, (g, gh) and (h, gh) are two pairs of neighbours. We call the unordered set (g, h, gh) a *generator triple*. Again we define this up to taking inverses and conjugation. Thus (g, h, gh) is the same generator triple as (g, h^{-1}, gh) and also (g, h, hg). To each pair of neighbours (g, h) there are two generator triples, namely (g, h, gh) and (g, h, gh^{-1}).

A convenient geometric model describing this set up is the *Farey graph*. Each vertex of this graph is an equivalence class of generators (that is a generator up to conjugacy and taking inverses). Two vertices of this graph are joined by an edge if and only if they are neighbours. Each vertex of the Farey graph is the endpoint of infinitely many edges. Every edge lies on exactly two cycles of length three, corresponding to the generator triples. Nielsen's result referred to above implies that this graph is connected.

The Farey graph may be embedded in the closed upper half plane as the 1-skeleton of the *Farey tessellation*. In this embedding every vertex is the rational number (or infinity) given by the slope of its lift to $\mathbb{C} - \mathbb{Z}[i]$ indicated above. Two vertices p/q and r/s are connected by an edge, which is a Euclidean semi-circle, if and and only if $ps - qr = \pm 1$ (including straight lines joining $\infty = 1/0$ to any integer $n = n/1$).

We make the identification between Teichmüller space of Σ and the upper plane in such a way that the the rational points of the boundary $\mathbb{R} \cup \{\infty\}$ correspond to the simple closed curves as described above. We also require that the action of the mapping class group of Σ (Teichmüller modular group) on the upper half plane is the standard action of $\mathrm{PSL}(2, \mathbb{Z}) = \Gamma$. Otherwise, we will not be concerned with exactly which marked surface correspond to which point in the upper half plane. Extending this identification to the whole of the boundary implies that the irrational points of $\mathbb{R}$ correspond to laminations with infinite leaves in the Thurston boundary, in the usual way.

2.2. Punctured torus bundles

A *pseudo-Anosov* element ϕ of the mapping class group is a hyperbolic element of $\mathrm{PSL}(2, \mathbb{Z})$. That is,

$$\phi = \begin{pmatrix} a & b \\ c & d \end{pmatrix}$$

where $\mathrm{tr}^2(\phi) = (a + d)^2 > 4$. The fixed points of ϕ are

$$v_{\pm} = \frac{a - d \pm \sqrt{(a+d)^2 - 4}}{2c}.$$

Because $(a+d)^2 > 4$ we see that $(a+d)^2 - 4$ cannot be a perfect square and so $\sqrt{(a+d)^2-4}$ is irrational. Hence these fixed points lie in the quadratic number field $\mathbb{Q}\left(\sqrt{(a+d)^2-4}\right)$. This shows that pseudo-Anosov diffeomorphisms preserve two laminations on Σ and both of these laminations have infinite leaves. Hence our definition of pseudo-Anosov agrees with the standard definition using projective measured laminations (see Theorem 2.5 of [Thu86b]).

We can decompose ϕ as $L^{m_k}U^{n_k}\cdots L^{m_1}U^{n_1}$ for integers m_j and n_j, where

$$L = \begin{pmatrix} 1 & 0 \\ 1 & 1 \end{pmatrix}, \quad U = \begin{pmatrix} 1 & 1 \\ 0 & 1 \end{pmatrix}.$$

If ϕ has this form, then the continued fraction expansions of the fixed points of ϕ have tails $[m_k, n_k, \ldots, m_1, n_1, m_k, n_k, \ldots]$ but have to be read in opposite directions (see [Ser85b]).

In [JM79] Jørgensen and Marden construct two examples of punctured torus bundles. Both these examples have $k = 1$, one has $m_1 = n_1 = 1$, the other has $m_1 = n_1 = 2$. In [AH97] Alestalo and Helling give the details of punctured torus bundles when $k = 1$ for more general m_1 and n_1. In [Koc99] Koch considered many examples of punctured torus bundles associated to more general pseudo-Anosov maps. In Section 5 we consider in detail the example where $k = 1$, $m_1 = 2$ and $n_1 = 1$.

The mapping class ϕ induces an automorphism ϕ_* of G. The matrix U corresponds to the Nielsen move $(g,h) \longmapsto (gh^{-1}, h)$ and the matrix L corresponds to the Nielsen move $(g,h) \longmapsto (g, g^{-1}h)$ (see page 66 of [McM96]). Therefore using the LU decomposition of ϕ given above we can write ϕ_* as a sequence of elementary Nielsen moves. Conversely, given an automorphism of G as a sequence of Nielsen moves we can write down the corresponding mapping class ϕ.

Let ϕ be a pseudo-Anosov diffeomorphism of the punctured torus Σ and let M be the *mapping torus* of ϕ. That is M is homeomorphic to $\Sigma \times [0,1]/\sim$ where $\sim$ identifies $(x,0)$ with $(\phi(x),1)$ for all points $x \in \Sigma$. The fundamental group of M is $H = \langle g, h, f \mid \phi_*(g) = fgf^{-1}, \phi_*(h) = fhf^{-1} \rangle$ where $G = \langle g, h \rangle$ is the fundamental group of Σ and ϕ_* is the automorphism of G induced by ϕ, as described above.

We now show that to any pseudo-Anosov map ϕ in $\mathrm{PSL}(2,\mathbb{Z})$ we may associate a group G^∞ which is a strong limit of quasi-Fuchsian groups. Later we indicate how to show G and G^∞ are the same group. The details of this process are given in Chapter 3 of McMullen [McM96]. We briefly summarise them here. Wada's computer programme OPTi [Wad] may be used to visualise this process.

Let $\phi \in \mathrm{PSL}(2,\mathbb{Z})$ be hyperbolic. As we saw above, ϕ has fixed points $v_\pm$. Consider the semicircle α in the upper half plane joining these two points. Because ϕ

preserves both α and the Farey tessellation $\mathscr{F}$ it is clear that the intersection pattern of α and $\mathscr{F}$ is preserved by ϕ. Since ϕ acts as a hyperbolic translation along α we see that this intersection pattern is periodic. Moreover, if $\phi = L^{m_k} U^{n_k} \cdots L^{m_1} U^{n_1}$ then this intersection pattern corresponds to exiting m_k successive triangles on the right then n_k on the left and so on, see [Ser85b], [Ser85a].

Let x_0 be a point of the interior of α. For each positive integer n let α_n be the arc joining $\phi^{-n}(x_0)$ to $\phi^n(x_0)$. Using Bers' simultaneous uniformization theorem [Ber60] there is a quasi-Fuchsian punctured torus group G_n so that two ends of G_n have hyperbolic structures corresponding to the points $\overline{\phi^{-n}(x_0)}$ and $\phi^n(x_0)$. As n tends to infinity, the groups G_n tend strongly towards G^∞, the group with end invariants ν_+ and ν_-. This strong convergence is a consequence of Thurston's double limit theorem Theorem 4.1 of [Thu86b] or Theorem 3.8 of [McM96]. For details of the strong convergence, see Section 3.5 of [McM96] where McMullen works in a Bers slice and then uses a sequence of changes of marking.

2.3. Ideal triangulations

We can associate an ideal triangulation of Σ to any generator triple. This is done as follows (see for example page 328 of [Bow97]). Every generator corresponds to a non-trivial, non-peripheral homotopy class $[\gamma]$ of simple closed curves on Σ. Given $\gamma \in [\gamma]$, there is a unique non-trivial homotopy class $[\gamma^*]$ of arcs with both endpoints on the boundary (that is, at the puncture) of Σ so that there exists γ^* in $[\gamma^*]$ disjoint from γ. Given a generator triple, say (g, h, gh) with associated classes $[\gamma]$, $[\alpha]$, $[\beta]$, we can find disjoint representatives γ^*, α^*, β^* of the three homotopy classes of arcs $[\gamma^*]$, $[\alpha^*]$, $[\beta^*]$. These form the edge of our ideal triangulation. Cutting the punctured torus along the arcs γ^* and α^* corresponding to a pair of neighbours gives an ideal quadrilateral. The two generator triples containing this pair of neighbours corresponds to choosing the third arc to be one of the two diagonals of this quadrilateral. Performing a Nielsen move replaces one of these diagonals with the other. Thus given two adjacent generator triples there are two ideal triangles for each triple. These four triangles form the faces of an ideal tetrahedron. Given a pseudo-Anosov diffeomorphism ϕ, by decomposing ϕ into a sequence of elementary Nielsen moves, we can use the process described above to construct a combinatorial ideal triangulation of the mapping torus M associated to ϕ.

The main problem addressed in this paper is to show that when we give M its unique hyperbolic structure then this combinatorial triangulation is realised as an ideal triangulation of M by ideal hyperbolic tetrahedra. In order to construct this hyperbolic ideal triangulation of M we must show that the orientation of these hyperbolic tetrahedra are consistent so that they fit together without overlap. We show that this indeed happens.

3. Ford domains for punctured torus bundles

In this section we give a structure theorem for the Ford domains of punctured torus bundles normalised so that ∞ is a parabolic fixed point. The normalisation is due to Jørgensen [Jør03] and our theorem is a consequence of his structure theorem for quasi-Fuchsian punctured torus groups (Theorem 1 of [Jør03]).

Given $g \in \mathrm{PSL}(2, \mathbb{C})$ not fixing ∞, consider the hemispheres in $\mathbb{H}^3$ centred at $g^{-1}(\infty)$, the *pole* of g, and those centred at $g(\infty)$, the pole of g^{-1}. Among these there is exactly one hemisphere centred at the pole of g which is mapped to the hemisphere centred at the pole of g^{-1} of the same (Euclidean) radius. These hemispheres are the *isometric spheres* of g and g^{-1} respectively, see [For29]. The *Ford domain* D for G is the intersection of the exteriors of the isometric spheres of all those elements of G not fixing ∞, see [For29]. The Ford domain is invariant under G_∞, the stabiliser of ∞ in G. The intersection of the Ford domain with some fundamental domain for G_∞ is a fundamental domain for G. The details for Fuchsian groups are given in [Leh66] (page 57-58) and [Bea83] (page 239), and may be extended to Kleinian groups in the obvious way.

Following Jørgensen (Section 2 of [Jør03]) we make the following normalisation for punctured torus groups. First, if $\mathrm{tr}(gh^{-1}g^{-1}h) = +2$ then $\langle g, h \rangle$ is elementary (see [Jør03] or Lemma 3.1 of [PS01]). Therefore, if $gh^{-1}g^{-1}h$ is parabolic then its trace must be -2. Jørgensen's normalises so that $gh^{-1}g^{-1}h$ fixes infinity and translates by 2. That is

$$k = gh^{-1}g^{-1}h = \begin{pmatrix} -1 & -2 \\ 0 & -1 \end{pmatrix}. \tag{3.1}$$

Two consequences of this normalisation (see equation (1) of [Jør03] and the sentence which follows this equation) are that the isometric sphere of a generator g has radius $1/|\mathrm{tr}(g)|$ and the poles of g and g^{-1} are related by

$$g(\infty) = 1 + g^{-1}(\infty).$$

(We choose to write g for Jørgensen's A and h for his B^{-1}. Thus $h(\infty) = 1 + h^{-1}(\infty)$ as well.)

The Ford domain is a locally finite polyhedron whose side pairing maps are as follows: Suppose that F_g is a face of D contained in the isometric sphere of g, then there is a face $F_{g^{-1}}$ contained in the isometric sphere of g^{-1}. Then g is a side pairing from F_g to $F_{g^{-1}}$. From these side pairings we can obtain edge and vertex cycles in the usual way: An edge e_0 of the Ford domain lies in the intersection of the isometric spheres of two elements of the group, g_0^{-1} and g_1. The image of e_0 under g_1, which we denote by e_1 lies in the intersection of the isometric spheres of g_1^{-1} and g_2. Repeating this process and using local finiteness we eventually find edges $e_2 = g_2(e_1), \ldots, e_{n-1} = g_{n-1}(e_{n-2}), e_0 = g_n(e_{n-1})$. The edges $e_0, \ldots, e_{n-1}$ comprise an *edge cycle*. The Ford domain and its images under $g_1, \ldots, g_n \circ \cdots \circ g_1$ glue together to cover a neighbourhood of e_0. Likewise, given a vertex v_0 we can find other vertices $v_1 = h_1(v_0), \ldots, v_m = h_m(v_0)$, which constitute a *vertex cycle*, so that the Ford domain and its images under $h_1, \ldots, h_m$ glue together to form a neighbourhood of v_0. In the Ford domains we will be considering here, all edge cycles have length 3 and all vertex cycles have length 4.

As Jørgensen observed (see the end of [Jør03] Section 2), the Ford domain is also invariant under $z \longmapsto z + 1$, the square root of the commutator, which sends the isometric sphere of a generator to the isometric sphere of its inverse. This map sends the face F_g to the face $F_{g^{-1}}$ for any generator g.

Theorem 3.1. *The Ford domain D of a group H fibring over the circle with fibre the once punctured torus, normalised as above, has the following structure:*

(i) *every face of D is contained in the isometric sphere of a generator;*

(ii) *every edge of D is contained in the intersection of the isometric spheres of a pair of neighbours;*

(iii) *every edge of D lies in a cycle of length 3 and this cycle corresponds to the pairwise intersection of the isometric spheres of a generator triple;*

(iv) *every vertex of D is the common intersection of the isometric spheres of a generator triple;*

(v) *if the isometric spheres of a generator triple intersect pairwise, then their common intersection is non-empty.*

Moreover, the faces and edges which arise are determined by the intersection of a fundamental arc of α with the Farey tessellation $\mathcal{F}$.

Proof. (Sketch) Let $H = \langle g, h, f \rangle$ be the fundamental group of a punctured torus bundle corresponding to the pseudo-Anosov map ϕ. With the normalisation given above, the Ford domain for H is the same as the Ford domain for its infinite cyclic cover

$G = \langle g, h \rangle$. In order to find a fundamental domain for these two groups we must intersect the Ford domain with a fundamental domain for the stabiliser of ∞. The stabilisers are a rank two parabolic group $H_\infty = \langle gh^{-1}g^{-1}h, f \rangle$ and a cyclic parabolic group $G_\infty = \langle gh^{-1}g^{-1}h \rangle$ respectively. Because of the symmetry of the Ford domain by f, the combinatorics of the faces must be periodic.

Our strategy for the proof will be to consider G^∞, the group on the boundary of quasi-Fuchsian space that is the limit of the quasi-Fuchsian groups G_n described in Section 2.2. Because of geometric convergence, the Ford domains for the quasi-Fuchsian groups G_n converge to the Ford domain of G^∞. We can then use Jørgensen's theorem to find the structure of the Ford domains for these quasi-Fuchsian groups and show that this structure persists in the limit (this result has been announced by Akiyoshi [Aki99]). In particular, this will imply that the Ford domain has a periodic structure and so is invariant under f. Since the limit sets of both G and G^∞ are the whole of S^2, Sullivan rigidity implies they are the same group (we could have also used Minsky's theorem [Min99] at this point instead).

We now complete our sketch of this result by outlining how the structure of the Ford domains change under the geometric convergence. Jørgensen has shown that the conditions of our Theorem hold for quasi-Fuchsian punctured torus groups. Conditions (i) and (iv) are characteristics 1 and 2 of Theorem 1 of [Jør03]; conditions (ii) and (iii) are essentially contained in the discussion of Figure 5 of [Jør03]. Finally (v) is the first assertion verified in the proof of Theorem 1 of [Jør03]. The last assertion is part of Theorem 3 of [Jør03]. He also shows that the intersection of the Ford domain with the ideal boundary is made up of exactly two simply connected components part 4 of [Jør03] Theorem 1. As we deform the quasi-Fuchsian group both of these components shrink to the empty set.

As Jørgensen observed, as we deform quasi-Fuchsian groups in the interior of quasi-Fuchsian space the only change to the combinatorial structure of the Ford domain D that can occur is that an ideal edge of D shrinks to a point and then is lifted to a vertex of D in the interior of $\mathbb{H}^3$. As this happens, a new face of D appears together with a new ideal edge and two new edges in $\mathbb{H}^3$ (see [ASWY00], especially Figure 9.1). The new faces which arise in this way are determined by the combinatorics of the intersection of the path in Teichmüller space joining the hyperbolic structures of the endpoints with Farey tessellation (Theorem 3 of [Jør03]). This path is an arc of the semicircle α defined in Section 2.2. As this process of adding more faces occurs (so the arc of α becomes longer and longer) the combinatorial structure of a fixed part of the Ford domain away from the ideal boundary remains the same. In this way any compact subset of the Ford domain of G^∞ is realised combinatorially as a subset of the Ford domain of some G_n. Hence the Ford domain of the limiting group G^∞ also satisfies conditions (i) to (v). Moreover, the combinatorics of its faces is given by the intersection of Farey tessellation with the semicircle joining $v_\pm$, the fixed points of ϕ.

The Ford domain of G also has these properties since, as indicated above, G and G^∞ are the same group. $\qquad\qquad\qquad\qquad\qquad\qquad\qquad\qquad\qquad\qquad\qquad\qquad\quad$ □

4. Cutting surfaces and tetrahedra

We now outline our method for producing cutting surfaces and decomposing punctured torus bundles into ideal tetrahedra. This follows the construction given in [PS01].

By Theorem 3.1, every edge cycle is contained in the pairwise intersection of the isometric spheres of a generator triple. Consider such an edge cycle and the associated generator triple (g, h, gh). In fact, there are two edge cycles, each of length three, corresponding to this generator triple. We will show that transverse to these six edges there is a canonical cutting surface whose quotient under H is an embedded punctured torus in M. This punctured torus is made up of two ideal triangles glued together with constant bending angle along their edges and is a fibre of M. Moreover, this triangulation is the same as the triangulation associated to (g, h, gh) described in Section 2.3. In this way we can construct a sequence of pleated punctured tori which leave any compact subset of $\hat{M} = \mathbb{H}^3/G$, see [PS01].

Choose a pair of neighbours g and h so that the intersection of the isometric spheres of g^{-1} and h^{-1} contains an edge e of the Ford domain D. Let $T(g, h)$ denote the ideal triangle in $\mathbb{H}^3$ with vertices at ∞, $g(\infty)$ and $h(\infty)$. The latter two points are the centres of the isometric spheres of g^{-1} and h^{-1}, respectively. The edge e is orthogonal to the plane containing $T(g, h)$, although the geodesic containing e may intersect this plane outside the triangle $T(g, h)$. By Theorem 3.1 the edge e lies in a cycle of length 3. The other members of this cycle are first $g^{-1}(e)$, contained in the intersection of the isometric spheres of $(g, h^{-1}g)$, and secondly $h^{-1}(e)$, contained in the intersection of the isometric spheres of $(h, g^{-1}h)$. These two edges are orthogonal to the planes containing the ideal triangles $g^{-1}T(g, h) = T(g^{-1}h, g^{-1})$, with vertices at ∞, $h^{-1}g(\infty)$ and $g(\infty)$, and $h^{-1}T(g, h) = T(h^{-1}, h^{-1}g)$, with vertices at ∞, $h(\infty)$ and $g^{-1}h(\infty)$, respectively. The triangle $T(g, h)$ intersects the Ford domain in a quadrilateral (denoted by E in Figure 1). Likewise, the intersection of $T(g^{-1}h, g^{-1})$ and $T(h^{-1}, h^{-1}g)$ with the Ford domain of G are quadrilaterals (denoted by C and A respectively). The pull backs of these three quadrilaterals then tile $T(g, h)$. In Figure 1 these quadrilaterals are E, $h(A)$ and $g(C)$.

Similarly, consider the ideal triangle $T(g^{-1}, h^{-1})$ whose vertices are ∞, $g^{-1}(\infty)$ and $h^{-1}(\infty)$, and its images $gT(g^{-1}, h^{-1}) = T(gh^{-1}, g)$ and $hT(g^{-1}, h^{-1}) = T(h, hg^{-1})$. These three triangles intersect the Ford domain in quadrilaterals B, F and D respec-

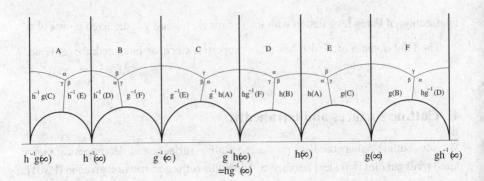

Figure 1: A cutting surface corresponding to the generator triple (g, h, gh^{-1}). In order to embed this into the manifold, bend along the bold lines. Compare this with Figure 1 of [Jør03].

tively. Figure 1 illustrates the tessellation of $T(g^{-1}, h^{-1})$ by the quadrilaterals B, $h^{-1}(D)$ and $g^{-1}(F)$. The six quadrilaterals A, B, C, D, E and F may be glued together to form a polygon inside the Ford domain of G, as indicated in Figure 1. (Compare this surface with Figure 1 of [Jør03] where X is our g and Y is our h^{-1}.) Using the side pairings for the Ford domain, this polygon becomes a punctured torus embedded in $M = \mathbb{H}^3/H$. By construction, this embedded punctured torus consists of two flat triangles (corresponding to $T(g, h)$ and $T(g^{-1}, h^{-1})$) glued together at constant angle along three disjoint geodesic arcs which begin and end at the puncture. We call such a surface a *cutting surface*. The way in which a cutting surface is embedded in the Ford domain makes it clear that the cutting surface is one of the fibres of M.

If g corresponds to $[\gamma] \in \pi_1(\Sigma)$ as in Section 2.3, then the arc joining ∞ to $g(\infty)$ is a lift of an arc in the homotopy class $[\gamma^*]$. Similarly for h and gh^{-1}. This shows that the triangulation of Σ arising in the cutting surface and the triangulation constructed in Section 2.3 are the same.

We remark that Figure 1 illustrates the generic way a cutting surface intersects the Ford domain, where the intersection of the cutting surface and the isometric spheres is contained in the triangles $T(g, h)$ and $T(g^{-1}, h^{-1})$. If the radius of the isometric sphere of g^{-1} is very much larger than that of h^{-1} it may happen that the intersection of the isometric spheres of g^{-1} and h^{-1} does not lie in $T(g, h)$. Even with more general configurations, such as this one, our construction will still work; the crucial point is that $T(g, h)$ and $T(g^{-1}, h^{-1})$ may be glued to form an embedded simplicial surface.

Given an edge e of the Ford domain for H (or G) we have constructed a cutting surface. We now repeat this construction for all edges of the Ford domain. Since there are only finitely many edges up to equivalence by H_∞, the stabiliser of ∞, we obtain finitely many cutting surfaces.

If we begin with the cutting surface we have already constructed and move along the boundary of the Ford domain in the direction of f, we next encounter a vertex cycle of the Ford domain. This vertex cycle has length four and, by Theorem 3.1, each vertex is contained in the common intersection of the isometric spheres of a generator triple. This triple is either the triple we started with, or else a triple corresponding to an adjacent triangle in the Farey tessellation (both possibilities occur, each for two of the vertices in the cycle). In the latter case, the new generator triple may be obtained from the initial triple by applying a Nielsen move. The other six edges incident to these vertices form two edge cycles of length three which are associated to this new generator triple (see Section 5 or page 196 of [PS01]). We repeat the process described above to obtain a cutting surface associated to this new generator triple. We shall see that the region between these two surfaces is an ideal tetrahedron.

Moving from the generator triple (g,h,gh) to the triple (g,h,gh^{-1}) may be interpreted as crossing the edge in the Farey tessellation joining g and h. Following [Min99], such an edge is called a *spanning edge*. If a generator is the endpoint of at least two spanning edges, then, again following [Min99], it is called a *pivot*. It is easy to see that a generator is a pivot if and only if the corresponding face in the Ford domain has at least 6 edges.

We now turn our attention from the cutting surfaces and investigate the regions between adjacent cutting surfaces. As indicated by the following result, these regions are ideal tetrahedra and the boundary of each of these tetrahedra consists of two pairs of ideal triangles arising from the cutting surfaces.

Proposition 4.1 (Proposition 3.12 of [PS01]). *Assume that both the generator triples* (g,h,gh) *and* (g,h,gh^{-1}) *give rise to edge cycles of the Ford domain (and hence cutting surfaces). Consider the ideal tetrahedron with vertices* ∞, $g(\infty)$, $h(\infty)$ *and* $gh(\infty)$. *Then the projection of this tetrahedron to* M *has as its boundary the cutting surfaces for the generator triples* (g,h,gh) *and* (g,h,gh^{-1}).

In the worked example in the next section we illustrate this process in action. For example, in Figure 4 consider the region between the dashed line, the cutting surface corresponding to (g,h,gh^{-1}) illustrated in Figure 1, and the solid line, the cutting surface corresponding to $(h,gh^{-1},hg^{-1}h)$. This consists of four triangles. The edges of these triangles are the vertical projections of the ideal hyperbolic triangles $T(g,h)$ etc. The regions bounded by these triangles and the Ford domain are polyhedra with three infinite faces (corresponding to quadrilaterals in the cutting surfaces) and three finite faces (corresponding to pieces of the isometric spheres associated to the vertices of the triangle). When we use the side identifications to glue these polyhedra together we obtain the ideal tetrahedron of Proposition 4.1 (in just the same way that the triangle $T(g,h)$ is formed by identifying sides of the three quadrilaterals A, C and E). For definiteness, consider the tetrahedron $\mathscr{T}$ with vertices ∞, $g(\infty)$, $h(\infty)$ and $gh^{-1}(\infty)$. The vertical projection of this tetrahedron is a triangle in Figure 4 with two dashed sides and one solid side. The dashed sides are the vertical projections of $T(g,h)$ and $T(g^{-1},h^{-1})$. As described above, after performing side pairings, the cutting surface associated to (g,h,gh^{-1}) may be identified with these two triangles. The solid side of $\mathscr{T}$ is the triangle $T(h,gh^{-1})$ and the base is $gT(h^{-1},g^{-1}h)$. The triangles $T(h,gh^{-1})$ and $T(h^{-1},g^{-1}h)$ correspond to the cutting surface of $(h,gh^{-1},hg^{-1}h)$. Thus the boundary of $\mathscr{T}$ comprises two adjacent cutting surfaces.

Observe that the four triangles between the vertical projections of a pair of adjacent cutting surfaces come in two pairs of congruent triangles. This corresponds to the symmetry of the Ford domain under the square root of the commutator and to the well known fact that opposite edges of an ideal hyperbolic tetrahedron have the same dihedral angle.

Proceeding in this manner for all the faces in the Ford domain we obtain a decomposition of M into ideal tetrahedra all of whose vertices lie at the cusp. We claim that this is combinatorially the same as the abstract triangulation described given on page 328 of [Bow97]. It is clear from our construction that these two triangulations are the same: to each generator triple there correspond a pair of triangles, moving to an adjacent generator triple produces a second pair of triangles and together these four triangles form the faces of one of the tetrahedra. Furthermore, the tetrahedra are indexed by the Nielsen moves in the decomposition of our pseudo-Anosov diffeomorphism ϕ. It only remains to check that our decomposition into ideal tetrahedra is consistent. In other words, we must show that the vertical projection of the tetrahedral decomposition of the Ford domain D gives a triangulation of the plane dual to the tessellation given by the faces of D (compare [ASWY00]). This would be trivial if each vertex of the triangulation (that is the centre of the associated isometric sphere) lay vertically below the interior of the corresponding face of D. This is not always the case. Since the faces of D lie in hemispheres whose centres all lie on the Riemann sphere, when travelling orthogonally to an edge of D, the order we meet the centres of the isometric spheres is the same as the order we meet the faces of the Ford domain. Using this fact it is then easy to see that the two triangulations of M are consistent.

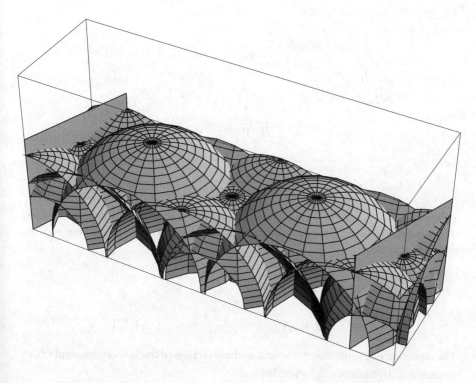

Figure 2: A fundamental domain for our example. This is obtained by intersecting the Ford domain with a fundamental parallelogram for the stabiliser of ∞. This figure was drawn by Tino Koch; see [Koc99].

5. An example

We now work through the details of a concrete example. This is the group for which the pseudo-Anosov diffeomorphism

$$\phi = \begin{pmatrix} 1 & 1 \\ 2 & 3 \end{pmatrix} = \begin{pmatrix} 1 & 0 \\ 2 & 1 \end{pmatrix} \begin{pmatrix} 1 & 1 \\ 0 & 1 \end{pmatrix} = L^2 U^1.$$

This may be interpreted as the Nielsen move $(g, h) \longmapsto (gh^{-1}, h)$ followed by the Nielsen move $(g, h) \longmapsto (g, g^{-1}h)$ twice (see page 66 of [McM96] or [AH97]). This has the effect of

$$(g, h) \longmapsto (gh^{-1}, h) \longmapsto (gh^{-1}, hg^{-1}h) \longmapsto (gh^{-1}, hg^{-1}hg^{-1}h).$$

Thus we have $\phi_*(g) = gh^{-1}$ and $\phi_*(h) = hg^{-1}hg^{-1}h$.

We also remark that the fixed points of ϕ are $(-1 \pm \sqrt{3})/2$. The continued fraction

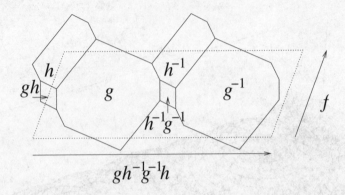

Figure 3: Vertical projection of a fundamental domain for the our example. Compare Abbildung 3 of [Koc99]. The labels g, h etc on the faces give the element of the group in whose isometric sphere the face lies.

expansions of these points are

$$\frac{\sqrt{3}-1}{2} = [0,2,1,2,1,\ldots], \quad -\frac{\sqrt{3}+1}{2} = -[1,2,1,2,1,\ldots].$$

The repeating ones and twos in the tails of these continued fractions correspond to this sequence of Dehn twists, see [Ser85b].

In Figures 2 and 3 we give the isometric and vertical projections of a fundamental domain for H. This consists of the intersection of the Ford domain with a parallel-ogram whose opposite sides are identified by f and $gh^{-1}g^{-1}h$. In Figure 3 we have drawn the vertical projection of the faces of the Ford domain contained in the isometric spheres of g, h, gh and their inverses.

In Figure 4 we have drawn the vertical projection of several faces of the Ford domain together with four cutting surfaces. The left and right ends of each cutting surface are identified by $gh^{-1}h^{-1}g$. The two dotted cutting surfaces are identified by f. Observe that the edges of the Ford domain are orthogonal to the cutting surfaces and their vertical projections form dual tessellations of the plane. This shows how M may be triangulated by three ideal tetrahedra.

We now describe in greater detail the transition from one cutting surface to the next. The first cutting surface we consider (shown as dotted in the figure) is associated to the generator triple (g,h,gh) and is made up of the triangles joining ∞ to (in order) $h^{-1}gh(\infty)$, $h^{-1}(\infty)$, $h^{-1}g^{-1}(\infty)$, $g^{-1}(\infty) = h^{-1}g^{-1}h(\infty)$, $h(\infty)$, $gh(\infty)$ and $g(\infty)$. Moving in the direction of f the edges between the isometric spheres of gh and h, gh and g meet. Similarly the edges between their the isometric spheres of their inverses also meet. The edges between the isometric spheres of h and $h^{-1}g^{-1}h$ and between

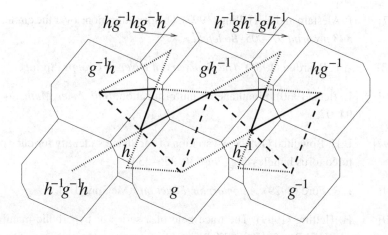

Figure 4: Vertical projection of the Ford domain and cutting surfaces for the our example.

their inverses each bifurcate. The six new edges are associated with the generator triple (g, h^{-1}, gh^{-1}). The associated cutting surface, illustrated by a dashed line, is made up of triangles joining ∞ to (in order) $h^{-1}g(\infty)$, $h^{-1}(\infty)$, $g^{-1}(\infty)$, $hg^{-1}(\infty) = g^{-1}h(\infty)$, $h(\infty)$, $g(\infty)$ and $gh^{-1}(\infty)$. It is this cutting surface which is illustrated in Figure 1. The two remaining cutting surfaces are associated to the generator triples $(h, gh^{-1}, hg^{-1}h)$, shown with a solid line, and $(gh^{-1}, hg^{-1}hg^{-1}h, hg^{-1}h) = (\phi_*(g), \phi_*(h), \phi_*(gh))$, shown with a dotted line again.

The four triangles between each pair of adjacent cutting surfaces may be glued together by the side pairings of the Ford domain to form a tetrahedron. Thus we may obtain a fundamental polyhedron made up of the three tetrahedra with the following vertices: first ∞, $g(\infty)$, $h(\infty)$, $gh(\infty)$; secondly ∞, $g(\infty)$, $h(\infty)$, $gh^{-1}(\infty)$ and finally ∞, $gh^{-1}(\infty)$, $h(\infty)$, $hg^{-1}h(\infty)$. This gives a tetrahedral decomposition of the associated 3-manifold.

References

[Aki99]   H. Akiyoshi (1999). On the Ford domain of once-punctured torus groups. In *Hyperbolic Spaces and Related Topics*, edited by *S. Kamiya, R.I.M.S. Kokyuroku* **1104**, Kyoto University, 109–121.

[ASWY00] H. Akiyoshi, M. Sakuma, M. Wada and Y. Yamashita (2000). Ford domains of punctured torus groups and two-bridge knot groups. In *Knot Theory, Proceedings of the workshop held in Toronto*, 14–71.

[AH97]   P. Alestalo and H. Helling (1997). On torus fibrations over the circle, *SFB 343 preprint 97–005, Bielefeld*.

[Bea83]   A.F. Beardon (1983). *The Geometry of Discrete Groups*, Springer.

[Ber60]   L. Bers (1960). Simultaneous uniformization, *Bull. Amer. Math. Soc.* **66**, 94–97.

[Bow97]   B.H. Bowditch (1997). A variation of McShane's identity for once punctured torus bundles, *Topology* **36**, 325–334.

[For29]   L.R. Ford (1929). *Automorphic Functions*, McGraw-Hill.

[Hel99]   H. Helling (1999). The trace field of a series of hyperbolic manifolds, *SFB 343 Preprint 99–072, Bielefeld*.

[JM79]   T. Jørgensen and A. Marden (1979). Two doubly degenerate groups, *Quart. J. Maths* **30**, 143–156.

[Jør03]   T. Jørgensen (2003). On pairs of once-punctured tori, *this volume*.

[Koc99]   T. Koch (1999). Hyperbolische einfach-punktierte Torus-Bündel, *Ph.D. Thesis, Bielefeld*.

[Leh66]   J. Lehner (1966). *A Short Course on Automorphic Functions*, Holt, Rinehart and Winston.

[McM96]   C.T. McMullen (1996). Renormalization and 3-Manifolds Which Fiber Over the Circle, *Ann. Math. Studies* **142**, Princeton.

[Min99]   Y.N. Minsky (1999). The classification of punctured-torus groups, *Ann. Math.* **149**, 559–626.

[Nie18]   J. Nielsen (1918). Die Isomorphismen der allgemeinen unendlichen Gruppe mit zwei Erzeugenden, *Math. Ann.* **78**, 385–397.

[Ota01]   J.-P. Otal (2001). *The Hyperbolization Theorem for Fibred 3-Manifolds*, SMF/AMS Texts and Monographs **7**.

[PS01]   J.R. Parker and B.O. Stratmann (2001). Kleinian groups with singly cusped parabolic fixed points, *Kodai Mathematical J.* **24**, 169–206.

[Ser85a]   C. Series (1985). The geometry of Markoff numbers, *Math. Intelligencer* **7**, 20–29.

[Ser85b]   C. Series (1985). The modular surface and continued fractions, *J. Lond. Math. Soc.* **31**, 69–80.

[Thu86b]   W.P. Thurston (1986). Hyperbolic structures on 3-manifolds II: Surface groups and 3-manifolds which fiber over the circle, *arXiv:math.GT/9801045.*

[Wad]   M. Wada. OPTi.
           http://vivaldi.ics.nara-wu.ac.jp/ wada/OPTi/

J.R. Parker

Department of Mathematical Sciences
University of Durham
Durham, DH1 3LE
England

J.R.Parker@durham.ac.uk

AMS Classification: 30F40, 51M10

Keywords: hyperbolic structure, punctured torus bundle

Kleinian Groups and Hyperbolic 3-Manifolds
Lond. Math. Soc. Lec. Notes **299**, 293–304

Y. Komori, V. Markovic & C. Series (Eds.)
Cambridge Univ. Press, 2003

On the boundary of the Earle slice for punctured torus groups

Yohei Komori

Abstract

We shall show that the Earle slice $\mathcal{E}$ for punctured torus groups is a Jordan domain, and every pleating ray in $\mathcal{E}$ lands at a unique boundary group whose pair of end invariants is equal to the pleating invariants of this ray. We will also study the asymptotic behavior of the boundary of $\mathcal{E}$.

1. Introduction

In [Min99], Minsky showed that any marked punctured torus group can be characterized by its pair of end invariants, where a punctured torus group is a rank two free Kleinian group whose commutator of generators is parabolic. To prove this result, called the *Ending Lamination Theorem*, he also proved another important result, called the *Pivot Theorem*, which controls thin parts of the corresponding hyperbolic manifold from the data of end invariants. As one of applications of these theorems, he showed that the Bers slice and the Maskit slice are Jordan domains.

In this paper we apply his results to the *Earle slice* which is a holomorphic slice of quasi-fuchsian space representing the Teichmüller space of once-punctured tori. This slice was considered originally by Earle in [Ear81], and its geometric coordinates, named *pleating coordinates* was studied by Series and the author in [KoS01]. By using rational pleating rays, the figure of the Earle slice $\mathcal{E}$ realized in the complex plane $\mathbb{C}$ was drawn by Liepa. (See Figure 1. In fact only the upper half of the Earle slice is shown, the picture being symmetrical under reflection in the real axis.) In this paper we will show that:

(i) $\mathcal{E}$ is a Jordan domain.

(ii) There is a right half plane which is contained in $\mathcal{E}$.

(iii) Every pleating ray in $\mathcal{E}$ lands at a unique boundary point of $\mathcal{E}$.

This paper is organized as follows. Section 2 is dedicated to the background material, especially the space of punctured torus groups. We introduce the definitions and basic facts about the Earle slice in Section 3. In Sections 4 and 5 we duplicate the definitions and arguments from [Min99] which are necessary to understand Minsky's Ending Lamination Theorem and Pivot Theorem. The previous three claims are proven in Sections 6, 7 and 8 respectively.

The author would like to express his special gratitude to Minsky's fundamental work [Min99] on punctured torus groups without which the present paper would not have been possible. He is also grateful to Hideki Miyachi and Caroline Series for fruitful discussions, and to the referee for helpful comments that greatly improved the exposition.

2. Punctured torus groups

Let S be an oriented once-punctured torus and $\pi_1(S)$ be its fundamental group. An ordered pair α, β of generators of $\pi_1(S)$ is called *canonical* if the oriented intersection number $i(\alpha, \beta)$ in S with respect to the given orientation of S is equal to $+1$. The commutator $[\alpha, \beta] = \alpha\beta\alpha^{-1}\beta^{-1}$ represents a loop around the puncture.

Define $\mathscr{R}(\pi_1(S))$ to be the set of $PSL_2(\mathbb{C})$-conjugacy classes of representations from $\pi_1(S)$ to $PSL_2(\mathbb{C})$ which take the commutator of generators to a parabolic element. Let $\mathscr{D}(\pi_1(S))$ denote the subset of $\mathscr{R}(\pi_1(S))$ consisting of conjugacy classes of discrete and faithful representations. Any representative of an element of $\mathscr{D}(\pi_1(S))$ is called a *marked punctured torus group*. Let $\mathscr{QF}$ denote the subset of $\mathscr{D}(\pi_1(S))$ consisting of conjugacy classes of representations ρ such that for the action of $\Gamma = \rho(\pi_1(S))$ on the Riemann sphere $\hat{\mathbb{C}}$ the region of discontinuity Ω has exactly two simply connected invariant components $\Omega_\pm$. The quotients $\Omega_\pm/\Gamma$ are both homeomorphic to S and inherit an orientation induced from the orientation of $\hat{\mathbb{C}}$. We choose the labeling so that Ω_+ is the component such that the homotopy basis of Ω_+/Γ induced by the ordered pair of marked generators $\rho(\alpha), \rho(\beta)$ of Γ is canonical. Any representative of an element of $\mathscr{QF}$ is called a *marked quasifuchsian punctured torus group*. Considering the algebraic topology, $\mathscr{D}(\pi_1(S))$ is closed in $\mathscr{R}(\pi_1(S))$ and $\mathscr{QF}$ is open in $\mathscr{D}(\pi_1(S))$ (see [MT98]). A quasifuchsian group Γ is called *Fuchsian* if $\Omega_\pm$ are round discs.

Recall that the set of measured geodesic laminations on a hyperbolic surface is independent of the hyperbolic structure. Denote by $PML(S)$ the set of projective measured laminations on S. Let $\mathscr{C}(S)$ denote the set of free homotopy classes of unoriented simple non-peripheral curves on S. After choosing a canonical basis (α, β) for $\pi_1(S)$, any element of $H_1(S,\mathbb{Z})$ can be written as $(p,q) = p[\alpha] + q[\beta]$ in the basis $([\alpha],[\beta])$ for $H_1(S,\mathbb{Z})$, and we associate to this element the slope $-q/p \in \hat{\mathbb{Q}} := \mathbb{Q} \cup \{\infty\}$, which describes an element of $\mathscr{C}(S)$. Considering projective classes of weighted counting measures, we can embed $\mathscr{C}(S)$ into $PML(S)$ whose image is called the set of projective *rational* laminations. Recall that $PML(S)$ can be identified with $\hat{\mathbb{R}}$, in such a way that projective rational laminations correspond to $\hat{\mathbb{Q}}$.

3. The Earle slice of punctured torus groups

The following theorem is an adaptation of a result of Earle in [Ear81].

Theorem 3.1. (See Theorem 1 in [Ear81] and Theorem 2.1 in [KoS01])
Let (α, β) be a canonical basis of $\pi_1(\mathscr{T}_1)$ where $\mathscr{T}_1$ is an analytically finite Riemann surface homeomorphic to S. Let θ be an involution of $\pi_1(\mathscr{T}_1)$ defined by $\theta(\alpha) = \beta$ and $\theta(\beta) = \alpha$. Then, up to conjugation in $\mathrm{PSL}_2(\mathbb{C})$, there exists a unique marked quasifuchsian group $\rho : \pi_1(\mathscr{T}_1) \to \Gamma$, such that:

1. *There is a conformal map $\mathscr{T}_1 \to \Omega_+/\Gamma$ inducing the representation ρ.*

2. *There is a Möbius transformation $\Theta \in \mathrm{PSL}_2(\mathbb{C})$ of order two which induces a conformal homeomorphism $\Omega_+ \to \Omega_-$ such that $\Theta(\gamma z) = \theta(\gamma)\Theta(z)$ for all $\gamma \in \Gamma$ and $z \in \Omega_+$.*

Theorem 3.1 implies that for any marked Riemann surface $(\mathscr{T}_1; \alpha, \beta)$ which is analytically finite and homeomorphic to S, there is a marked quasifuchsian group $\Gamma = \langle A, B \rangle$ such that $(\mathscr{T}_1, \alpha, \beta)$ is conformal to $(\Omega_+/\Gamma; A, B)$ and $(\Omega_-/\Gamma; B, A)$. This theorem shows that the Earle slice is a holomorphic embedding of the Teichmüller space $\mathrm{Teich}(S)$ of once-punctured tori into $\mathscr{QF}$. The embedding depends only on the choice of the involution θ of $\pi_1(\mathscr{T}_1)$; in fact we can take any involution of $\pi_1(\mathscr{T}_1)$ which is induced from an orientation reversing diffeomorphism of $\mathscr{T}_1$ (see [Ear81]). We call the image of $\mathrm{Teich}(S)$ in $\mathscr{QF}$, the *Earle slice* of $\mathscr{QF}$. This slice can be thought of as a holomorphic extension of the rhombus line in the *Fuchsian locus* $\mathscr{F}$ into $\mathscr{QF}$ (see [KoS01]), hence it may be called the *rhombic* Earle slice. Next we show how to realize the Earle slice in $\mathbb{C}$.

Theorem 3.2. (See Theorem 3.1 in [KoS01])
Let $\rho : \pi_1(\mathscr{T}_1) \to \mathrm{PSL}_2(\mathbb{C})$ be a marked quasifuchsian punctured torus group in the Earle slice. Then, after conjugation by $\mathrm{PSL}_2(\mathbb{C})$ if necessary, we can take representatives of $A = \rho(\alpha), B = \rho(\beta)$ in $SL(2,\mathbb{C})$ of the form $A_d, B_d \in SL(2,\mathbb{C}), \quad d \in \mathbb{C}^+ := \{d \in \mathbb{C} | Re\, d > 0\}$, where

$$A_d = \begin{pmatrix} \frac{d^2+1}{d} & \frac{d^3}{2d^2+1} \\ \frac{2d^2+1}{d} & d \end{pmatrix}, B_d = \begin{pmatrix} \frac{d^2+1}{d} & -\frac{d^3}{2d^2+1} \\ -\frac{2d^2+1}{d} & d \end{pmatrix}.$$

The parameter $d \in \mathbb{C}^+$ is uniquely determined by the conjugacy class of ρ.

The map $\varphi : \mathbb{C}^+ \to \mathscr{R}(\pi_1(S))$ defined by $d \mapsto (A_d, B_d)$ is a holomorphic injection and we can realize the Earle slice in $\mathbb{C}^+$. We denote the corresponding region in $\mathbb{C}^+$ by $\mathscr{E}$. Then $\mathscr{E}$ has the following symmetries.

Proposition 3.3. (See Proposition 3.3, 3.4 and 3.6 in [KoS01])
The positive real line $\mathbb{R}^+$ corresponds to the Fuchsian locus of $\mathscr{E}$, the rhombus line. Moreover there exist two involutions of $\mathscr{E}$; a holomorphic involution $\sigma(d) = 1/(2d)$ and an anti-holomorphic involution $\iota(d) = \bar{d}$ where $\bar{d}$ denotes the complex conjugate of d.

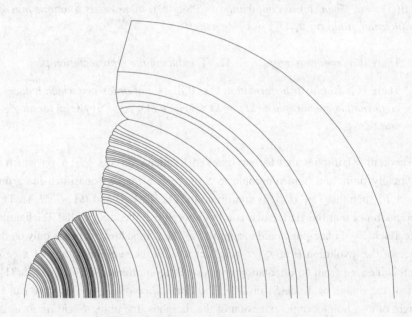

Figure 1: The upper half of the Earle slice.

Finally we review the notion of pleating rays (see [KS, KoS01]). For a quasifuchsian punctured torus group Γ, let $\mathscr{C}/\Gamma$ be the convex core of $\mathbb{H}^3/\Gamma$; equivalently $\mathscr{C}$ is the hyperbolic convex hull of the limit set Λ of Γ in $\mathbb{H}^3$. The boundary $\partial\mathscr{C}/\Gamma$ of $\mathscr{C}/\Gamma$ has two connected components $\partial\mathscr{C}^\pm/\Gamma$, each homeomorphic to S. These components are each pleated surfaces whose pleating loci carry the bending measure whose projective classes we denote $pl^\pm(\Gamma)$.

For $x, y \in PML(S) = \hat{\mathbb{R}}$, The (x,y)-*pleating ray* in $\mathscr{E}$ is the set defined by $\mathscr{P}(x,y) := \{d \in \mathscr{E} \mid pl^+(d) = x, pl^-(d) = y\}$. Since the boundary components $\partial \mathscr{C}^\pm$ are conjugate under the involution Θ for groups in $\mathscr{E}$, we have

Proposition 3.4. (See Proposition 6.6 and Corollary 6.14 in [KoS01])
$\mathscr{P}(x, 1/x) \neq \emptyset$ *provided* $x \neq \pm 1$, *and* $\mathscr{P}(x,y) = \emptyset$ *otherwise.*

In particular,

Theorem 3.5. (See Theorem 5.11 in [KoS01])
The set of rational pleating rays $\mathscr{P}(x, 1/x)$ $(x \in \hat{\mathbb{Q}} \setminus \{\pm 1\})$ *are dense in* $\mathscr{E}$.

This allows us to draw the picture shown in Figure 1. The positive real axis represents Fuchsian groups with rhombic symmetry, and only the upper half of the Earle slice is shown, the picture being symmetrical under reflection in the real axis.

4. Minsky's Ending Lamination Theorem

We associate to a marked punctured torus group an ordered pair of *end invariants* (ν_-, ν_+), each lying in $\overline{\mathbb{H}}^2 := \mathbb{H}^2 \cup \hat{\mathbb{R}}$. Let $\rho : \pi_1(S) \to \mathrm{PSL}_2(\mathbb{C})$ denote a marked punctured torus group and $N := \mathbb{H}^3 / \rho(\pi_1(S))$ be its associated manifold. Then by Bonahon's theorem of geometric tameness (see [MT98]), N is homeomorphic to $S \times \mathbb{R}$. Let us name the ends e_+ and e_-. We choose the labeling as follows (see [Min99]). Let the orientation $S \times \{1\}$ agree with the orientation of S. Orient $S \times (-1,1)$ by the orientation of $S \times \{1\}$ and its inward-pointing vector. The orientation of $\mathbb{H}^3$ induces the orientation of N. Then up to homotopy there exists uniquely an orientation preserving homeomorphism between N and $S \times (-1,1)$ which induces the representation ρ. Let e_+ be the end of N whose neighborhoods are neighborhoods of $S \times \{1\}$ under this identification. Let Ω denote the (possibly empty) domain of discontinuity of $\Gamma = \rho(\pi_1(S))$ and $\overline{N}$ denote the quotient $(\mathbb{H}^3 \cup \Omega)/\Gamma$. Any component of the boundary Ω/Γ is reached by going to one of the ends e_+ or e_-, and this divides it into two disjoint pieces Ω_+/Γ and Ω_-/Γ. There are three possibilities for each of these boundaries, corresponding to three types of end invariants (here let s denote either $+$ or $-$):

(i) If Ω_s is a topological disc, then Ω_s/Γ is a marked punctured torus, and so Ω_s/Γ is represented by a point ν_s in the Teichmüller space of S; that is, a unique point $\nu_s \in \mathbb{H}^2$ such that the one-point compactifications of Ω_+/Γ and Ω_-/Γ are equivalent as marked Riemann surfaces to the marked flat tori $\mathbb{C}/(\mathbb{Z} \cdot 1 + \mathbb{Z} \cdot \nu_+)$ and $\mathbb{C}/(\mathbb{Z} \cdot \overline{\nu}_- + \mathbb{Z} \cdot 1)$ respectively.

In particular, $v_+ = v_-$ if and only if Γ is a Fuchsian group.

(ii) If Ω_s is an infinite union of round discs, then Ω_s/Γ is a thrice-punctured sphere, obtained from the corresponding boundary of $S \times (-1, 1)$ by deleting a simple closed curve γ_s. In this case $v_s \in \hat{\mathbb{Q}}$ denotes the slope of γ_s. The conjugacy class of γ_s in Γ is parabolic in this case.

(iii) If Ω_s is empty, then we can find a sequence of simple closed curves $\{\gamma_n\}$ in S whose geodesic representatives γ_n^* in N are eventually contained in any neighborhood of e_s, and the slopes of γ_n converge in $\hat{\mathbb{R}}$ to a unique irrational number. This limiting irrational slope is denoted by v_s and is called an *ending lamination*.

To a marked punctured torus group $\rho : \pi_1(S) \rightarrow \mathrm{PSL}_2(\mathbb{C})$ one may associate the ordered pair of end invariants (v_-, v_+) lying in $\overline{\mathbb{H}}^2 \times \overline{\mathbb{H}}^2 \setminus \Delta$, where Δ denotes the diagonal of $\hat{\mathbb{R}} \times \hat{\mathbb{R}}$. Minsky's Ending Lamination Theorem is

Theorem 4.1. (See Theorems A and B in [Min99])
The map
$$v : \mathscr{D}(\pi_1(S)) \rightarrow \overline{\mathbb{H}}^2 \times \overline{\mathbb{H}}^2 \setminus \Delta$$
defined by $\rho \mapsto (v_-, v_+)$ *is bijective.* v *is not continuous while its inverse* v^{-1} *is continuous.*

5. Minsky's Pivot Theorem

Next we review Minsky's Pivot Theorem which is a key idea to prove the Ending Lamination Theorem 4.1, and is also a main tool for our results in this paper.

First we define the *Farey triangulation* of the upper half plane $\mathbb{H}^2$ as follows. For any two rational numbers written in lowest terms as p/q and r/s, say they are *neighbors* if $|ps - qr| = 1$. Allow also the case $\infty = 1/0$. Joining any two neighbors by a hyperbolic geodesic, we obtain a triangulation of $\mathbb{H}^2$ invariant under the natural action of $\mathrm{PSL}_2(\mathbb{Z})$.

Next we recall the notion of *pivots* for marked punctured torus groups. Let (v_-, v_+) be the end-invariant pair of a marked punctured torus group $\rho : \pi_1(S) \rightarrow \mathrm{PSL}_2(\mathbb{C})$. Letting s denote $+$ or $-$, define a point $\alpha_s \in \hat{\mathbb{R}}$ to be closest to v_s in the following sense. If $v_s \in \hat{\mathbb{R}}$, define $\alpha_s = v_s$. If $v_s \in \mathbb{H}^2$, let $\alpha_s \in \mathscr{C}(S)$ represent a geodesic of shortest length in the hyperbolic structure corresponding to v_s. More precisely, if v_s is contained in a Farey triangle Δ, we divide up Δ into six regions by the axes of its reflection symmetries. Then each vertex $u \in \mathscr{C}(S)$ of Δ has minimal hyperbolic length in the hyperbolic structure corresponding to v_s when v_s is in the pair of regions that meet u. Now define $E = E(\alpha_-, \alpha_+)$ to be the set of edges of the Farey graph which separate α_- from α_+ in $\mathbb{H}^2$. Let P_0 denote the set of vertices of $\mathscr{C}(S)$ which belong to at least 2 edges in E. We call these vertices *internal pivots* of ρ. The edges of E admit a natural order where $e < f$ if e separates the interior of f from α_-, and this induces an ordering on P_0. The full pivot sequence P of ρ is obtained by appending to the beginning of P_0 the vertex α_- if $\alpha_- \in \mathscr{C}(S)$, and appending to the end of P_0 the vertex α_+ if $\alpha_+ \in \mathscr{C}(S)$.

Now we can state the Pivot Theorem. Let $\gamma(\alpha) \in \pi_1(S)$ represent $\alpha \in \mathscr{C}(S) = \hat{\mathbb{Q}}$, and for $\rho \in \mathscr{D}(\pi_1(S))$, let $\lambda_\rho(\alpha) := l_\rho(\alpha) + i\theta_\rho(\alpha)$ denote the complex translation length of $\rho(\gamma(\alpha)) \in \mathrm{PSL}_2(\mathbb{C})$ where $l_\rho(\alpha)$ is the translation length of $\rho(\gamma(\alpha))$ along its axis and $\theta_\rho(\alpha)$ is its rotation.

For each $\beta \in \mathscr{C}(S)$, fix an element of $\mathrm{PSL}_2(\mathbb{Z})$ such that β is taken to ∞ by this element. Then this element takes the set of neighbors of β to $\mathbb{Z}$. Such a transformation is unique up to integer translation. Let $v_+(\beta)$ and $v_-(\beta)$ denote the points of $\overline{\mathbb{H}}^2$ to which $v_\pm \in \overline{\mathbb{H}}^2$ are taken by this transformation. Minsky's Pivot Theorem is

Theorem 5.1. (See Theorem 4.1 in [Min99])
There exist positive constants ε, c_1 such that for any marked punctured torus group ρ,

1. *If $l_\rho(\beta) \leq \varepsilon$ then β is a pivot of ρ.*

2. *Let α be a pivot of ρ. If we take a branch of $\lambda_\rho(\alpha)$ satisfying $-\pi < Im\,\lambda_\rho(\alpha) \leq \pi$, then*

$$d_{\mathbb{H}^2}\Big(\frac{2\pi i}{\lambda_\rho(\alpha)}, v_+(\alpha) - \overline{v_-(\alpha)} + i\Big) < c_1$$

where $d_{\mathbb{H}^2}(\cdot,\cdot)$ denotes the hyperbolic metric on $\mathbb{H}^2$.

6. $\mathscr{E}$ is a Jordan domain

Now we can show that $\mathscr{E}$ is a Jordan domain. First we consider the relation between the closure of the Earle slice $\varphi(\mathscr{E})$ in $\mathscr{QF}$ and the closure of $\mathscr{E}$ in the d-plane.

Lemma 6.1. *If non-zero $d \in \mathbb{C}$ is on the imaginary axis of the d-plane, $A_d B_d$ or $A_d B_d^{-1}$ is elliptic.*

Proof. From the trace equations $\mathrm{Tr}\,A_d B_d = 2 + \frac{1}{d^2}$ and $\mathrm{Tr}\,A_d B_d^{-1} = 2(2d^2 + 1)$, we can easily check this claim. $\square$

Proposition 6.2.

1. The closure $\bar{\mathcal{E}}$ of $\mathcal{E}$ in $\mathbb{C}^+$ is homeomorphic to the closure $\overline{\varphi(\mathcal{E})}$ of $\varphi(\mathcal{E})$ in $\mathscr{D}(\pi_1(S))$ under φ.

2. The closure of $\mathcal{E}$ in $\hat{\mathbb{C}}$ is equal to $\bar{\mathcal{E}} \cup \{0, \infty\}$.

Proof.

(i) φ is a homeomorphism from $\mathbb{C}^+$ to its image under φ, and $\varphi(\mathbb{C}^+) \cap \mathscr{D}(\pi_1(S))$ is closed in $\mathscr{D}(\pi_1(S))$ by Lemma 6.1.

(ii) From the above Lemma 6.1 and the fact that $\mathcal{E}$ contains the positive real line $\mathbb{R}^+$, we can check the claim.

$\square$

Now we have the following diagram:

$$
\begin{array}{ccc}
\mathbb{C}^+ & \xrightarrow{\;\varphi\;} & \mathscr{R}(\pi_1(S)) \\
\uparrow & & \uparrow \\
\mathcal{E} & \xrightarrow{\;\varphi\;} & \mathscr{D}(\pi_1(S)) \xrightarrow{\;\nu\;} \overline{\mathbb{H}}^2 \times \overline{\mathbb{H}}^2 \setminus \Delta
\end{array}
$$

Next we consider the restriction of ν to the Earle slice $\varphi(\mathcal{E})$ in $\mathscr{QF}$.

Proposition 6.3. $\nu \circ \varphi(\mathcal{E}) = \{(\nu_-, \nu_+) \in \mathbb{H}^2 \times \mathbb{H}^2 \,|\, \nu_- \overline{\nu_+} = 1\}$

Proof. Recall that for a marked quasifuchsian groups $\Gamma = \langle A, B \rangle$ in the Earle slice $\mathcal{E}$, $(\Omega_+/\Gamma; A, B)$ is conformal to $(\Omega_-/\Gamma; B, A)$. But for each $\nu_+ \in \mathbb{H}^2$, there is a unique $\nu_- \in \mathbb{H}^2$ such that the marked flat tori $\mathbb{C}/(\mathbb{Z} \cdot 1 + \mathbb{Z} \cdot \nu_+)$ and $\mathbb{C}/(\mathbb{Z} \cdot \bar{\nu}_- + \mathbb{Z} \cdot 1)$ are equivalent to the one-point compactifications of Ω_+/Γ and Ω_-/Γ respectively as marked Riemann surfaces. The proposition follows from the fact that $\mathbb{C}/(\mathbb{Z} \cdot 1 + \mathbb{Z} \cdot \tau)$ is conformal to $\mathbb{C}/(\mathbb{Z} \cdot \frac{1}{\tau} + \mathbb{Z} \cdot 1)$. $\square$

Therefore its closure in $\overline{\mathbb{H}}^2 \times \overline{\mathbb{H}}^2 \setminus \Delta$ can be written as

Corollary 6.4. $\overline{\nu \circ \varphi(\mathcal{E})} = \{(\nu_-, \nu_+) \in \overline{\mathbb{H}}^2 \times \overline{\mathbb{H}}^2 \setminus \Delta \,|\, \nu_- \overline{\nu_+} = 1\}$. *In particular, it is homeomorphic to the closed disc minus two boundary points.*

Next result is an application of Minsky's Theorems 4.1 and 5.1.

Proposition 6.5. *If a sequence of points* (v_-^i, v_+^i) *in* $\overline{v \circ \varphi(\mathscr{E})}$ *goes to the point* $(1,1)$ *in* $\hat{\mathbb{R}} \times \hat{\mathbb{R}}$, *then* $d_i = (v \circ \varphi)^{-1}((v_-^i, v_+^i))$ *converges to* 0 *in the d-plane. Similarly if* (v_-^i, v_+^i) *goes to* $(-1,-1)$, *then* d_i *diverges to infinity.*

Proof. Suppose first that $(v_-^i, v_+^i) \to (1,1)$. There is a unique element $A \in \mathrm{PSL}_2(\mathbb{Z})$ satisfying $A(1) = \infty$ and $A(-1) = 1/2$. Let $v_\pm^i(1)$ denote the points of $\overline{\mathbb{H}}^2$ to which $v_\pm^i$ are taken by A. $v_+^i(1)$ and $v_-^i(1)$ are related by $v_-^i(1) = 1 - v_+^i(1)$ from the relation in Corollary 6.4.

First we show that for sufficiently large i, $1 \in \hat{\mathbb{Q}}$ becomes a pivot for the representation ρ_i whose pair of end invariants is (v_-^i, v_+^i). When $Im\ v_+^i(1) \to \infty$, then $Im\ v_-^i(1) \to \infty$ by the relation $v_-^i(1) = 1 - v_+^i(1)$. The hyperbolic length $l_\pm^i(1)$ of the geodesic corresponding to the slope $1 \in \hat{\mathbb{Q}}$ becomes short on the boundary tori $\Omega_\pm / \rho_i(\pi_1(S))$. Then by Bers' inequality $1/l_{\rho_i}(1) \geq \frac{1}{2}(1/l_+^i(1) + 1/l_-^i(1))$, the length $l_{\rho_i}(1)$ of the geodesic in $\mathbb{H}^3/\rho_i(\pi_1(S))$ corresponding to $1 \in \hat{\mathbb{Q}}$ is also short, hence by the Pivot Theorem 5.1(1), $1 \in \hat{\mathbb{Q}}$ is a pivot for ρ_i. When $Im\ v_+^i(1)$ remains bounded and hence $Re\ v_+^i(1) \to \pm\infty$, then $Re\ v_-^i(1) \to \mp\infty$ and in this case, by definition, $1 \in \hat{\mathbb{Q}}$ is also a pivot for ρ_i.

Hence by the Pivot Theorem 5.1(2), the complex translation length $\lambda_{\rho_i}(1)$ satisfying $-\pi < Im\ \lambda_{\rho_i}(1) \leq \pi$ goes to 0. This implies that $\mathrm{Tr}\,\rho_i(\gamma(1))$ goes to 2. From the equality $\mathrm{Tr}\,\rho_i(\gamma(1)) = \mathrm{Tr}\,A_{d_i}B_{d_i}^{-1} = 2(2d_i^2 + 1)$, d_i goes to 0.

The remaining case that $(v_-^i, v_+^i) \to (-1,-1)$ can be proved by the same argument. $\square$

Theorem 6.6. *The restriction of* v^{-1} *to* $\overline{v \circ \varphi(\mathscr{E})}$ *is a homeomorphism from* $\overline{v \circ \varphi(\mathscr{E})}$ *to* $\overline{\varphi(\mathscr{E})}$.

Proof. Because $v^{-1}(\overline{v \circ \varphi(\mathscr{E})})$ is closed in $\mathscr{D}(\pi_1(S))$ by Proposition 6.5, it must be equal to the closure $\overline{\varphi(\mathscr{E})}$ of $\varphi(\mathscr{E})$ in $\mathscr{D}(\pi_1(S))$. By the same reason the restriction of v^{-1} to $\overline{v \circ \varphi(\mathscr{E})}$ is a homeomorphism. $\square$

Now the next result is an immediate corollary of Theorem 6.6 and Proposition 6.2.

Corollary 6.7.

(i) *The boundary of* $\mathscr{E}$ *in* $\mathbb{C}^+$ *consists of two open Jordan arcs terminating* 0 *and* ∞.

(ii) *The boundary of* $\mathscr{E}$ *in* $\hat{\mathbb{C}}$ *is a Jordan curve. Therefore* $\mathscr{E}$ *is a Jordan domain.*

7. Asymptotic behavior of the boundary $\partial\mathscr{E}$

First we show that $\mathscr{E}$ is flat at the origin.

Theorem 7.1. *In the d-plane, there exists an open round disc B in $\mathscr{E}$ whose closure is tangent to the imaginary axis at 0.*

Proof. First we fix a branch of the complex length function $\lambda_d(1)$ on $\mathscr{E}$ by the condition that it is real valued on the positive real line $\mathbb{R}^+$. We remark that $Re\ \lambda_d(1) = l_d(1) > 0$ on $\mathscr{E}$, hence $\lambda(\mathscr{E}) := \{\lambda(d) \in \mathbb{C} | d \in \mathscr{E}\}$ is contained in the right half λ-plane $\mathbb{C}^+$.

Next we extend this branch to a neighborhood of 0 in the d-plane. The equality $\mathrm{Tr} A_d B_d^{-1} = 2\cosh\frac{\lambda_d(1)}{2} = 2(2d^2 + 1)$ implies that $d = \sinh\frac{\lambda_d(1)}{4}$, hence the branch $\lambda_d(1)$ can be extended conformally in a a neighborhood U of 0 in $\mathbb{C}$. Especially by taking U sufficiently small, we may assume that $|Re\ \lambda_d(1)|$ and $|Im\ \lambda_d(1)|$ are both small. Then by the Pivot Theorem 5.1(1), $1 \in \hat{\mathbb{Q}}$ is a pivot for any points in $U \cap \mathscr{E}$.

Now for $k > 0$ let L_k denote the horizontal line in $\mathbb{H}^2$ passing through ik; that is, let $L_k := \{z(s) \in \mathbb{H}^2 \mid z(s) = s + ik \quad (s \in \mathbb{R})\}$. By Theorem 6.6, $(A \circ v_+)^{-1}(z(s))$ goes to 0 as $|s| \to \pm\infty$. In particular, there exists $r_1 > 0$ such that $(A \circ v_+)^{-1}(z(s)) \in U \cap \mathscr{E}$ for $|s| > r_1$. Here $A \in \mathrm{PSL}_2(\mathbb{Z})$ satisfies $A(1) = \infty$ and $A(-1) = 1/2$.

On the other hand, by the Pivot Theorem 5.1(2),

$$d_{\mathbb{H}^2}\Big(\frac{2\pi i}{\lambda_{(A\circ v_+)^{-1}(z(s))}(1)}, 2s - 1 + i(2k+1)\Big) < c_1$$

for $|s| > r_1$ which implies that the curve $\{\lambda_{(A\circ v_+)^{-1}(z(s))}(1)\}_{s\in\mathbb{R}}$ is tangent to the imaginary axis at 0. Therefore in $\lambda(U \cap \mathscr{E})$, we can take a small open round disc tangent to the imaginary axis at 0. Take B as the image of this disc under the conformal map $d = \sinh(\frac{\lambda}{4})$ around 0. $\qquad\square$

Now we have the following result for the asymptotic behavior of the boundary $\partial\mathscr{E}$.

Corollary 7.2. *In the d-plane there exists a right half region contained in $\mathscr{E}$.*

Proof. Take the image of the round disc B in Theorem 7.1 under the conformal involution $\sigma(d) = 1/(2d)$ of $\mathscr{E}$. $\qquad\square$

Remark 7.3. By using the Pivot Theorem 5.1, we can show that $\mathscr{E}$ is not a quasi-disk (see [Miy]). Miyachi recently showed a more strong result; for the Bers slice, the Maskit slice and the Earle slice of punctured torus groups, every boundary point corresponding to a cusp group is an inward-pointing cusp (see [Miy]).

8. End invariants and pleating invariants

In [KoS01], we showed that every rational pleating ray $\mathcal{P}(x, 1/x)$ $(x \in \hat{\mathbb{Q}} \setminus \{\pm 1\})$ lands at a unique point $c_x \in \partial \mathcal{E}$ representing a cusp group whose pair of end invariants is equal to $(x, 1/x)$ (see Theorem 5.1 in [KoS01]). We extend this result to irrational cases.

Theorem 8.1. *For $x \in \hat{\mathbb{R}} \setminus \{\pm 1\}$, every pleating ray $\mathcal{P}(x, 1/x)$ lands to the boundary group whose pair of end invariants is $(x, 1/x)$.*

Proof. We may assume that x is irrational. From Theorem 6.16 in [KoS01], $\mathcal{P}(x, 1/x)$ is a simple arc with one boundary point on $\mathbb{R}^+$. If we follow $\mathcal{P}(x, 1/x)$ to the other boundary point, $l_d(x)$ goes to 0 by Proposition 6.15 in [KoS01], hence it can't accumulate to points in $\mathcal{E}$ by the Limit Pleating Theorem 5.1 in [KS]. Since $\partial \mathcal{E}$ consists of two Jordan arcs by Corollary 6.7, cusps are dense in $\partial \mathcal{E}$ and the accumulation set I_x of $\mathcal{P}(x, 1/x)$ is a closed interval in $\partial \mathcal{E}$. Therefore I_x must be a unique point whose pair of end invariants is $(x, 1/x)$. $\qquad\Box$

References

[Ear81] C.J. Earle (1981). Some intrinsic coordinates on Teichmüller space, *Proc. Amer. Math. Soc.* **83**, 527–531.

[KS] L. Keen and C. Series. Pleating invariants for punctured torus groups, *to appear Topology.*

[KoS01] Y. Komori and C. Series (2001). Pleating coordinates for the Earle embedding, *Ann. Fac. Sci. Toulouse Math.* , 69–105.

[Min99] Y.N. Minsky (1999). The classification of punctured-torus groups, *Ann. Math.* **149**, 559–626.

[Miy] H. Miyachi. Cusps in complex boundaries of one-dimensional Teichmüller spaces, *to appear Conform. Geom. Dyn.*

[MT98] K. Matsuzaki and M. Taniguchi (1998). *Hyperbolic manifolds and Kleinian groups*, Oxford Mathematical Monograph.

Yohei Komori

Department of Mathematics
Osaka City University
Sugimoto, Sumiyoshi-ku
Osaka, 558-8585
Japan

komori@sci.osaka-cu.ac.jp

AMS Classification: 30F40, 32G05

Keywords: Kleinian group, pleating coordinates, Teichmuller space

Part III
Related topics

Kleinian Groups and Hyperbolic 3-Manifolds   Y. Komori, V. Markovic & C. Series (Eds.)
Lond. Math. Soc. Lec. Notes **299**, 307–341   Cambridge Univ. Press, 2003

Variations on a theme of Horowitz

James W. Anderson

Abstract

Horowitz [Hor72] showed that for every $n \geq 2$, there exist elements $w_1, \ldots, w_n$ in $F_2 = \text{free}(a, b)$ which generate non-conjugate maximal cyclic subgroups of F_2 and which have the property that $\text{trace}(\rho(w_1)) = \cdots = \text{trace}(\rho(w_n))$ for all faithful representations ρ of F_2 into $SL_2(\mathbb{C})$. Randol [Ran80] used this result to show that the length spectrum of a hyperbolic surface has unbounded multiplicity. Masters [Mat00] has recently extended this unboundness of the length spectrum to hyperbolic 3-manifolds. The purpose of this note is to present a survey of what is known about characters of faithful representations of F_2 into $SL_2(\mathbb{C})$, to give a conjectural topological characterization of such n-tuples of elements of F_2, and to discuss the case of faithful representations of general surface groups and 3-manifold groups.

1. Introduction, history, and motivation

The purpose of this survey is to explore the following question, asked during the Special Session on Geometric Function Theory, held in Hartford, Connecticut, during the 898th meeting of the AMS in March, 1995 (for a list of all the questions asked during that session, we refer the reader to Basmajian [Bas97]):

15. According to a theorem of Horowitz (see Horowitz [Hor72], Randol [Ran80]), there exist pairs of closed curves on a closed [orientable] surface S [with negative Euler characteristic] for which the lengths of the geodesics in the respective homotopy classes are equal for any hyperbolic structure on S. These constructions all involve writing down a pair of words in the fundamental group for S and then applying trace identities to show that the words have the same trace, independent of the representation into $PSL(2, \mathbb{R})$. Find a topological characterization of such a pair of curves.

We begin with some definitions, and make the observation that while we generally state the definitions in terms of the free group of rank two F_2, the definitions and many of the observations hold true for a general finitely generated group G. Also, while we generally restrict our attention to faithful representations of a group G into $SL_2(\mathbb{C})$, the assumption of faithfulness is not necessary. In fact, the results of Horowitz [Hor72] and Ginzburg and Rudnick [GR98] hold for all representations of F_2 into $SL_2(\mathbb{C})$.

Let $F_2 = \text{free}(a, b)$ be the free group of rank 2, and let $\mathscr{F}(F_2)$ denote the space of all faithful representations of F_2 into $SL_2(\mathbb{C})$. The topology on $\mathscr{F}(F_2)$ is given by realizing it as a subset of $SL_2(\mathbb{C}) \times SL_2(\mathbb{C}) = \text{Hom}(F_2, SL_2(\mathbb{C}))$, the space of all representations of F_2 into $SL_2(\mathbb{C})$, by associating the representation $\rho \in \mathscr{F}(F_2)$ with the point $(\rho(a), \rho(b))$ in $SL_2(\mathbb{C}) \times SL_2(\mathbb{C})$. Note that $\mathscr{F}(F_2)$ is dense in $SL_2(\mathbb{C}) \times SL_2(\mathbb{C})$. We denote by $\mathscr{D}F(F_2)$ the subspace of $\mathscr{F}(F_2)$ consisting of all those faithful representations ρ of F_2 into $SL_2(\mathbb{C})$ whose image $\rho(F_2)$ is a discrete subgroup of $SL_2(\mathbb{C})$.

For an element w of F_2, the *character associated to w* is the function $\chi[w] : \mathscr{F}(F_2) \to \mathbb{C}$ given by setting $\chi[w](\rho) = \text{trace}(\rho(w))$, where $\text{trace}(A)$ is the usual trace of the 2×2 matrix A. Note that by this definition, an element w of F_2 and its inverse w^{-1} determine equal characters $\chi[w] = \chi[w^{-1}]$, since $\text{trace}(A) = \text{trace}(A^{-1})$ for a 2×2 matrix A with determinant 1. Direct calculation also establishes the identities $\chi[g] = \chi[h \cdot g \cdot h^{-1}]$, since $\text{trace}(A) = \text{trace}(B \cdot A \cdot B^{-1})$ for 2×2 matrices A and B with determinant 1, and $\chi[g \cdot h] = \chi[g]\chi[h] - \chi[g \cdot h^{-1}]$, since $\text{trace}(A \cdot B) = \text{trace}(A)\text{trace}(B) - \text{trace}(A \cdot B^{-1})$ for 2×2 matrices A and B with determinant 1.

Hence, if we let $\mathscr{C}(F_2)$ denote the set of conjugacy classes of maximal cyclic subgroups of F_2, the first two of the three identities just described yield that there is a well-defined map from $\mathscr{C}(F_2)$ to the set of characters, by taking the character of a generator. In this language, the purpose of this note is to describe the extent to which this map is not injective, and to describe means of determining when different elements of $\mathscr{C}(F_2)$ give rise to the same character.

An element w of F_2 is *maximal* if it generates a maximal cyclic subgroup of F_2, and hence is not a proper power of another element of F_2. An element w of F_2 is *primitive* if there exists a free basis S for F_2 containing w. Note that primitive elements are necessarily maximal, though not conversely. For a general group G (admitting a faithful representation into $SL_2(\mathbb{C})$), the notion of maximality of an element of G still holds, namely, that an element of G is maximal if it is not a proper power of another element of G, though the notion of primitivity is restricted to elements of free groups. We are able to restrict our attention to maximal elements, since for a maximal element w of G, the character $\chi[w^n]$ is a polynomial in $\chi[w]$, see Section 2; in fact, there exists a polynomial $\tau_n(x)$, independent of w, so that $\chi[w^n] = \tau_n(\chi[w])$.

A *character class* in F_2 is the collection of all maximal cyclic subgroups of F_2 which give rise to the same character; that is, two maximal cyclic subgroups $\langle w \rangle$ and $\langle u \rangle$ of F_2 belong to the same character class if and only if $\chi[w](\rho) = \chi[u](\rho)$ for all $\rho \in \mathscr{F}(F_2)$. The *stable multiplicity* $\mathrm{mult}(w)$ of a maximal element w of F_2 is the number of conjugacy classes in its character class. It was shown by Horowitz, Theorem 8.1 of [Hor72], that the stable multiplicity of an element of F_2 is always finite.

The starting point for our discussion is the following result of Horowitz:

Theorem 1.1 (Example 8.2 of Horowitz [Hor72]). *Let F_2 be a free group on two generators. For each $m \geq 1$, there exist elements $w_1, \ldots, w_m$ of F_2 which generate pairwise non-conjugate maximal cyclic subgroups of F_2 and which satisfy $\chi[w_1] = \cdots = \chi[w_m]$. That is, the stable multiplicity of maximal elements of F_2 is unbounded.*

Note that this result of Horowitz does not apply directly to lengths of closed curves on surfaces or in 3-manifolds, since most surfaces and 3-manifolds have fundamental groups that are not free of rank two, but rather is a statement about characters of representations of F_2 into $\mathrm{SL}_2(\mathbb{C})$. This will be discussed in more detail in Section 5. For the time being, we focus our attention on algebraic properties of representations of F_2.

In the same paper, Horowitz also gives the following necessary condition for two elements of F_2 to have the same character.

Theorem 1.2 (Lemma 6.1 of Horowitz [Hor72]). *Let U, U^* be elements in the free group $F_2 = \mathrm{free}(a, b)$ on two generators a and b of the form*

$$U = a^{\alpha_1} \cdot b^{\beta_1} \cdot a^{\alpha_2} \cdot b^{\beta_2} \cdots a^{\alpha_s} \cdot b^{\beta_s},$$

$$U^* = a^{\alpha_1^*} \cdot b^{\beta_1^*} \cdot a^{\alpha_2^*} \cdot b^{\beta_2^*} \cdots a^{\alpha_t^*} \cdot b^{\beta_t^*},$$

where $s, t > 0$ and $\alpha_1, \ldots, \alpha_s, \beta_1, \ldots, \beta_s, \alpha_1^*, \ldots, \alpha_t^*, \beta_1^*, \ldots, \beta_t^*$ are non-zero integers. If $\chi[U] = \chi[U^*]$, then $s = t$. Also, the numbers $|\alpha_1^*|, \ldots, |\alpha_s^*|$ are a rearrangement of the numbers $|\alpha_1|, \ldots, |\alpha_s|$, and the numbers $|\beta_1^*|, \ldots, |\beta_s^*|$ are a rearrangement of the numbers $|\beta_1|, \ldots, |\beta_s|$.

We note here that given an element w of $F_2 = \mathrm{free}(a, b)$, Theorem 1.2 gives an inefficient algorithm for determining all elements u of F_2 for which $\chi[w] = \chi[u]$, namely by considering all elements u of F_2 constructed by permuting the exponents of a and b in w, as well as changing their signs. As it is known that Theorem 1.2 is not optimal, and as there is not obvious direct generalization of Theorem 1.2 to other groups, we continue the search for better necessary conditions for two elements of F_2 to determine the same character.

310   J. W. Anderson

One corollary of Theorem 1.2 is the following result, which we highlight due to the role that it plays later.

Theorem 1.3 (Theorem 7.1 of Horowitz [Hor72]). *Let u be an element of a free group F (of any countable rank). If $\chi[u] = \chi[c^m]$, where c is a primitive element of F, then u is conjugate to $c^{\pm m}$.*

These results of Horowitz give necessary conditions for two elements of F_2 to give rise to the same character. Before stating what is known in terms of partial converses to Theorem 1.2, we need the following. There is an involution I on $F_2 = \text{free}(a,b)$, defined as follows. First consider the automorphism $J : F_2 \to F_2$ defined by setting $J(a) = a^{-1}$ and $J(b) = b^{-1}$ and then extending so that J is an automorphism of F_2.

Define the involution $I : F_2 \to F_2$ by $I(w) = J(w^{-1}) = (J(w))^{-1}$. We refer to I as the *canonical involution* for F_2 with respect to the generators a and b. Note that I is not an automorphism of F_2 but rather is an anti-automorphism, with $I(w \cdot u) = I(u) \cdot I(w)$ and $I(w^{-1}) = (I(w))^{-1}$. It is not difficult to see that I is character preserving, in the sense that for any element w of F_2, we always have that $\chi[w] = \chi[I(w)]$. This and other properties of the involution I are discussed in more detail in Section 3.

As a partial answer to the original question, we note the following Proposition, which is an easy exercise using the uniqueness of normal forms in F_2, see for example Lyndon and Schupp [LyS77].

Proposition 1.4. *Let $w = a^{n_1} \cdot b^{m_1} \cdots a^{n_p} \cdot b^{m_p}$ be an element of $F_2 = \text{free}(a,b)$, with $n_1,\ldots,n_p$ and $m_1,\ldots,m_p$ all non-zero. Then, w is conjugate to $I(w)$ if and only if there exists c so that $n_k = n_l$ for $k+l \equiv c \pmod{p}$ and $m_k = m_l$ for $k+l \equiv c-1 \pmod{p}$.*

In particular, if $p \geq 3$ and if either the n_k are distinct or the m_k are distinct, then w and $I(w)$ generate non-conjugate maximal cyclic subgroups of F_2. In the case $p = 2$, for $w = a^{n_1} \cdot b^{m_1} \cdot a^{n_2} \cdot b^{m_2}$ with n_1, m_1, n_2, and m_2 distinct integers, we have that w and $I(w)$ generate non-conjugate maximal cyclic subgroups of F_2.

So, in a loose sense, for most maximal elements w of F_2, there is another element u, necessarily maximal, so that w and u generate non-conjugate cyclic subgroups of F_2 and so that $\chi[w] = \chi[u]$.

We now refine our terminology. Say that a maximal element w of F_2 (or more precisely, the conjugacy class of maximal cyclic subgroups of F_2 generated by w) is *pseudo-simple* if for any element u with $\chi[u] = \chi[w]$, we have that u is conjugate either to $w^{\pm 1}$ or to $I(w)^{\pm 1}$. With this language, the stable multiplicity of a pseudo-simple element of F_2 is at most 2, since we allow the possibility that w and $I(w)$ are conjugate.

Further, say that a maximal element w of F_2 (or more precisely, the conjugacy class of maximal cyclic subgroups of F_2 generated by w) is *simple* if for any element u with $\chi[u] = \chi[w]$, we have that u is conjugate to $w^{\pm 1}$. For example, primitive elements in free groups are simple, by Theorem 1.3. In particular, if w is simple, then $I(w)$ is conjugate to w. The stable multiplicity of a simple element of F_2 is 1.

An element w of F_2 is *strictly pseudo-simple* if it is pseudo-simple but not simple. In particular, if w is strictly pseudo-simple, then w and $I(w)$ are not conjugate. The stable multiplicity of a strictly pseudo-simple element w of F_2 is exactly 2, with the two conjugacy classes in its character class being represented by $\langle w \rangle$ and $\langle I(w) \rangle$.

Ginzburg and Rudnick [GR98] prove the following. Given an element $w = a^{n_1} \cdot b^{m_1} \cdots a^{n_p} \cdot b^{m_p}$, consider the two p-tuples of exponents $\mathbf{n} = (n_1, \ldots, n_p)$ and $\mathbf{m} = (m_1, \ldots, m_p)$. Say that $\mathbf{n}$ is *non-singular* if $n_k \neq \sum_{j \in S} n_j$ for all $1 \leq k \leq p$ and for all subsets $S \subset \{1, \ldots, p\}$, $S \neq \{k\}$. (In particular, note that if $\mathbf{n}$ is non-singular, then all the n_j are distinct and also $\sum_{j \in S} n_j \neq 0$ if S is non-empty.) Say that the element w is *non-singular* if both p-tuples of its exponents $\mathbf{n}$ and $\mathbf{m}$ are non-singular.

Theorem 1.5 (Theorem 1.1 of Ginzburg and Rudnick [GR98]). *If w is a non-singular element of F_2, then w is strictly pseudo-simple.*

(The terminology in this note differs slightly from Ginzburg and Rudnick [GR98], who use simple where we use pseudo-simple.) Moreover, Ginzburg and Rudnick [GR98] also refine the statement of Theorem 1.2 for a non-singular element w of F_2.

Theorem 1.6 (Corollary 3.1 of Ginzburg and Rudnick [GR98]). *Let U, U^* be non-singular elements in the free group $F_2 = \mathrm{free}(a,b)$ on two generators a and b of the form*

$$U = a^{\alpha_1} \cdot b^{\beta_1} \cdot a^{\alpha_2} \cdot b^{\beta_2} \cdots a^{\alpha_s} \cdot b^{\beta_s},$$

$$U^* = a^{\alpha_1^*} \cdot b^{\beta_1^*} \cdot a^{\alpha_2^*} \cdot b^{\beta_2^*} \cdots a^{\alpha_t^*} \cdot b^{\beta_t^*},$$

where $s, t > 0$ and $\alpha_1, \ldots, \alpha_s, \beta_1, \ldots, \beta_s, \alpha_1^*, \ldots, \alpha_t^*, \beta_1^*, \ldots, \beta_t^*$ are non-zero integers. If $\chi[U] = \chi[U^*]$, then $s = t$. Moreover, either the numbers $\alpha_1^*, \ldots, \alpha_s^*$ are a rearrangement of the numbers $\alpha_1, \ldots, \alpha_s$, and the numbers $\beta_1^*, \ldots, \beta_s^*$ are a rearrangement of the numbers $\beta_1, \ldots, \beta_s$, or else the numbers $\alpha_1^*, \ldots, \alpha_s^*$ are a rearrangement of the numbers $-\alpha_1, \ldots, -\alpha_s$, and the numbers $\beta_1^*, \ldots, \beta_s^*$ are a rearrangement of the numbers $-\beta_1, \ldots, -\beta_s$.

Theorem 1.2 and Theorem 1.6 share a common approach to their proof. Namely, they are both proven starting from the observation that if w and u are elements of F_2 with $\chi[w] = \chi[u]$, then for any family $\mathcal{P}$ of representations in $\mathcal{F}(F_2)$, we have that trace($\rho(w)$) = trace($\rho(u)$) for all $\rho \in \mathscr{P}$. Any identities satisfied by all representations in $\mathscr{F}(F_2)$ must be satisfied by all the representations in $\mathscr{P}$, and so the identities arising from analyzing the representations in $\mathscr{P}$ give conditions that yield necessary conditions for all representations in $\mathscr{F}(F_2)$.

Horowitz [Hor72] considers the collection $\mathscr{P}$ of representations $\mathscr{P} = \{\rho\}$ defined by

$$\rho(a) = \begin{pmatrix} \lambda & t \\ 0 & \lambda^{-1} \end{pmatrix} \text{ and } \rho(b) = \begin{pmatrix} \mu & 0 \\ t & \mu^{-1} \end{pmatrix}$$

for complex numbers λ, μ, and t. Ginzburg and Rudnick [GR98] consider the collection $\mathscr{R}$ of representations $\mathscr{R} = \{\rho\}$ defined by

$$\rho(a) = \begin{pmatrix} a & 0 \\ 0 & a^{-1} \end{pmatrix} \text{ and } \rho(b) = \begin{pmatrix} 1 & x \\ 1 & x+1 \end{pmatrix} \begin{pmatrix} b & 0 \\ 0 & b^{-1} \end{pmatrix} \begin{pmatrix} 1 & x \\ 1 & x+1 \end{pmatrix}^{-1}$$

for complex numbers a, b, and x.

In fact, Horowitz analyzes the leading term of the Fricke polynomial of $\chi[w]$, described in Section 2, expressed as a polynomial in $\chi[a \cdot b]$, with coefficients in $\mathbb{Z}[\chi[a], \chi[b]]$, evaluated at the representations in $\mathscr{P}$. Ginzberg and Rudnick analyze the coefficient of $\chi[a \cdot b]$ in this expansion, evaluated at the representations in $\mathscr{R}$.

2. Fricke polynomials

One of the main results in Horowitz [Hor72] was to give a proof of the following claim of Fricke (see p. 338 and 366 of Fricke and Klein [FK97]).

Theorem 2.1 (Theorem 3.1 of Horowitz [Hor72]). *Let F_n be a free group on n generators $a_1, \ldots, a_n$. If u is an arbitrary element of F_n, then the character $\chi[u]$ of u can be expressed as a polynomial with integer coefficients in the $2^n - 1$ characters $\chi[a_{i_1} \cdot a_{i_2} \cdots a_{i_k}]$, where $1 \leq k \leq n$ and $1 \leq i_1 < i_2 < \cdots < i_k \leq n$.*

We refer to this polynomial as the *Fricke polynomial* for u. One of the keys to the proof of Theorem 2.1 is the identity $\chi[w \cdot u] = \chi[w]\chi[u] - \chi[w \cdot u^{-1}]$, which follows immediately from the analogous identity for traces of 2×2 matrices, as well as the other basic identities already mentioned, that $\chi[w] = \chi[w^{-1}]$ and that $\chi[w] = \chi[u \cdot w \cdot u^{-1}]$; see Section 1.

One consequence of Theorem 2.1 is the following construction of Fricke. Let w and u be any pair of elements of F_2 for which $\chi[w] = \chi[u]$, and let $p = p(a,b)$

be any element of $F_2 = \text{free}(a,b)$. Since $\chi[p]$ is expressible as a polynomial $\chi[p] = P(\chi[a], \chi[b], \chi[a \cdot b])$, we see that

$$\chi[p(w,u)] = P(\chi[w], \chi[u], \chi[w \cdot u]) = P(\chi[u], \chi[w], \chi[u \cdot w]) = \chi[p(u,w)],$$

where the middle equality is a consequence of the assumption that $\chi[w] = \chi[u]$ and the fact that $\chi[w \cdot u] = \chi[u \cdot w]$.

This leads to the following definition. Let w and u be elements of F_2 for which $\chi[w] = \chi[u]$, and let F be the subgroup of F_2 generated by w and u. There is an automorphism σ on F, the *switching automorphism*, defined by setting $\sigma(w) = u$ and $\sigma(u) = w$ and then extending σ to be an automorphism of F. The discussion above yields that σ is a character preserving automorphism on the subgroup F of F_2, which in general does not extend to an automorphism of all of F_2.

Theorem 2.1 can be thought of as the analogue for these characters of the result that the Teichmüller space of an orientable surface S of negative Euler characteristic and finite analytic type can be parametrized by the lengths of a fixed finite set of simple closed curves on the surface. For more information on this, we refer the interested reader to Abikoff [Abi80], Schmutz Schaller [Sch99], and Hamenstädt [Ham01].

Let $\mathscr{B}_n$ be the ring of polynomials with integer coefficients in the $2^n - 1$ indeterminates $x_{i_1 i_2 \cdots i_k}$, where $1 \leq k \leq n$ and $1 \leq i_1 < i_2 < \cdots < i_k \leq n$. Theorem 2.1 can also be interpreted as describing a map $\Theta : \mathscr{C}(F_n) \to \mathscr{B}_n$, by taking the character of a generator.

One question as yet unresolved is to determine the image $\Theta(\mathscr{C}(F_n))$ in $\mathscr{B}_n$. It is an easy observation that Θ is not surjective, even for $n = 1$. To see this, define a family $\tau_n(s)$, $n \geq 0$, of polynomials by setting $\tau_0(s) = 2$, $\tau_1(s) = s$, and $\tau_{n+1}(s) = s\,\tau_n(s) - \tau_{n-1}(s)$. The $\tau_n(s)$ are Chebychev polynomials of the second kind. Using the above identity for $\chi[w \cdot u]$, we see that $\chi[w^n] = \tau_n(\chi[w])$. This is discussed by Horowitz [Hor72], Section 2, and was exploited to great effect by Jørgensen [Jør82].

Let $\mathscr{I}_n$ be the ideal in $\mathscr{B}_n$ consisting of those polynomials which are identically 0 under the substitution

$$x_{i_1 i_2 \cdots i_k} = \chi[a_{i_1} \cdot a_{i_2} \cdots a_{i_k}].$$

(The polynomials in $\mathscr{I}_n$ are the obstruction to the uniqueness of the Fricke polynomial of a word in F_n.) Horowitz considered the question of determining the structure of $\mathscr{I}_n$. He showed, see Theorem 4.1 of [Hor72], that $\mathscr{I}_1$ and $\mathscr{I}_2$ are both the trivial ideal, so that the character of elements of $F_1 = \text{free}(a_1)$ and of $F_2 = \text{free}(a_1, a_2)$ are represented by unique polynomials.

In the case $n = 3$, though, he shows that the ideal $\mathscr{I}_3$ is non-zero. Specifically, let

$\mathbf{x} = (x_1, x_2, x_3, x_{12}, x_{13}, x_{23})$, and set

$$k_1(\mathbf{x}) = x_{12}x_3 + x_{13}x_2 + x_{23}x_1$$

and

$$k_0(\mathbf{x}) = x_1^2 + x_2^2 + x_3^2 + x_{12}^2 + x_{13}^2 + x_{23}^2 - x_1x_2x_{12} - x_1x_3x_{13} - x_2x_3x_{23} + x_{12}x_{13}x_{23} - 4.$$

Then, $\mathscr{I}_3$ is the principal ideal in $\mathscr{B}_3$ generated by $k(\mathbf{x}, x_{123}) = x_{123}^2 - k_1(\mathbf{x})x_{123} + k_0(\mathbf{x})$; this is derived from the character relation

$$\chi[a_1 \cdot a_2 \cdot a_3] = \chi[a_1]\chi[a_2 \cdot a_3] \quad + \quad \chi[a_2]\chi[a_1 \cdot a_3] + \chi[a_3]\chi[a_1 \cdot a_2]$$
$$- \quad \chi[a_1]\chi[a_2]\chi[a_3] - \chi[a_1 \cdot a_3 \cdot a_2],$$

which is derivable from the basic identities for characters discussed at the beginning of this Section, and the consequent identity for $\chi[u \cdot v \cdot w]\chi[u \cdot w \cdot v]$. In contrast to this, Whittemore [Whi73b] showed that $\mathscr{I}_n$ is not a principal ideal for $n \geq 4$.

Another reason for Whittemore's interest is the following question, as described in [Whi73a]. Following Artin, define the *braid group* B_n to be the group of automorphisms of the free group $F_n = \text{free}(a_1, \dots, a_n)$ generated by the automorphisms β_k of F_n, $1 \leq k \leq n-1$, defined by: $\beta_k(a_k) = a_{k+1}$, $\beta_k(a_{k+1}) = a_{k+1} \cdot a_k \cdot a_{k+1}^{-1}$, and $\beta_k(a_j) = a_j$ for $j \neq k, k+1$. It is known that every knot group G (where a *knot group* is the fundamental group of $\mathbf{S}^3 - K$ for a knot K) can be obtained from F_n by identifying the generators of F_n with their images under an element β_G of B_n.

Using Theorem 2.1, we can realize the set $\text{Hom}(F_n, \text{SL}_2(\mathbb{C}))$ of all representations of F_n into $\text{SL}_2(\mathbb{C})$ with a subset $\mathscr{T}_n$ of $\mathbb{C}^{2^n-1}$ by taking an element ρ of $\text{Hom}(F_n, \text{SL}_2(\mathbb{C}))$ to the point

$$(\chi[a_1](\rho), \chi[a_2](\rho), \dots, \chi[a_{i_1} \cdot a_{i_2} \cdots a_{i_k}](\rho), \dots, \chi[a_1 \cdot a_2 \cdots a_n](\rho))$$

of $\mathbb{C}^{2^n-1}$, for all $2^n - 1$ possible values of $i_1, \dots, i_k$ satisfying $1 \leq k \leq n$ and $1 \leq i_1 < i_2 < \cdots < i_k \leq n$.

Magnus conjectured that the points of $\mathscr{T}_n$ corresponding to a knot group G are exactly the fixed points in $\mathscr{T}_n$ of the automorphism of $\mathscr{T}_n$ induced by β_G. In Theorem 1 of [Whi73a], Whittemore determined the points of $\mathscr{T}_2$ corresponding to the representations of the group G of Listing's knot, given by the presentation

$$G = \langle a, b \mid b^{-1} \cdot a^{-1} \cdot b \cdot a \cdot b^{-1} \cdot a \cdot b \cdot a^{-1} \cdot b^{-1} \cdot a = 1 \rangle.$$

Let $\mathscr{A}_n$ denote the group of automorphisms of the quotient ring $\mathscr{B}_n / \mathscr{I}_n$, and let $\text{Out}(F_n)$ denote the group of outer automorphism classes of F_n. Each automorphism of

F_n induces in a natural way an element of $\mathscr{A}_n$. Horowitz then argues that this induces a natural isomorphism between $\text{Out}(F_n)$ and $\mathscr{A}_n$ for $n \geq 3$. (We refer the interested reader to the discussion in [Hor75] preceeding Corollary 1 for a more detailed treatment.)

Consequently, it was suggested $\text{Out}(F_n)$ might be profitably studied by analyzing the structure of $\mathscr{A}_n$. However, it is unclear to what extent this programme was carried out, and it is unclear that it would significantly add to the current state of the knowledge of the structure of $\text{Out}(F_n)$, though some further work on this general question has been done by Magnus [Mag80], González-Acu na and Montensinos-Amilibia [GM93], and Humphries [Hum01].

3. Properties of the involution I

The purpose of this Section is to explore some of the properties of the canonical involution I and the automorphism J on $F_2 = \text{free}(a, b)$.

The first observation is that the property of an element w of F_2 being conjugate to $I(w)$ is independent of the choice of generating set for F_2. In fact, up to an inner automorphism of F_2, the two operations of changing generators and applying the canonical involution (with respect to the appropriate set of generators) commute. This is an easy application of Nielsen transformations. For a discussion of Nielsen transformations, see Lyndon and Schupp [LyS77].

The second observation is that both the canonical involution I and the automorphism J of F_2 are character preserving. One proof of this begins with the following Lemma, due originally to Jørgensen [Jør78]. (There are other proofs, for instance the proof given by Ginzburg and Rudnick [GR98].)

Lemma 3.1 (Section 4 of Jørgensen [Jør78]). *Let A and B be two elements of* $SL_2(\mathbb{C})$. *Then, there exists an element E of* $SL_2(\mathbb{C})$ *so that $E \cdot A \cdot E^{-1} = A^{-1}$ and* $E \cdot B \cdot E^{-1} = B^{-1}$.

Moreover, given the geometric description of E (for instance, in the case that A and B are hyperbolic elements of $SL_2(\mathbb{C})$ with distinct fixed points, E is the half-turn whose axis is the common perpendicular to the axes of A and B), it is easy to see that E varies continuously with A and B. We note here that, as has been observed and exploited by Jørgensen and others, see in particular Jørgensen [Jør78], Jørgensen and Sandler [JS93], and Pignataro and Sandler [PS74], an element w of F_2 is equal to $I(w)$ if and only if w is a palindrome in a and b.

Combining Lemma 3.1 with the facts that conjugation and inversion are both trace preserving, we see that the automorphism J, and hence the canonical involution I, are both character preserving.

The third observation is that most of the examples and constructions found to date regarding elements of F_2 which generate non-conjugate maximal cyclic subgroups of F_2 and which give rise to the same character can largely be captured by the action of character preserving involutions, either the canonical involution I, the automorphism J, or the switching automorphism σ of some subgroup (as described in Section 2).

Consider, for example, the following elements of $F_2 = \text{free}(a,b)$, due originally to Horowitz [Hor72]. Given an infinite-tuple $(\varepsilon_1, \varepsilon_1, \ldots, \varepsilon_n, \ldots)$, where each $\varepsilon_n = \pm 1$, we get a rooted binary tree T of elements of F_2. Namely, set $w_0 = a$ and for $m \geq 1$ set

$$
\begin{aligned}
w_m(\varepsilon_1, \varepsilon_2, \ldots, \varepsilon_m) &= w_{m-1}(\varepsilon_1, \varepsilon_2, \ldots, \varepsilon_{m-1})^{-\varepsilon_m} \cdot b^{2m} \cdot \\
&\quad w_{m-1}(\varepsilon_1, \varepsilon_2, \ldots, \varepsilon_{m-1})^{\varepsilon_m} \cdot b^{2m-1} \cdot \\
&\quad w_{m-1}(\varepsilon_1, \varepsilon_2, \ldots, \varepsilon_{m-1})^{-\varepsilon_m} \cdot b^{2m} \cdot \\
&\quad w_{m-1}(\varepsilon_1, \varepsilon_2, \ldots, \varepsilon_{m-1})^{\varepsilon_m}.
\end{aligned}
$$

Note that for $m \geq 0$, there are 2^m elements of *depth m* in T, namely the elements $w_m(\varepsilon_1, \varepsilon_2, \ldots, \varepsilon_m)$ for the 2^m choices of $\varepsilon_k = \pm 1$ for $1 \leq k \leq m$. Horowitz proves that the 2^m elements $w_1, \ldots, w_{2^m}$ generate pairwise non-conjugate maximal cyclic subgroups of F_2 and that $\chi[w_1] = \cdots = \chi[w_{2^m}]$. His proof of the former part of the statement is just an application of the existence of unique normal forms for elements in free groups. His proof of the latter part is a direct calculation, using the family of representations described at the end of Section 1. We give here an alternative proof of the second part of his statement, using a slightly difficult argument.

The basic fact we need is the following.

Lemma 3.2. *Let $F_2 = \text{free}(a,b)$ be the free group on a and b, and let T be the tree described above. Then,*

$$
I(w_m(\varepsilon_1, \varepsilon_2, \ldots, \varepsilon_m)) = w_m(-\varepsilon_1, -\varepsilon_2, \ldots, -\varepsilon_m).
$$

Proof. The proof of the Lemma is by induction. We begin with the calculation of $I(w_1(\varepsilon_1))$. Note that

$$
w_1(\varepsilon_1) = a^{-\varepsilon_1} \cdot b^2 \cdot a^{\varepsilon_1} \cdot b \cdot a^{-\varepsilon_1} \cdot b^2 \cdot a^{\varepsilon_1}
$$

and that

$$
\begin{aligned}
I(w_1(\varepsilon_1)) &= I(a^{-\varepsilon_1} \cdot b^2 \cdot a^{\varepsilon_1} \cdot b \cdot a^{-\varepsilon_1} \cdot b^2 \cdot a^{\varepsilon_1}) \\
&= a^{\varepsilon_1} \cdot b^2 \cdot a^{-\varepsilon_1} \cdot b \cdot a^{\varepsilon_1} \cdot b^2 \cdot a^{-\varepsilon_1} = w_1(-\varepsilon_1),
\end{aligned}
$$

as desired.

Suppose now that

$$
I(w_{m-1}(\varepsilon_1,\varepsilon_2,\dots,\varepsilon_{m-1})) = w_{m-1}(-\varepsilon_1,-\varepsilon_2,\dots,-\varepsilon_{m-1}),
$$

and consider $I(w_m(\varepsilon_1,\varepsilon_2,\dots,\varepsilon_m))$. Using that I is an anti-automorphism and the inductive hypothesis, we see that

$$
\begin{aligned}
I(w_m(\varepsilon_1,\varepsilon_2,\dots,\varepsilon_m)) &= I(w_{m-1}(\varepsilon_1,\varepsilon_2,\dots,\varepsilon_{m-1}))^{\varepsilon_m} \cdot b^{2m} \cdot \\
&\quad I(w_{m-1}(\varepsilon_1,\varepsilon_2,\dots,\varepsilon_{m-1}))^{-\varepsilon_m} \cdot b^{2m-1} \cdot \\
&\quad I(w_{m-1}(\varepsilon_1,\varepsilon_2,\dots,\varepsilon_{m-1}))^{\varepsilon_m} \cdot b^{2m} \cdot \\
&\quad I(w_{m-1}(\varepsilon_1,\varepsilon_2,\dots,\varepsilon_{m-1}))^{-\varepsilon_m} \\
&= w_{m-1}(-\varepsilon_1,-\varepsilon_2,\dots,-\varepsilon_{m-1})^{\varepsilon_m} \cdot b^{2m} \cdot \\
&\quad w_{m-1}(-\varepsilon_1,-\varepsilon_2,\dots,-\varepsilon_{m-1})^{-\varepsilon_m} \cdot b^{2m-1} \cdot \\
&\quad w_{m-1}(-\varepsilon_1,-\varepsilon_2,\dots,-\varepsilon_{m-1})^{\varepsilon_m} \cdot b^{2m} \cdot \\
&\quad w_{m-1}(-\varepsilon_1,-\varepsilon_2,\dots,-\varepsilon_{m-1})^{-\varepsilon_m} \\
&= w_m(-\varepsilon_1,-\varepsilon_2,\dots,-\varepsilon_m),
\end{aligned}
$$

as desired. $\qquad\square$

Note that the two elements $w_m(\varepsilon_1,\varepsilon_2,\dots,\varepsilon_m)$ and $w_m(-\varepsilon_1,-\varepsilon_2,\dots,-\varepsilon_m)$ lie in different branches of the tree rooted at w_0. In fact, we can apply this Lemma to the subtree rooted at any element $w = w_k(\varepsilon_1,\varepsilon_2,\dots,\varepsilon_k)$, which corresponds in this construction to the subgroup of F_2 generated by w and b, with its canonical involution defined in terms of w and b. By considering all such subtrees and their relative canonical involutions, we see that all the elements in this tree below w_0 and of the same depth must have equal characters, as they are related by this collection of involutions.

We can recast this construction slightly as follows. Set $v_m(a,b) = a^{-1} \cdot b^{2m} \cdot a \cdot b^{2m-1} \cdot a^{-1} \cdot b^{2m} \cdot a$. Then,

$$
w_1(\varepsilon_1) = v_1(a^{\varepsilon_1},b) \quad \text{and} \quad w_1(-\varepsilon_1) = v_1(a^{-\varepsilon_1},b^{-1})^{-1} = I(v_1(a^{\varepsilon_1},b)).
$$

In general,

$$
w_m(\varepsilon_1,\dots,\varepsilon_m) = v_m(w_{m-1}(\varepsilon_1,\dots,\varepsilon_{m-1})^{\varepsilon_m},b)
$$

and

$$I(w_m(\varepsilon_1,\ldots,\varepsilon_m)) = v_m(w_{m-1}(-\varepsilon_1,\ldots,-\varepsilon_{m-1})^{-\varepsilon_m},b^{-1})^{-1}$$
$$= w_m(-\varepsilon_1,\ldots,-\varepsilon_m).$$

Suppose however that we consider the action on $w_m(\varepsilon_1,\ldots,\varepsilon_m)$ of the involution I^*, where we consider $w_m(\varepsilon_1,\ldots,\varepsilon_m)$ as a word in $w_{m-1}(\varepsilon_1,\ldots,\varepsilon_{m-1})$ and b and I^* is the canonical involution for this generators. Then,

$$I^*(w_m(\varepsilon_1,\ldots,\varepsilon_m)) = w_m(\varepsilon_1,\ldots,\varepsilon_{m-1},-\varepsilon_m).$$

There are several other constructions of n-tuples of words in $F_2 = \text{free}(a,b)$ generating non-conjugate maximal cyclic subgroups of F_2 whose characters are equal.

One is due to Buser [Bus92]. Set $v(a,b) = b \cdot a^{-1} \cdot b^{-1} \cdot a \cdot b$, set $W_1(\varepsilon_1)(a,b) = v(a^{\varepsilon_1},b^{\varepsilon_1})$, and for $m \geq 2$ inductively define

$$W_m(\varepsilon_1,\ldots,\varepsilon_m)(a,b) = W_{m-1}(\varepsilon_1,\ldots,\varepsilon_{m-1})(a^{\varepsilon_m},v(a^{\varepsilon_m},b^{\varepsilon_m})),$$

where again each $\varepsilon_k = \pm 1$, so that there are 2^m words at the mth step of the construction. As with Horowitz's construction, these 2^m words generate non-conjugate maximal cyclic subgroups of F_2 and give rise to the same character.

Since J is an automorphism, $J(a^\varepsilon) = a^{-\varepsilon}$, and since $J(v(a^\varepsilon,b^\varepsilon)) = v(a^{-\varepsilon},b^{-\varepsilon})$, we have that

$$J(W_m(\varepsilon_1,\ldots,\varepsilon_m)(a,b)) = W_{m-1}(\varepsilon_1,\ldots,\varepsilon_{m-1})(J(a^{\varepsilon_m}),v(J(a^{\varepsilon_m}),J(b^{\varepsilon_m})))$$
$$= W_{m-1}(\varepsilon_1,\ldots,\varepsilon_{m-1})(a^{-\varepsilon_m},v(a^{-\varepsilon_m},b^{-\varepsilon_m}))$$
$$= W_m(\varepsilon_1,\ldots,\varepsilon_{m-1},-\varepsilon_m)(a,b).$$

We note here that in his discussion, Buser also gives a very nice geometric description of this construction.

Masters [Mat00] uses a slightly different approach. Let $\{p_n\}$, $\{q_n\}$, and $\{k_n\}$ be sequences of positive integers, for $n \geq 1$, and set

$$W_n(x,y) = (x^{p_n-1+q_n}y^{-q_n})^{k_n}\, x\, (x^{p_n-1+q_n}y^{-q_n})\, x^{-1}$$

and

$$\overline{W}_n(x,y) = x\, (x^{p_n-1+q_n}y^{-q_n})^{k_n}\, x^{-1}\, (x^{p_n-1+q_n}y^{-q_n}).$$

Note that if I is the canonical involution for the free group generated by x and y, we have that

$$I(W_n(x,y)) = (x^{-1} y^{-q_n}) \overline{W}_n(x,y) (x^{-1} y^{-q_n})^{-1},$$

and so $\chi[W_n(x,y)] = \chi[\overline{W}_n(x,y)]$.

In this case, the nodes in the tree are ordered pairs of elements. Consider the words

$$w_{1,1} = W_1(a,b)$$

and

$$w_{1,2} = \overline{W}_1(a,b).$$

By the argument in the previous paragraph, $\chi[w_{1,1}] = \chi[w_{1,2}]$, and the root of the tree is the ordered pair $(w_{1,1}, w_{1,2})$. The left branch from the root corresponds to the ordered pair

$$(w_{2,1}, w_{2,2}) = (W_2(w_{1,1}, w_{1,2}), \overline{W}_2(w_{1,1}, w_{1,2})),$$

and the right branch from the root corresponds to the ordered pair

$$(w_{2,3}, w_{2,4}) = (W_2(w_{1,2}, w_{1,1}), \overline{W}_2(w_{1,2}, w_{1,1})).$$

The canonical involution yields that the two words in each ordered pair have the same character, and the switching automorphism relative to the generators for the root of the tree interchange the two branches.

To generate the binary tree, we iterate this construction: each node v in the tree is marked by an ordered pair of elements of F_2 of the same character; if the depth of v is m (where here the root has depth 1), one of the two branches of depth $m + 1$ descending from v is marked by W_{m+1} and $\overline{W}_{m+1}$ applied to the pair of elements marking v, while the other branch descending from v is marked by first applying the switching automorphism to the pair of words marking v and then applying W_{m+1} and $\overline{W}_{m+1}$ to these words. Again, all the words of the same depth have the same character. (We note that Masters considers only a part of this tree, as he uses only m elements of depth m and not all 2^m.) The reason for choosing the sequences of exponents is to ensure that, when the free groups are realized inside a 3-manifold group, the m considered non-conjugate elements in the free groups remain non-conjugate in the ambient 3-manifold group.

Pignataro and Sandler [PS74] also construct a binary tree in which each node is marked by an ordered pair of elements of $F_2 = \text{free}(a,b)$. The tree is rooted at F_2. Consider the word $W_0(a,b) = a \cdot b^2 \cdot a^{-1}$, and set $W(a,b) = W_0(a,b) \cdot J(W_0(a,b)) \cdot W_0(a,b)^{-1} \cdot J(W_0(a,b))^{-1}$. (Substituting in $W_0(a,b)$

into the expression for $W(a,b)$ gives that $W(a,b) = a \cdot b^2 \cdot a^{-2} \cdot b^{-2} \cdot a^2 \cdot b^{-2} \cdot a^{-2} \cdot b^2 \cdot a$, and it is important for their analysis that $W(a,b)$ is a palindrome in a and b.)

Suppose that a node v is marked by the ordered pair (U,V). The node on the left branch descending from v is then marked by the ordered pair $(W(U,V), W(U,V^{-1}))$; since

$$W(U,V) = W_0(U,V) \cdot J(W_0(U,V)) \cdot W_0(U,V)^{-1} \cdot J(W_0(U,V))^{-1}$$

and

$$
\begin{aligned}
W(U,&V^{-1})\\
&= W_0(U,V^{-1}) \cdot J(W_0(U,V^{-1})) \cdot W_0(U,V^{-1})^{-1} \cdot J(W_0(U,V^{-1}))^{-1}\\
&= W_0(U,V)^{-1} \cdot J(W_0(U,V))^{-1} \cdot W_0(U,V) \cdot J(W_0(U,V)),
\end{aligned}
$$

we see that $W(U,V^{-1})$ is conjugate to $W(U,V)$ and hence the two elements marking the node v give rise to the same character, for all nodes except the root of the tree.

The node on the right branch descending from v is marked by the ordered pair $(W(V,U), W(V,U^{-1}))$, which is obtained from the ordered pair marking the node on the left branch by applying the switching automorphism relative to U, V to the first element in the pair and by applying the switching automorphism and the automorphism J, both relative to U, V, to the second element in the pair. However, since $W(U,V)$ is a palindrome, the action of J is the same as inversion.

4. The main structural conjecture

The following is an attempt to formulate a loose conjecture to describe when elements give rise to the same character in F_2:

Conjecture 4.1. Let F_2 be the free group of rank two, and suppose that there are elements w and u of F_2 which generate non-conjugate maximal cyclic subgroups of F_2 and whose associated characters $\chi[w]$ and $\chi[u]$ are equal. Then, there exists a binary tree T of subgroups of F_2 with the following properties:

(i) each node v of the tree is a free subgroup of F_2 of rank two;

(ii) the branches denote proper inclusion, so that if a branch descends from a node V to a node V', then V' is a proper subgroup of V, where we think of the tree as being arranged vertically, with the root at the top;

(iii) for each node V of T, there is a character preserving involution I_V on V which interchanges the two branches descending from V;

(iv) there are nodes V_w and V_u containing w and u, respectively, which have the same depth in T and which are related by the action of the character preserving involutions I_V for nodes V in the tree above V_w and V_u.

Roughly speaking, the constructions of Horowitz, Buser, Masters, and Pignataro and Sandler all fall within the scope of the conjecture.

In the case of Horowitz's construction, the free subgroups in the tree are the subgroups generated by $w_m(\varepsilon_1, \ldots, \varepsilon_m)$ and b, and the involutions are the canonical involutions with respect to these generators.

In the case of Buser's construction, the free subgroups in the tree are the subgroups generated by the $W_m(\varepsilon_1, \ldots, \varepsilon_m)$ and a, and the involutions are the automorphisms J with respect to these generators.

In the case of Masters' construction, the free subgroups in the tree are generated by the pairs of elements marking the nodes in the tree, and the involutions are the switching automorphisms with respect to these generators. The main difficulty here is that the two elements marking the root may not generate a free group.

In the case of Pignataro and Sandlers' construction, the free subgroups are the tree are generated by the pairs of elements marking the nodes in the tree, and the involutions are the switching automorphisms on the ordered pairs marking the nodes. Here, though, for the conjecture to apply, we would need to take the tree in the conjecture to be the tree starting from one of the nodes of depth one in Pignataro and Sandler construction as described in the previous Section.

5. Connections to lengths of curves

We now consider in more detail the connection between discrete, faithful representations of a group G into $SL_2(\mathbb{C})$ and lengths of curves in hyperbolic 2- and 3-manifolds. We begin by resolving a slight ambiguity, as the fundamental groups of hyperbolic 2- and 3-manifolds are discrete subgroups of $PSL_2(\mathbb{C})$ (or $PSL_2(\mathbb{R})$, in the case of surfaces), and not of $SL_2(\mathbb{C})$. Let $P : SL_2(\mathbb{C}) \to PSL_2(\mathbb{C})$ be the quotient map.

It is well known (see for instance Kra [Kra85] and the references contained therein) that a discrete, faithful representation $\widehat{\rho}$ of a finitely generated group G into $PSL_2(\mathbb{C})$ lifts to a discrete, faithful representation ρ of G into $SL_2(\mathbb{C})$ (by which we mean that $\widehat{\rho} = P \circ \rho$) if G contains no 2-torsion. Conversely, if G is a finitely generated group containing no 2-torsion and if ρ is a faithful representation of G into $SL_2(\mathbb{C})$, then the composition $\widehat{\rho} = P \circ \rho$ is necessarily a faithful representation of G into $PSL_2(\mathbb{C})$, as the image $\rho(G)$ of G in $\mathrm{SL}_2(\mathbb{C})$ cannot contain the non-trivial element of the kernel of P, namely $-\mathrm{id}$. (However, *a priori* there still may be a several-to-one correspondence between representations into $\mathrm{SL}_2(\mathbb{C})$ and representations into $\mathrm{PSL}_2(\mathbb{C})$, as there may be distinct representations ρ_1 and ρ_2 of G into $\mathrm{SL}_2(\mathbb{C})$ for which $P \circ \rho_1 = P \circ \rho_2$.)

So, given a discrete, faithful representation ρ of a finitely generated group G with no 2-torsion into $\mathrm{SL}_2(\mathbb{C})$, we can compose with P to obtain a discrete, faithful representation $\widehat{\rho} = P \circ \rho$ of G into $\mathrm{PSL}_2(\mathbb{C})$, which then gives rise to an orientable hyperbolic 3-manifold, namely the quotient $\mathbb{H}^3/\widehat{\rho}(G)$. (We make here the convention that when G is the fundamental group of a surface, we consider discrete, faithful representations ρ of G into $\mathrm{SL}_2(\mathbb{R})$, with quotient surface $\mathbb{H}^2/\widehat{\rho}(G)$, unless explicitly stated otherwise.) (In the cases of interest to us here, the group G will be the fundamental group of an orientable surface of negative Euler characteristic or of a compact hyperbolizable 3-manifold, and will in fact be torsion-free.)

Let A be a loxodromic (or hyperbolic) element of $\mathrm{PSL}_2(\mathbb{C})$, so that A is conjugate to $z \mapsto \lambda^2 z$ for some λ^2 in $\mathbb{C}$ with $|\lambda^2| > 1$. The number λ^2 is the *multiplier* of the loxodromic element A. Note that the multiplier of a loxodromic element of $\mathrm{PSL}_2(\mathbb{C})$ determines the trace of its lift to $\mathrm{SL}_2(\mathbb{C})$ up to sign, as there are two possible lifts of A to $\mathrm{SL}_2(\mathbb{C})$, with traces $\pm(\lambda + \lambda^{-1})$. The *axis* $\mathrm{axis}(A)$ of A is the hyperbolic line in $\mathbb{H}^3$ joining its two fixed points; A acts as translation along its axis. The *translation distance* of A along $\mathrm{axis}(A)$, defined to be the hyperbolic distance between x and $A(x)$ for any point x on $\mathrm{axis}(A)$, is $\ln(|\lambda^2|)$.

Let Γ be a discrete torsion-free subgroup of $\mathrm{PSL}_2(\mathbb{C})$. There is a one-to-one correspondence between free homotopy classes of closed curves in $\mathbb{H}^3/\Gamma$ (or in $\mathbb{H}^2/\Gamma$, in the case that Γ lies in $\mathrm{PSL}_2(\mathbb{C})$) and conjugacy classes of maximal cyclic subgroups of Γ. For a maximal loxodromic element A of Γ, the axis of A projects to a closed geodesic of length $\ln(|\lambda^2|)$ in the quotient manifold $\mathbb{H}^3/\Gamma$. Among all closed curves in the free homotopy class determined by A, the projection of the axis of A has minimal length. We define the length of the free homotopy class of curves determined by A, or equivalently of the conjugacy class of maximal cyclic subgroups of Γ determined by A, to be the length of this geodesic.

For a maximal parabolic element A of Γ, the axis of A is not defined, and there are closed curves in the free homotopy class of A whose lengths go to 0. We define the length of the free homotopy class of curves determined by A, or equivalently of the conjugacy class of maximal cyclic subgroups of Γ determined by A, to be 0. (There are no elliptic elements of Γ, by assumption.)

For a finitely generated group G and an element ρ of $\mathscr{F}(G)$ with discrete image, the *length spectrum* of $\widehat{\rho}(G)$ (where $\widehat{\rho} = P \circ \rho$), or of its quotient manifold

$\mathbb{H}^3/\widehat{\rho}(G)$, is the set of lengths of closed geodesics in $\mathbb{H}^3/\widehat{\rho}(G)$, counted with multiplicity. (Actually, in the case of interest to us here, since we have a representation of G into $\mathrm{PSL}_2(\mathbb{C})$, we have the *marked length spectrum*, which we can think of as the map from G into $\mathbb{R}$ obtained by composing $\widehat{\rho}$ with the function from $\widehat{\rho}(G)$ giving the length of a conjugacy class of maximal cyclic subgroups of $\widehat{\rho}(G)$, using the correspondence described in the previous paragraphs. For closed orientable surfaces equipped with a metric of constant negative curvature, the marked length spectrum contains sufficient information to completely determine the geometry of the surface. The marked length spectrum has been studied by a number of authors; we refer the interested reader to Croke [Cro90] or Otal [Ota90] for more information about the behavior of the length spectra of surfaces.)

We pause here to note the following. In recent years, there has been a great deal of interest in determining the exact behavior of the number $\mathcal{N}(\ell)$ of closed geodesics of length at most ℓ in a hyperbolic n-manifold, or n-orbifold, which is known to be asymptotically $\mathcal{N}(\ell) \sim \frac{1}{(n-1)\ell}e^{(n-1)\ell}$, as well as the statistics of their distribution. We will not explore this connection here, other than to say that arithmetic and non-arithmetic hyperbolic n-manifolds behave differently when viewed by $\mathcal{N}(\ell)$. For further information, we refer the interested reader to Schmutz [Sch96], Luo and Sarnak [LuS94], Marklof [Maf96], and Bolte [Bol93], and to the references contained therein.

Let ρ be an element of $\mathscr{F}(G)$ with discrete image. If two maximal elements w and u of G satisfy $\mathrm{trace}(\rho(w)) = \mathrm{trace}(\rho(u))$ with $\rho(w)$ (and hence $\rho(u)$) loxodromic, then $\widehat{\rho}(w)$ and $\widehat{\rho}(u)$ correspond to closed geodesics of equal length in the quotient manifold $\mathbf{H}^3/\widehat{\rho}(G)$, where $\widehat{\rho} = P \circ \rho$. This follows immediately, since the trace of an element in $\mathrm{SL}_2(\mathbb{C})$ determines the multiplier of the corresponding element in $\mathrm{PSL}_2(\mathbb{C})$, which in turn determines the length of the closed geodesic in the quotient manifold. Specifically, if $\widehat{\rho}(w)$ is loxodromic with multiplier λ^2, then $c = \mathrm{trace}(\widehat{\rho}(w)) = \pm(\lambda + \lambda^{-1})$, and so $\lambda^2 = \frac{1}{2}(c^2 - 2 \pm c\sqrt{c^2 - 4})$, where the sign of the $\pm$ is chosen so that $|\lambda^2| > 1$.

In particular, if G is any finitely generated group and if w and u are two elements of G which generate non-conjugate maximal cyclic subgroups and which satisfy $\chi[w] = \chi[u]$, then $\mathrm{trace}(\rho(w)) = \mathrm{trace}(\rho(u))$ for all $\rho \in \mathscr{F}(G)$, and so the lengths of the free homotopy classes determined by w and u are equal in $\mathbb{H}^3/\widehat{\rho}(G)$ (where $\widehat{\rho} = P \circ \rho$) (or in $\mathbb{H}^2/\widehat{\rho}(G)$, in the case that ρ is a representation into $\mathrm{SL}_2(\mathbb{R})$) for all representations ρ in $\mathscr{F}(G)$ with discrete image. So, finding pairs of closed curves on S whose geodesic representatives have the same hyperbolic length over all hyperbolic structures on S is equivalent to the problem of finding pairs of elements in G that generate non-conjugate maximal cyclic subgroups of G and that give rise to the same character over the space of faithful representations of G into $\mathrm{SL}_2(\mathbb{C})$. We refer the interested reader to Leininger [Lei01], particularly Section 3, for a more detailed discussion of this point.

Randol proved the following result for the length spectrum of a surface.

Theorem 5.1 (Main result of Randol [Ran80]). *Let S be an orientable surface of negative Euler characteristic. Then, the length spectrum of S has unbounded multiplicity.*

We pause here to make the following aside. Randol's theorem, Theorem 5.1, arose out of his interest in earlier work of Guillemin and Kazhdan [GK80], who prove the following. Let M be a closed surface with a metric of negative curvature and simple length spectrum; here, by *simple length spectrum*, we mean that there do not exist closed geodesics on M such that the ratio of their lengths is a rational number. Let Δ be the Laplace-Beltrami operator on $C^\infty(M)$. If there are functions q_1 and q_2 in $C^\infty(M)$ for which the operators $\Delta + q_1$ and $\Delta + q_2$ have coincident spectra, then $q_1 \equiv q_2$. (We note that this result has been generalized to compact negatively curved Riemannian manifolds by Croke and Sharafutdinov [CS98], to whom we refer the interested reader for more information.) In this language, Theorem 5.1 implies that surfaces with a constant negative curvature metric never satisfy this condition of simple length spectrum.

Of course, when discussing the spectrum of the Laplace-Beltrami operator on a hyperbolic surface, it would be remiss to not mention the Selberg trace formula. We refer the interested reader to the paper of McKean [McK72] and the books of Hejhal [Hej] for a more detailed discussion of the trace formula.

Masters proved the following result for the length spectrum of a hyperbolic 3-manifold.

Theorem 5.2 (Theorem 1.2 of Masters [Mat00]). *Let N be a hyperbolic 3-manifold with non-elementary fundamental group. Then, the length spectrum of N has unbounded multiplicity.*

Both Randol and Masters used the earlier work of Horowitz in their proofs. The main difficulty in both cases, more pronounced for 3-manifolds than for surfaces, is not the construction a free subgroup F_2 of the fundamental group G, but rather is to control the problem of elements in F_2 being non-conjugate in F_2 but becoming conjugate in G. For surfaces, the easiest way to get around this difficulty is to make use of the fact that the fundamental group of an orientable surface of negative Euler characteristic contains a large number of nicely behaved free subgroups of rank two. The nicest behaved such subgroups are the malnormal free subgroups. Recall that a subgroup H of a group G is *malnormal* if $gHg^{-1} \cap H = \{1\}$ for all $g \in G - H$. In particular, if F_2 is a malnormal subgroup of G, then elements of F_2 are conjugate in G if and only if they are conjugate in F_2. Hence, one approach to handling the case of a general group G is to construct malnormal free subgroups of G of rank two, and then apply the results from the preceeding Sections.

The fundamental group of an orientable surface S of negative Euler characteristic contains a large number of non-conjugate malnormal free subgroups of rank 2. Some can be constructed geometrically. For example, every pair of pants decomposition of S, of which there are infinitely many (if S is not itself a pair of pants) gives a number of embedded copies of a pair of pants in S, and the fundamental group of each such pair of pants is a malnormal free subgroup of rank 2 of $\pi_1(S)$. (Here, a *pair of pants* is topologically a thrice-punctured sphere, though conformally there are four types: a sphere with 3 points removed, with 2 points and 1 disc removed, with 1 point and 2 discs removed, and with 3 discs removed.) There are also malnormal subgroups of $\pi_1(S)$ corresponding to each embedded torus with one point or disc removed in S. This means that in order to characterize elements of $\pi_1(S)$ with the same character, it becomes necessary to characterize all malnormal free subgroups of $\pi_1(S)$, and even then there are elements with the same character that arise from other constructions, as will be described below.

We may also take a larger embedded subsurface of S whose fundamental group injects into $\pi_1(S)$. For example, if S is closed and we take the standard presentation

$$G = \langle a_1, b_1, \ldots, a_p, b_p \mid [a_1, b_1] \cdots [a_p, b_p] = 1 \rangle,$$

for $G = \pi_1(S)$, then the subgroup $\langle a_1, \ldots, a_p \rangle$ is malnormal and free. This subgroup is the fundamental group of the subsurface constructed by taking a regular neighborhood of $a_1 \cup \cdots \cup a_p$ in S. In fact, this is the subgroup used by Randol [Ran80].

Note that Theorem 5.1 can also be extended to surfaces of infinite type, as such surfaces contain many malnormal free subgroups, again arising from embedded copies of a pair of pants or a torus with one puncture or hole.

In attempting to generalize this method to the fundamental group of a 3-manifold M, we run into the difficulty that the construction of malnormal free subgroups of 3-manifold groups is much more difficult than the construction of such subgroups for surface groups.

Masters resolves this difficulty in the proof of Theorem 5.2 by choosing the elements in the free subgroup carefully and showing directly that they are not conjugate in $\pi_1(M)$, using number theory and a careful choice of the exponents p_n, q_n, and k_n, as described in Section 3.

It is possible to obtain a separate proof of part of Theorem 5.2 in the case of convex co-compact hyperbolic 3-manifolds, a class which includes closed hyperbolic 3-manifolds, using the following Theorem of I. Kapovich, avoiding number theory. (This approach does use different machinery, namely the fact that convex co-compact Kleinian groups are word hyperbolic in the sense of Gromov.)

Theorem 5.3 (Theorem C of Kapovich [Kap99]). *Let G be a torsion-free word hyperbolic group and let Γ be a non-elementary (i.e. not cyclic) subgroup of G. Then there exists a subgroup H of Γ such that H is free of rank 2 which is quasiconvex and malnormal in G.*

Malnormality is a strong condition to impose on a free subgroup F of a group G. There is a less exact but nonetheless still effective method, due to Pignataro and Sandler, which addresses the issue of when non-conjugate elements of F become conjugate in G, which avoids malnormal subgroups. The following Lemma is adapted from an argument given in the proof of Theorem 1 of Pignataro and Sandler [PS74].

Lemma 5.4. *Let G be a finitely generated group without torsion and without $\mathbb{Z} \oplus \mathbb{Z}$ subgroups, and suppose that there exists a discrete faithful representation ρ_0 of G into $\mathrm{SL}_2(\mathbb{C})$. Then, there exists a constant $K > 0$ so that the following holds: for any faithful (but not necessarily discrete) representation ρ of G into $\mathrm{SL}_2(\mathbb{C})$ and for any free subgroup F of rank 2 in G, the inclusion map from the collection $\mathscr{C}(\rho(F))$ of conjugacy classes of maximal cyclic subgroups of $\rho(F)$ to the collection $\mathscr{C}(\rho(G))$ of conjugacy classes of maximal cyclic subgroups of $\rho(G)$ is at most K-to-1.*

Proof. First, we can assume without loss of generality that $P \circ \rho_0(G)$ is a purely loxodromic, geometrically finite subgroup of $\mathrm{PSL}_2(\mathbb{C})$. [If $P \circ \rho_0(G)$ is not geometrically finite, then let M be a *compact core* for $\mathbb{H}^3/(P \circ \rho_0(G))$. We can uniformize M as $(\mathbb{H}^3 \cup \Omega(\Gamma))/\Gamma$ for a purely loxodromic, geometrically finite subgroup of $\mathrm{PSL}_2(\mathbb{C})$. Since Γ is necessarily isomorphic to $P \circ \rho_0(G)$, we can write $\Gamma = P \circ \rho_1(G)$ for a discrete faithful representation ρ_1 of G into $\mathrm{SL}_2(\mathbb{C})$, and then replace ρ_0 with ρ_1.] Set $\Gamma = P \circ \rho_0(G)$.

Since there are no cusps by assumption, there is a one-to-one correspondence between the collection $\mathscr{C}(\rho_0(G))$ of conjugacy classes of maximal cyclic subgroups of $\rho_0(G)$ and the collection of closed geodesics in the hyperbolic 3-manifold $\mathbb{H}^3/\Gamma$. Let $\mathrm{CC}(\mathbb{H}^3/\Gamma)$ be the *convex core* of the hyperbolic 3-manifold $\mathbb{H}^3/\Gamma$, and note that $\mathrm{CC}(\mathbb{H}^3/\Gamma)$ contains all of the closed geodesics in $\mathbb{H}^3/\Gamma$.

Note that $P \circ \rho_0(F)$ is a purely loxodromic, geometrically finite subgroup of Γ. Let $\pi : \mathbb{H}^3/(P \circ \rho_0(F)) \to \mathbb{H}^3/\Gamma$ be the covering map. Since the convex core of $\mathbb{H}^3/(P \circ \rho_0(F))$ is compact, its image under π is compact as well. Since there is a positive lower bound on the injectivity radius of $\mathbb{H}^3/\Gamma$, there is some $K > 0$ so that π is at most K-to-1.

We can reinterpret this geometric fact as saying that the map from the collection $\mathscr{C}(F)$ of maximal cyclic subgroups of F to the collection $\mathscr{C}(G)$ of conjugacy classes of maximal cyclic subgroups of G is at most K-to-1. Hence, any faithful representation ρ of G into $SL_2(\mathbb{C})$ has the same property. $\square$

Note that this argument can be made to work for a group G containing $\mathbb{Z} \oplus \mathbb{Z}$ subgroups, by carefully analyzing the behavior of the covering map at the cusps.

Underlying all of this discussion is the fact that free subgroups of rank 2 are very common in any group that admits a faithful representation into $SL_2(\mathbb{C})$, as such groups satisfy the Tits alternative: for any two elements A and B of $SL_2(\mathbb{C})$ of infinite order and with disjoint fixed point sets, there are integers n and m so that $\langle A^n, B^m \rangle$ is free of rank 2. We may then apply Lemma 5.4 to these subgroups. In particular, this implies that the characterization of pairs of elements of G with equal characters is extremely complicated.

We note that more is known about 2-generator subgroups of Kleinian groups. For instance, Ratcliffe [Rat87] shows that for a torsion-free, two generator, discrete subgroup Γ of either $SL_2(\mathbb{C})$ or of $PSL_2(\mathbb{C})$, either Γ is free abelian of rank two, $\mathbb{H}^3/\Gamma$ has finite volume, or Γ is free of rank two. Reid [Rei92] has shown that there are infinitely many closed 2-generator hyperbolic 3-manifolds which have a proper finite sheeted cover which is also 2-generator, which is behavior that is very unlike the surface case.

We now expand our horizons. Let S be an orientable surface of negative Euler characteristic, and let $\mathscr{T}(S)$ denote the Teichmüller space of hyperbolic structures on S. Let $\mathscr{C}(S)$ denote the set of free homotopy classes of homotopically non-trivial closed curves on S. There is a natural map $\mathscr{L} : \mathscr{C}(S) \times \mathscr{T}(S) \to [0, \infty)$, given by setting

$$\mathscr{L}([c], g) = \text{length}_g([c]),$$

where $\text{length}_g([c])$ is defined to be the infimum of the lengths of the closed curves on S in the free homotopy class $[c]$ determined by c, measured using the hyperbolic structure g on S.

As has already been noted, since an element of $\pi_1(S)$ and its inverse correspond to the same curve on S with opposite orientations, there is a one-to-one correspondence between the collection $\mathscr{C}(\pi_1(S))$ of conjugacy classes of maximal cyclic subgroups of $\pi_1(S)$ and the collection $\mathscr{C}(S)$ of free homotopy classes of closed curves on S. Theorem 1.5 and Proposition 1.4 can be thought of as evidence for the view that for fixed $g \in \mathscr{T}(S)$, the function $\mathscr{L}(\cdot, g) : \mathscr{C}(S) \to [0, \infty)$ often has multiplicity at least two, with two representative conjugacy classes generated by w and $I(w)$.

As noted by Randol [Ran80], the Bumpy Metric theorem (see Abraham [Abr70], Anosov [Ano82]) implies that, if we expand the second factor of the domain to be the space $\mathscr{R}(S)$ of all Riemannian metrics on S, then the function

$$\mathscr{L}(\cdot,g) : \mathscr{C}(S) \to [0,\infty)$$

for fixed $g \in \mathscr{R}(S)$ is generically injective.

Hence, there is something non-generic about the hyperbolic metrics on a surface, and it would be nice to have a conjecture that captures this non-genericity. Note that it cannot be as simple as saying that hyperbolic metrics are exactly the metrics g for which the function $\mathscr{L}(\cdot,g)$ on $\mathscr{C}(S)$ has unbounded multiplicity, by the following example due to Buser [private communication]. Let S be a closed orientable surface of genus 2, let c be a simple closed separating curve on S, let U be an open regular neighborhood of c, and consider a metric g on S that is hyperbolic on one component of $S - U$ and not hyperbolic on the other component. The hyperbolic component of the surface then contributes to the unboundedness of the multiplicity of the length spectrum of S, and the metric on the other component can be chosen to be anything.

So, consider the action of $\mathrm{Diff}(S)$ on $\mathscr{R}(S)$ by pullback. Let

$$G_S = \{f \in \mathrm{Diff}(S) \,|\, f^*(\mathscr{T}(S)) = \mathscr{T}(S)\}$$

be the collection of all diffeomorphisms of S that pull hyperbolic metrics back to hyperbolic metrics. It is immediate that G_S is a subgroup of $\mathrm{Diff}(S)$, by elementary properties of pullback.

Question 5.5. Does there exist a diffeomorphism f of S so that $f^*(\mathscr{T}(S))$ is a proper subset of $\mathscr{T}(S)$?

Say that a metric $g \in \mathscr{R}(S)$ is *wacky* if the map $\mathscr{L}(\cdot,g) : \mathscr{C}(S) \to [0,\infty)$ has unbounded multiplicity. For example, every hyperbolic metric is wacky, while a generic metric is not wacky. Let $\mathscr{W}(S)$ be the collection of all wacky metrics on S, and consider the group

$$H_S = \{f \in \mathrm{Diff}(S) \,|\, f^*(\mathscr{W}(S)) = \mathscr{W}(S)\}.$$

The following conjecture attempts to capture what is special about hyperbolic metrics in this context.

Conjecture 5.6. Let S be an orientable surface of negative Euler characteristic. Then, G_S is a maximal connected subgroup of H_S.

6. Character preserving automorphisms

Let $\mathrm{Aut}(G)$ denote the group of all automorphisms $\varphi : G \to G$ of G, and let $\mathrm{Inn}(G)$ denote the subgroup of $\mathrm{Aut}(G)$ consisting of the inner automorphisms $\varphi_g : G \to G$, given by $\varphi_g(h) = g \cdot h \cdot g^{-1}$ for $g \in G$. Let

$$\text{Aut}_\chi(G) = \{\varphi \in \text{Aut}(G) \mid \chi[g] = \chi[\varphi(g)] \text{ for all } g \in G\}$$

be the group of *character preserving automorphisms*. Note that $\text{Inn}(G) \subset \text{Aut}_\chi(G)$, by the basic properties of trace. As the constructions described in Section 3 and the conjecture given in Section 4 rely on the fact that J is a character preserving automorphism of F_2, and in some sense is the only one defined on all of F_2, we need to understand the group $\text{Aut}_\chi(G)$.

This group has been completely determined for free groups by Horowitz.

Theorem 6.1 (Theorem 1 of Horowitz [Hor75]). *Let F_n be the free group of rank n. If $n \geq 3$, we have that $\text{Aut}_\chi(F_n) = \text{Inn}(F_n)$. If $n = 2$, we have that $\text{Aut}_\chi(F_2) = \langle \text{Inn}(F_2), J \rangle$, where J is the automorphism defined in Section 1.*

It is known that automorphisms of the free group F_2 of rank two are all geometric, in that if we realize F_2 as the fundamental group of a punctured torus T, then every automorphism is induced by the action of a homeomorphism of T. (However, this is no longer true if we realize F_2 as the fundamental group of a thrice-punctured sphere.) Moreover, given any two elements w and u of $F_2 = \text{free}(a,b)$, the homomorphism $\varphi : F_2 \to F_2$ defined by $\varphi(a) = w$ and $\varphi(b) = u$ is an automorphism if and only if the commutator $[w,u]$ is conjugate to $[a,b]$. However, very few automorphisms of F_p for $p \geq 3$ are geometric, see Gersten [Ger83].

Let $G_p = \langle a_1, b_1, \ldots, a_p, b_p \mid [a_1, b_1] \cdots [a_p, b_p] = 1 \rangle$ be the standard presentation of the fundamental group of the closed orientable surface S_p of genus $p \geq 2$. We consider the question of determining $\text{Aut}_\chi(G_p)$. It is a result of Nielsen [Nie27] that all of the automorphisms of G_p are geometric.

In the case $p = 2$, there is an analogue on S_2 of the involution J on F_2, namely the *hyperelliptic involution*. This is a conformal involution of S_2. For more information about the hyperelliptic involution, we refer the interested reader to Farkas and Kra [FaK80].

By work of Haas and Susskind [HS89], the hyperelliptic involution has the following characterization. For $p \geq 2$, let S_p be the closed orientable surface of genus p, and let f be an orientation-preserving homeomorphism of S_p with the property that for every simple closed curve α on S_p, $f(\alpha)$ is freely homotopic to either α or $-\alpha$ (where $-\alpha$ is the curve α with the opposite orientation). Then, either f is homotopic to the identity, or $p = 2$ and f is homotopic to the hyperelliptic involution. Conversely, on a closed orientable surface S_2 of genus two, the hyperelliptic involution J preserves the free homotopy class of every simple closed curve, and reverses the orientation of the curve if and only if the curve is non-separating. So, in terms of the standard presentation for G_2 given above, we see that $J(a_k) = a_k^{-1}$ and $J(b_k) = b_k^{-1}$. In particular, since J preserves the length of every simple closed curve on S_2, we have that J is character preserving on G_2.

We begin with the following Lemma, which is the analogue for G_p of Theorem 1.3. We note that a different proof of this Lemma is given by McShane [McS93].

Lemma 6.2. *Let G_p be the fundamental group of the closed orientable surface S_p of genus $p \geq 2$. Let $g \in G_p$ be a maximal element that represents a simple closed curve on S_p. Then, g determines $\chi[g]$; that is, if there exists a maximal element $h \in G_p$ with $\chi[h] = \chi[g]$, then h is conjugate to $g^{\pm 1}$.*

Proof. First, we restrict attention to the discrete, faithful representations of G_p into $SL_2(\mathbb{R})$, so that we get hyperbolic structures on S_p by taking the quotient $\mathbb{H}^2$ by $P \circ \rho(G_p)$. (Here, we are using the fact that since G_p has no 2-torsion, $\rho(G_p)$ in $SL_2(\mathbb{R})$ is isomorphic to $P \circ \rho(G_p)$ in $PSL_2(\mathbb{R})$.)

For each hyperbolic structure on S_p, the length of a closed geodesic on S_p determines the character of the corresponding element of G_p, and vice versa, by the discussion in Section 5. In particular, equal characters for two elements of G_p imply that the corresponding closed geodesics on S_p have equal lengths, independent of the hyperbolic structure on S_p.

If c is a homotopically non-trivial non-simple closed curve on S_p, there is a uniform positive lower bound of $2\ln(1 + \sqrt{2})$ for the length for the closed geodesic homotopic to c over all hyperbolic structures on S_p, see Hempel [Hem84]. However, if c' is a homotopically non-trivial simple closed curve, there is no positive minimum length for the closed geodesic homotopic to c' over all hyperbolic structures on S_p. In fact, there exist hyperbolic structures on S_p for which the length of the closed geodesic homotopic to c' goes to 0. Hence, since g represents a simple closed curve on S_p, h must also represent a simple closed curve on S_p.

Now, we are reduced to considering two simple closed curves on S_p so that the lengths of their corresponding closed geodesics are equal, independent of the hyperbolic structure on S_p. If they intersect, then the Collar Lemma, see for instance Buser [Bus92], implies that the length of one goes to infinity as the length of the other goes to 0. If they are disjoint, we may use Fenchel-Nielsen coordinates, see for instance Abikoff [Abi80], to see that the length of one can be made to go to 0 without changing the length of the other. Hence, we see that the curves must coincide, which is equivalent to saying that g and h are conjugate up to inverse, as desired.   □

Theorem 6.3. *Let G_p be the fundamental group of the closed orientable surface S_p of genus $p \geq 2$. For $p \geq 3$, we have that $\mathrm{Aut}_\chi(G_p) = \mathrm{Inn}(G_p)$. For $p = 2$, we have that*

$\mathrm{Aut}_\chi(G_2) = \langle \mathrm{Inn}(G_2), J \rangle$, *where the involution J of G_2 arises from the hyperelliptic involution on S_2.*

Proof. As in the case of free groups, the basic properties of trace yield immediately that $\mathrm{Inn}(G_p) \subset \mathrm{Aut}_\chi(G_p)$. For surfaces of genus 2, the hyperelliptic involution is an isometry for every hyperbolic structure on S, and so preserves the lengths of closed geodesics and hence also preserves characters. This shows that $\langle \mathrm{Inn}(G_2), J \rangle \subset \mathrm{Aut}_\chi(G_2)$

Now, let φ be an element of $\mathrm{Aut}_\chi(G_p)$. Let $g \in G_p$ be any element that represents a homotopically non-trivial simple closed curve on S_p. By Lemma 6.2, we see that $\varphi(g)$ must be conjugate to g. In particular, the automorphism φ of G_p corresponds to a homeomorphism f_φ of S_p that takes each simple closed geodesic to itself, possibly reversing the orientation of the geodesic.

We apply Theorem 1 of McShane [McS93] to see that this homeomorphism f_φ of S_p must be homotopic to an isometric map. (If we knew that φ was induced by an orientation-preserving homeomorphism of S_p, then we could apply the result of Haas and Susskind described above.) In the case $p = 2$, the only self-maps of S_p that are isometries of every hyperbolic structure are the identity and the hyperelliptic involution J. In the case $p \geq 3$, the only self-map of S_p that is an isometry of every hyperbolic structure on S_p is the identity, as desired.   □

7. Variants

There has been a wide variety of work in related areas by a number of authors.

Jørgensen [Jør82], [Jør00], [Jør76] has studied various aspects, properties, and applications of trace identities in $SL_2(\mathbb{C})$ and $PSL_2(\mathbb{C})$.

Sandler [San98] extended results of the sort discussed in this survey to certain families of faithful representations of F_2 into $SU(2, 1)$, with similar applications to the length spectra of certain complex hyperbolic manifolds.

Thompson [Tho89] showed for each $n \geq 2$, there exists a field k and a subgroup G_n of $SL_2(k)$ which contains a free group F_n of rank n, so that two elements of F_n give rise to the same character if and only if they are conjugate in G_n. Moreover, the field k is explicitly constructed as the algebraic closure of a finitely generated extension field of the rationals $\mathbb{Q}$.

Traina [Tra80] gives an explicit though complicated expression for the Fricke polynomial for an element of $F_2 = \mathrm{free}(a, b)$.

Baribaud [Bar99] studied the lengths of closed geodesics on a pair of pants. She defined a parameter, the *number of strings*, and gave a complete description of those closed geodesics which have the shortest length given their number of strings, for those geodesics with an odd numbers of strings.

Magnus [Mag75] considered this question for other groups. For example, consider the group G with the presentation

$$G = \langle a, b \,|\, W^k = 1 \rangle,$$

where W is a freely reduced word in a and b, and $k > 1$. Then, a necessary condition for G to have a faithful representation into $\mathrm{PSL}_2(\mathbb{C})$ is that if U is an element of G with the same Fricke polynomial as W, then U is conjugate to $W^{\pm 1}$.

McShane [McS98] (see also Bowditch [Bow96]) showed that for any hyperbolic structure on a punctured torus T, the equality

$$\sum_{\gamma} \frac{1}{1 + \exp(|\gamma|)} = \frac{1}{2}$$

holds, where the sum ranges over all simple closed geodesics γ on T and where $|\gamma|$ is the length of the closed geodesic γ. Bowditch [Bow97] has generalized this equality to hyperbolic once-punctured torus bundles.

This series can also be generalized as follows, see McShane [McS98]. Let M be a convex surface without boundary and with a hyperbolic structure of finite area and a cusp x. Then, the equality

$$\sum \frac{1}{1 + \exp(\frac{1}{2}(|\alpha| + |\beta|))} = \frac{1}{2}$$

holds, where the sum is over all pairs α and β of simple closed geodesics which bound a pair of pants containing the cusp x.

Pignataro and Sandler [PS74] use techniques similar to those described in this note to generalize earlier work of Jørgensen and Sandler [JS93]. Let S be an orientable surface of negative Euler characteristic and let c, c' be two closed curves on S, neither homotopic to a peripheral curve, that intersect essentially. For each hyperbolic structure on S, let c and c' also refer to the closed geodesics on S with the given hyperbolic structure that lie in the free homotopy classes determined by c and c'. Then, for any hyperbolic structure on S, let x be a point of intersection of two closed geodesics c and c'. (Note that, even though c and c' will vary on S as the hyperbolic structure varies, there is always a point of intersection corresponding to x for the corresponding closed geodesics with the new hyperbolic structure.) Then, for any $m \geq 2$, there are closed curves $w_1, \ldots, w_m$ passing through x with $\text{length}_g(w_1) = \cdots = \text{length}_g(w_m)$ for every hyperbolic structure g on S.

There is a necessary condition in terms of homology for two elements of a surface group to have the same character.

Proposition 7.1 (Corollary 3.4 of Leininger [Lei01]). *Let G be the fundamental group of an orientable surface S. Let w and u be elements of G with $\chi[w] = \chi[u]$. Then, w and u may be oriented so that they represent the same class in $G/[G,G] = H_1(S, \mathbb{Z})$.*

To close this Section, there is a folklore conjecture, that two closed curves on a surface S of equal length over the Teichmüller space of S can be characterized by their intersection numbers with simple closed curves. Specifically, given two closed curves w and u on an orientable surface S of negative Euler characteristic, let $\text{i}(w, u)$ denote their geometric intersection number, which is equal to the minimum number of intersection points of w' and u', where w' is freely homotopic to w and u' is freely homotopic to u. Equivalently, define $\text{i}(w, u)$ to be the number of (necessarily transversal) intersection points of the geodesic representatives of w and u for any hyperbolic structure on S.

The strong form of this conjecture has recently been resolved in the negative by Leininger [Lei01]. The following Proposition is essentially a consequence of the Collar Lemma; for a complete proof, see [Lei01]. Also see [Lei01] for an explicit example of two elements w and u of $\pi_1(S)$ for which $\text{i}(w, c) = \text{i}(u, c)$ for all simple closed curves c on S but $\chi[w] \neq \chi[u]$, and for a more detailed discussion of this question.

Proposition 7.2 (Corollary 5.4 of Leininger [Lei01]). *Given an orientable surface S of negative Euler characteristic, let w and u be closed curves on S for which $\text{length}_g(w) = \text{length}_g(u)$ for all hyperbolic structures g on S. Then, $\text{i}(w, c) = \text{i}(u, c)$ for all simple closed curves c on S.*

8. Questions and conjectures

There are a number of other questions that can be asked. We present a few of them here.

- The first concerns the relationship between elements of F_2 that are simple in the sense of having stable multiplicity 1 and those that are simple in the sense that they correspond to a simple closed curve on the punctured torus S, when we realize F_2 as the fundamental group of S.

There is an algorithm, see Series [Ser85a], for determining when a closed curve on a punctured torus S, given as a word $w = a^{n_1} \cdot b^{m_1} \cdots a^{n_k} \cdot b^{m_k}$ in $F_2 = \text{free}(a,b)$ $= \pi_1(S)$, is a simple curve. This algorithm involves constructing a finite collection of nested free subgroups of F_2. However, it seems that the character preserving automorphisms do not shed any light on this question of determining simplicity of curves. Even though w is conjugate to $I(w)$ for every simple closed curve w, there are also non-simple closed curves for which w is conjugate to $I(w)$.

Consider the element $w = a^p \cdot b^y \cdot a^p \cdot b^q \cdot a^x \cdot b^q$ in $F_2 = \text{free}(a,b)$, where p, q, x, and y are arbitrary distinct non-zero integers. For most values of p, q, x, and y, w does not represent a simple curve on S, by the algorithm [Ser85a]. However, for any choice of p, q, x, and y, $I(w) = b^y \cdot a^x \cdot b^q \cdot a^p \cdot b^y \cdot a^p$ is conjugate (by $a^p \cdot b^y \cdot a^p$) to w; preliminary computer calculations in this case support the conjecture that w is then a simple element of F_2, and hence has $\text{mult}(w) = 1$. The following Conjecture attempts to make this link precise.

Conjecture 8.1. Let w be an element of F_2. If w is conjugate to $I(w)$, then $\text{mult}(w) = 1$.

It is a difficult question to characterize those elements of F_2 with stable multiplicity 1. One possibility is the following. Let $w = w(a,b)$ be an element of $F_2 = \text{free}(a,b)$. Say that w is *prime* if w does not admit a non-trivial decomposition as $w(a,b) = w'(u,v)$, where $u = u(a,b)$ and $v = v(a,b)$ are elements of F_2. (We require that at least one of u and v be non-trivial in F_2, that is, not a primitive element of F_2.) If w is not prime, then say that w is *composite*.

Conjecture 8.2 (Ginzburg and Rudnick [GR98]). Let w be a prime element of F_2. Then, w is pseudo-simple.

Conjecture 8.3. Let w be a prime element of F_2, and suppose there exists an element u of F_2 for which $\chi[w] = \chi[u]$. Then, u is prime, and u is conjugate to $w^{\pm 1}$.

The difficulty now becomes characterizing which elements of F_2 are prime.

One small piece of evidence for Conjecture 8.2 is the following construction. Let w be a composite element of F_2, so that we may write w as a word $w = w'(u,v)$, where $u = u(a,b)$ and $v = v(a,b)$ are non-trivial words in F_2. Let I be the canonical involution on F_2, and let I' be the canonical involution on the subgroup $F = \text{free}(u,v)$ of F_2. Then, $\chi[w'] = \chi[I'(w')]$, but in general, one expects that w' and $I'(w')$ are not conjugate in F, and hence not in F_2, depending on the specifics of the expressions of w, u, and v. Moreover, in this case one expects that, when expressed in terms of a and b, $I'(w'(u(a,b),v(a,b)))$ is not conjugate to either w or $I(w)$, and so the stable multiplicity of w is then at least three.

There is a related question, due to Riven:

Question 8.4. Let u and w be elements of F_2 so that $\chi[w] = \chi[u]$. Does there exist a generating set $\{x,y\}$ for F_2 so that $u^{\pm 1}$ is conjugate to $I(w)$, where I is the canonical involution with respect to the generating set $\{x,y\}$?

It could also be asked whether the question of the existence of such a generating set is or is not decidable. I would like to thank the first referee for bringing this question to my attention.

There is a topological interpretation of Conjecture 4.1. Let S be the punctured torus, and consider F_2 as $\pi_1(S)$. Let g_1 and g_2 be two elements of F_2 corresponding to closed curves on S not homotopic to the puncture on S. Then, $H = \langle g_1, g_2 \rangle$ is free of rank two and of infinite index in $\pi_1(S)$, unless g_1 and g_2 generate $\pi_1(S)$. As all automorphisms of F_2 are realized by homeomorphisms of S, we can phrase Conjecture 4.1 in terms of a tree of covers of S, where each node is a torus with a hole or a pair of pants, and of homeomorphisms of the node surfaces that realize the respective automorphisms.

- (Asked by U. Hamenstädt at the Workshop on Kleinian Groups and Hyperbolic 3-Manifolds, held at the University of Warwick, September 2001) Is there a connection between the stable multiplicity of a closed curve on a surface and the number of its self-intersections?

Let S be an orientable surface of negative Euler characteristic. Given a closed curve c on S, let $\mathrm{mult}(c)$ denote its stable multiplicity and let $\mathrm{self\_int}(c)$ denote the number of its self intersections, defined to be the minimum of the number of self intersections of any closed curve freely homotopic to c. Note that $\mathrm{self\_int}(c)$ is independent of the hyperbolic structure on S.

Basmajian [Bas93] showed, see Corollary 1.2 of [Bas93], that for each $k \geq 1$, there exists a constant M_k, depending only on k and satisfying $\lim_{m \to \infty} M_k = \infty$, so that if $\mathrm{self\_int}(c) = k$, then $\mathrm{length}_g(c) \geq M_k$ for every hyperbolic structure g on S.

So, for a closed curve c on S with $\mathrm{self\_int}(c) = k$, choose a hyperbolic structure z on S which minimizes $\mathrm{length}_g(c)$ as g ranges over $\mathcal{T}(S)$. Since $\lim_{k \to \infty} M_k = \infty$, there exists K so that $M_k > \mathrm{length}_z(c)$ for all $k \geq K$. Hence, if c' is another closed curve on S and $\mathrm{self\_int}(c') \geq K$, then c and c' must determine distinct character classes, since there is a hyperbolic structure on S, namely z, for which the closed geodesics freely homotopic to c and c' must have different lengths. However, this argument has the flaw that it is not uniform in the self intersection number of c, but relies on first determining the minimal length of c over all hyperbolic structures on S.

On a punctured torus, Conjecture 8.1 and the algorithm for simplicity which is described in Series [Ser85a] imply that there exist closed curves c_n on S for which self_int$(c_n) \to \infty$ but mult$(c_n) = 1$ for all n. So, the following question remains unresolved: does a bound on self_int(c) give a bound on mult(c)?

- This whole paper has been concerned with determining when there are elements w and u of F_2, or of a finitely generated group G, for which $\chi[w] - \chi[u] = 0$. Are there other functions, perhaps variants of McShane's identity, as discussed in Section 7, that hold for characters?

In general, there cannot exist w and u for which $\chi[w] + \chi[u] = 0$. Such pairs of elements would correspond to closed curves of equal length on the quotient manifold but would not be detected by the methods that have been discussed in this note, as their characters are not equal. Let G be a finitely generated group with the property that every point in $\mathrm{Hom}(G, \mathrm{SL}_2(\mathbb{C}))$ is an accumulation point of $\mathscr{F}(G)$; in particular, $\mathscr{F}(G)$ is dense in $\mathrm{Hom}(G, \mathrm{SL}_2(\mathbb{C}))$. Free groups of finite rank and fundamental groups of closed orientable surfaces are examples of such groups. If there were elements w and u of G for which $\chi[w] = -\chi[u]$, then for every odd $m \geq 1$ we would have that $\chi[w^m] = -\chi[u^m]$. We could choose m large enough so that $\langle w^m, u^m \rangle$ is a free group of rank two. By the assumption on G, there would exist a sequence of representations $\{\rho_n\}$ in $\mathscr{F}(G)$ converging to the element of $\mathrm{Hom}(G, \mathrm{SL}_2(\mathbb{C}))$ taking every element of G to the identity. In particular, both $\{\rho_n(w^m)\}$ and $\{\rho_n(u^m)\}$ would converge to the identity, at which point $\chi[w^m] = \chi[u^m] = 2$, as both would be equal to $\chi[\mathrm{id}]$, a contradiction. (This argument is adapted from an argument due to Horowitz [Hor75].)

- Are there analogous results for the length spectra of more general classes of spaces?

In this note, we have discussed this question for hyperbolic 2- and 3-manifolds. Leininger [Lei01] discusses and answers this question for certain classes of path metrics on surfaces, specifically the singular Euclidean metrics. However, the question of whether analogous results hold, for instance, for pleated surfaces, or singular hyperbolic surfaces, or for 3-dimensional hyperbolic cone manifolds, is still open.

Acknowledgements

I would like to thank Chris Croke, Ruth Gornet, Peter Buser, Ursula Hamenstädt, and Chris Leininger for helpful conversations over the life of this work. I would also like to thank Chris Leininger for pointing out several mistakes in an earlier version of the manuscript, and both referees for their helpful comments, which greatly helped to improve the quality of the exposition.

References

[Abi80] W. Abikoff (1980). *Topics in the real analytic theory of Teichmüller space (Lec. Notes Math.* **820***),* Springer-Verlag.

[Abr70] R. Abraham (1970). Bumpy metrics. In *Global Analysis, Proc. Sympos. Pure Math.* **XIV**, Berkeley, CA, 1968, Amer. Math. Soc., 1–3.

[Ano82] D.V. Anosov (1982). Generic properties of closed geodesics (Russian), *Izv. Akad. Nauk SSSR Ser. Math.* **46**, 675–709, 896.

[Bar99] C.M. Baribaud (1999). Closed geodesics on pairs of pants, *Israel J. Math.* **109**, 339–347.

[Bas93] A. Basmajian (1993). The stable neighborhood theorem and lengths of closed geodesics, *Proc. Amer. Math. Soc.* **119**, 217–224.

[Bas97] A. Basmajian (1997). Selected problems in Kleinian groups and hyperbolic geometry, *Contemp. Math.* **211**, Amer. Math. Soc..

[Bol93] J. Bolte (1993). Periodic orbits in arithmetical chaos on hyperbolic surfaces, *Nonlinearity* **6**, 935–951.

[Bow96] B.H. Bowditch (1996). A proof of McShane's identity via Markoff triples, *Bull. Lond. Math. Soc.* **28**, 73–78.

[Bow97] B.H. Bowditch (1997). A variation of McShane's identity for once-punctured torus bundles, *Topology* **36**, 325–334.

[Bus92] P. Buser (1992). Geometry and spectra of compact Riemann Surfaces, *Progr. Math.* **106**, Birkhäuser.

[Cro90] C.B. Croke (1990). Rigidity for surfaces of non-positive curvature, *Comment. Math. Helv.* **65**, 150–169.

[CS98]   C.B. Croke and V.A. Sharafutdinov (1998). Spectral rigidity of a compact negatively curved manifold, *Topology* **37**, 1265–1273.

[FaK80]   H. Farkas and I. Kra (1980). *Riemann surfaces (Graduate Texts in Mathematics 71)*, Springer-Verlag.

[FK97]   R. Fricke and F. Klein (1897). *Vorlesungen über die Theorie der Automorphen Funktionen, volume 1*, B. G. Teubner.

[Ger83]   S.M. Gersten (1983). Geometric automorphisms of a free group of rank at least three are rare, *Proc. Amer. Math. Soc.* **89**, 27–31.

[GK80]   V. Guillemin and D. Kazhdan (1980). Some inverse spectral results for negatively curved 2-manifolds, *Topology* **19**, 301–312.

[GM93]   F. González-Acu na and J.M. Montesinos-Amilibia (1993). On the character variety of group representations in SL$(2, \mathbb{C})$ and PSL$(2, \mathbb{C})$, *Math. Zeit.* **214**, 627–652.

[GR98]   D. Ginzburg and Z. Rudnick (1998). Stable multiplicities in the length spectrum of Riemann surfaces, *Israel J. Math.* **104**, 129–144.

[Ham01]   U. Hamenstädt (2001). New examples of maximal surfaces, *L'Enseignement Math.* **47**, 65–101.

[Hej]   D.A. Hejhal. The Selberg Trace Formula for PSL$_2(\mathbb{R})$.
            (1976). Volume 1, *Lec. Notes Math.* **548**, Springer-Verlag.
            (1983). Volume 2, *Lec. Notes Math.* **1001**, Springer-Verlag.

[Hem84]   J. Hempel (1984). Traces, lengths, and simplicity of loops on surfaces, *Topology Appl.* **18**, 153–161.

[Hor72]   R.D. Horowitz (1972). Characters of Free Groups Represented in the Two-Dimensional Special Linear Group, *Comm. Pure Appl. Math* **25**, 635–649.

[Hor75]   R.D. Horowitz (1975). Induced automorphisms on Fricke characters of free groups, *Trans. Amer. Math. Soc.* **208**, 41–50.

[HS89]   A. Haas and P. Susskind (1989). The geometry of the hyperelliptic involution in genus two, *Proc. Amer. Math. Soc.* **105**, 159–165.

[Hum01]   S. Humphries (2001). Action of braid groups on compact spaces with chaotic behavior, *preprint*.

[Jør76]   T. Jørgensen (1976). On discrete groups of Möbius transformations, *Amer. J. Math.* **98**, 739–749.

[Jør78]   T. Jørgensen (1978). Closed geodesics on Riemann surfaces, *Proc. Amer. Math. Soc.* **72**, 140–142.

[Jør82]   T. Jørgensen (1982). Traces in 2-generator subgroups of $SL_2(\mathbb{C})$, *Proc. Amer. Math. Soc.* **84**, 339–343.

[Jør00]   T. Jørgensen (2000). Composition and length of hyperbolic motions, *Contemp. Math.* **256**, 211–220.

[JS93]   T. Jørgensen and H. Sandler (1993). Double points on hyperbolic surfaces, *Proc. Amer. Math. Soc.* **119**, 893–896.

[Kap99]   I. Kapovich (1999). A non-quasiconvexity embedding theorem for hyperbolic groups, *Math. Proc. Cam. Phil. Soc.* **127**, 461–486.

[Kra85]   I. Kra (1985). On Lifting Kleinian Groups to $SL(2,\mathbb{C})$. In *Differential Geometry and Complex Analysis, edited by I. Chavel and H. M. Farkas*, Springer-Verlag, 181–193.

[Lei01]   C.J. Leininger (2001). Equivalent curves in surfaces, *to appear, Geometriae Dedicata.*

[LuS94]   F. Luo and P. Sarnak (1994). Number variance for arithmetic hyperbolic surfaces, *Comm. Math. Phys.* **161**, 419–432.

[LyS77]   R.C. Lyndon and P.E. Schupp (1977). Combinatorial group theory, *Ergebnisse der Mathematik und ihrer Grenzgebiete* **89**, Springer-Verlag.

[Mag75]   W. Magnus (1975). Two generator subgroups of $PSL_2(\mathbb{C})$, *Nachr. Akad. Wiss. Göttingen Math. – Phys. Kl. II* **7**, 81–94.

[Mag80]   W. Magnus (1980). Rings of Fricke characters and automorphism groups of free groups, *Math. Zeit.* **170**, 91–103.

[Maf96]   J. Marklof (1996). On multiplicities of length spectra of arithmetic hyperbolic three-orbifolds, *Nonlinearity* **9**, 517–536.

[Mat00]   J.D. Masters (2000). Length multiplicities of hyperbolic 3-manifolds, *Israel J. Math.* **119**, 9–28.

[McK72]   H.P. McKean (1972). Selberg's trace formula as applied to a compact Riemann surface, *Comm. Pure Appl. Math* **25**, 225–246.

[McS93]   G. McShane (1993). Homeomorphisms which preserve simple geodesics, *preprint*.

[McS98]   G. McShane (1998). Simple geodesics and a series constant over Teichmüller space, *Invent. Math.* **132**, 607–632.

[Nie27]   J. Nielsen (1927). Untersuchungen zur Topologie der geschlossenen zwei-seitigen Flächen, *Acta Math.* **50**, 189–358.

[Ota90]   J.-P. Otal (1990). Le spectre marqué des longueurs des surfaces á courbure né gative, *Ann. Math.* **131**, 151–162.

[PS74]   T. Pignataro and H. Sandler (1974). Families of closed geodesics on hyperbolic surfaces with common self-intersections, *Contemp. Math.* **169**, 481–489.

[Ran80]   B. Randol (1980). The length spectrum of a Riemann surface is always of unbounded multiplicity, *Proc. Amer. Math. Soc.* **78**, 455–456.

[Rat87]   J.G. Ratcliffe (1987). Euler characteristics of 3-manifold groups and discrete subgroups of $SL(2,\mathbb{C})$, *J. Pure Appl. Algebra* **44**, 303–314.

[Rei92]   A.W. Reid (1992). Some remarks on 2-generator hyperbolic 3-manifolds. In *Discrete groups and geometry, edited by W.J. Harvey and C. Maclachlan (LMS Lecture Notes 173)*, Cambridge University Press, 209–219.

[San98]   H. Sandler (1998). Trace equivalence in $SU(2,1)$, *Geom. Dedicata* **69**, 317–327.

[Sch96]   P. Schmutz (1996). Arithmetic groups and the length spectrum of Riemann surfaces, *Duke Math. J.* **84**, 199–215.

[Sch99]   P. Schmutz Schaller (1999). Teichmüller space and fundamental domains for Fuchsian groups, *L'Enseignement Math.* **45**, 169–187.

[Ser85a] C. Series (1985). The geometry of Markoff numbers, *Math. Intelligencer* **7**, 20–29.

[Tho89]   J.G. Thompson (1989). Fricke, free groups, and SL_2. In *Group Theory: proceedings of the Singapore Group Theory Conference, June 8–19, 1987, edited by K.H. Cheng and Y.K. Leong*, de Gruyter, 207–214.

[Tra80]   C.R. Traina (1980). Trace polynomial for two generator subgroups of $SL(2,\mathbb{C})$, *Proc. Amer. Math. Soc.* **79**, 369–372.

[Whi73a] A. Whittemore (1973). On representations of the group of Listing's knot by subgroups of $SL_2(\mathbb{C})$, *Proc. Amer. Math. Soc.* **40**, 378–382.

[Whi73b] A. Whittemore (1973). On special linear characters of free groups of rank $n \geq 4$, *Proc. Amer. Math. Soc.* **40**, 383–388.

James W. Anderson

Faculty of Mathematical Studies
University of Southampton
Southampton, SO17 1BJ
England

J.W.Anderson@maths.soton.ac.uk

AMS Classification: 57M50, 30F40, 20H10, 51M25

Keywords: Fricke polynomial, length, character, Teichmüller space

Kleinian Groups and Hyperbolic 3-Manifolds   Y. Komori, V. Markovic & C. Series (Eds.)
Lond. Math. Soc. Lec. Notes **299**, 343–362   Cambridge Univ. Press, 2003

Complex angle scaling

D. B. A. Epstein, A. Marden and V. Markovic

Abstract

We introduce the method of complex angle scaling to study deformations of hyperbolic structure. We show how the method leads to a construction of counterexamples to the equivariant $K = 2$ conjecture for hyperbolic convex hulls.

1. Introduction

Our purpose is to give an exposition of the theory of *complex angle scaling*.[1] Complex angle scaling mappings are quasiconformal homeomorphisms or locally injective quasiregular developing mappings of $\mathbb{H}^2$ into or onto $\mathbb{S}^2$. In many cases, complex angle scalings are *continous* analogues of the *complex earthquakes* described by McMullen [McM98]. The mappings are also closely related to "quakebends" introduced in [EM87]; these are mappings $\mathbb{H}^2 \to \mathbb{H}^3$.

In the form described here our theory applies to finite or infinite discrete measured laminations in hyperbolic 2-space $\mathbb{H}^2$. For certain parameter values they can be defined to be quasiconformal on all $\mathbb{S}^2$ and they even have a natural extension to quasiconformal mappings of $\mathbb{H}^3$.

In contrast, McMullen's complex earthquakes are defined on a finite area hyperbolic surface S as maps $\mathscr{ML}(S) \to \textbf{Teich}(S)$. However for finite laminations on S, namely those which lift to infinite discrete laminations in $\mathbb{H}^2$, both give exactly the same boundary values on $\partial \mathbb{H}^2$ and therefore determine the same deformation of a fuchsian group covering S. Complex angle scalings do not naturally extend to the lifts of general measured laminations on S.

We will apply our theory to prove that Thurston's equivariant $K = 2$ conjecture does not hold. We have proven this in [EMM01]; the proof presented here is expressed in terms of angle scaling, which is how we discovered it. The angle scaling method also yields a new and perhaps more natural proof of Sullivan's Theorem 5.1 that the Teichmüller distance between the conformal structure on a simply connected region

---

[1] We want to thank the referee for helpful comments.

and on the convex hull boundary that faces it differs by a universally bounded amount. This requires an elaboration of the theory presented here.

For the full details of the work discussed, we refer to [EMM01] and the work in progress [EMM02].

2. Complex angle scaling

2.1. The mappings on a wedge

We begin with the wedge

$$W = \{z \in \mathbb{C} : 0 < \arg z < \alpha < \pi\}.$$

The *complex angle scaling map* on W is defined in terms of the complex parameter $t = u + iv \in \mathbb{C}$ as

$$f_+(z) = e^{t\theta}z = re^{u\theta}e^{i(1+v)\theta}, \quad \text{for } z = re^{i\theta} \in W.$$

The boundary values of f_+ are $f_+(\zeta) = \zeta$, if $\zeta \geq 0$, and $f_+(\zeta) = e^{t\alpha}\zeta$, if $\arg\zeta = \alpha$. In parallel we define

$$f_-(z) = e^{-t\theta}z = re^{-u\theta}e^{i(1-v)\theta}, \quad \text{for } z \in W.$$

We list the properties of these mappings:

- For all $-1 < \text{Im}(t) \leq +1$, the map f_+ is a K_t-quasiconformal homeomorphism[2] and for $-1 < \text{Im}(t)$ is K_t-quasiregular[3] where

$$K_t = \frac{1 + |\kappa(t)|}{1 - |\kappa(t)|}, \quad \kappa(t) = \frac{-t}{2i+t} \cdot \frac{z}{\bar{z}}. \tag{2.1}$$

- For all $-1 \leq \text{Im}(t) = v < +1$, the map f_- is K_t-quasiconformal and for $\text{Im}(t) < +1$ is K_t-quasiregular where

$$K_t = \frac{1 + |\kappa(t)|}{1 - |\kappa(t)|}, \quad \kappa(t) = \frac{t}{2i-t}. \tag{2.2}$$

---

[2] A K-quasiconformal mapping is an orientation preserving homeomorphism whose maximal dilatation (or distortation) is $\leq K$.

[3] A K-quasiregular mapping has the form $G \circ F$ where F is K-quasiconformal and G is holomorphic on the range of F; it is locally injective if and only if G is.

- The boundary values of f_+, f_- on each edge of the wedge W is the restriction of a Möbius transformation.

- f_+ and f_- commute with any Möbius transformation that maps W onto itself (that is, any hyperbolic transformation with fixed points at the vertices 0 and ∞).

For larger angles, we must think of the image of W as spread over $\mathbb{S}^2$.

Important special cases are:

(i) $\text{Re}\, t = u = 0$. Then $f_+(z) = e^{iv}z = re^{i(1+v)\theta}$. This is a smoothed[4] version of *grafting*. The wedge W of angle α is mapped onto a wedge of angle $(1+v)\alpha$.

(ii) $t = i$; $v = 1$. f_+ is called the *angle doubling map* whereas f_- is called the *angle collapsing map*.

(iii) $\text{Im}(t) = v = 0$. Then $f_+(z) = e^{u\theta}z$. This is a smoothed version of an earthquake. The wedge W is mapped onto itself but it is stretched (or contracted if $u < 0$) continuously from its right edge to its left.

Among all quasiconformal mappings of W, f_+, f_- are extremal for their boundary values [Str76]. When $t = i$, $f_+(z)$ is 2-quasiconformal in W.

2.2. Extension of the action to space

Here we will work in the upper halfspace model of $\mathbb{H}^3$ which we will parameterize over the complex plane as

$$\mathbb{H}^3 = \{(z, s) : z \in \mathbb{C}, s > 0\}.$$

Let $W^* = \{(z, s) : z \in W, s > 0\} \subset \mathbb{H}^3$ denote the wedge over W. The mappings f_+, f_- have natural continuous extensions respectively as

$$f_+ : (z, s) \mapsto (ze^{t\theta}, s|e^{t\theta}|), \quad f_- : (z, s) \mapsto (ze^{-t\theta}, s|e^{-t\theta}|).$$

Their boundary values on the faces of W^* are the extensions to space of the Möbius transformations that preserve the corresponding edges of W. The extensions to space of f_+, f_- also commute with the extensions to W^* of the Möbius transformations that preserve W. They are 3-dimensional quasiconformal when they are homeomorphisms.

---

[4] Here smoothed means C^0, not C^∞.

2.3. Action on a pair of symmetric wedges

Consider now the situation that we have the symmetric wedges in $\mathbb{R}$,

$$W_+ = \{z : 0 \leq \beta < \arg z < \alpha + \beta \langle \pi \}; \quad W_- = \{z : -\alpha - \beta < \arg z < -\beta \leq 0\}.$$

Define f_+ in W_+ by conjugating W_+ to angle scaling on our standard W above by the rotation $z \mapsto e^{-i\beta}z$.

Let $\overline{W}$ denote the reflected image of the standard wedge W in $\mathbb{R}$. Define f_- in W_- by conjugating it via the rotation $z \mapsto e^{i\beta}z$ to the angle scaling map $z \mapsto e^{-t\theta}z$, $z \in \overline{W}$.

Interpolating by the identity and a loxodromic Möbius transformation, we obtain the following continuous map $E_t(z) : \mathbb{C} \to \mathbb{C}$:

- $E_t(z) = f_+(z)$ if $z \in W_+$,

- $E_t(z) = f_-(z)$, if $z \in W_-$,

- $E_t(z) = z$ for z in the component of $\mathbb{C} \setminus W_+ \cup W_-$ containing the positive real axis,

- $E_t(z) = e^{t\alpha}z$ for z in the component of $\mathbb{C} \setminus W_+ \cup W_-$ containing the negative real axis.

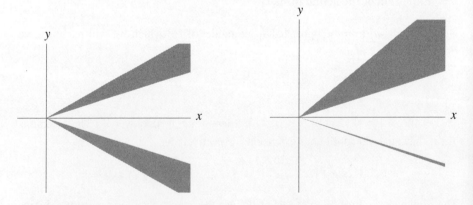

Figure 1: The lefthand picture shows two symmetric wedges of angle $\alpha = 0.2$ radians. The righthand picture shows the result of angle scaling by $t \in \mathbb{C}$ lying on the positive y-axis, with a value just less than $t = i$. So the angle of the upper wedge is almost doubled, and the angle of the lower wedge is almost collapsed. The corners in the shaded wedges are due to clipping in the production of the pictures.

Note that the E_t-image of the negative real axis is inclined at angle $\pi + \nu\alpha$ to the positive real axis.

The map as described on $\mathbb{C}$ extends to a continuous map of all $\mathbb{H}^3$. It will become quite messy as $|\mathrm{Im}(t)|$ becomes large.

The construction we have described is the basis of our angle scaling method.

2.4. Crescents associated with a finite lamination

By a *crescent* we mean an open region bounded by two circular arcs. A wedge is a special case. Any two crescents with the same interior vertex angle are Möbius equivalent in such a way that a designated initial edge of one is sent to a designated initial edge of the other.

Suppose Λ is a finite or infinite discrete geodesic lamination in $\mathbb{H}^2$. An *associated set of crescents* is a set of crescents in one-to-one correspondence with the leaves of Λ with the two properties

- Each leaf $\ell \in \Lambda$ corresponds to a crescent in $\mathbb{H}^2$ whose vertices are the endpoints of ℓ,

- The interiors of the crescents are mutually disjoint.

If Λ is invariant under a fuchsian group, we require that the set of associated crescents be invariant as well.

In turn, associated with the crescents, is the positive measured lamination

$$(\Lambda, \mu) = \{(\ell, \alpha_\ell)\}, \quad 0 < \alpha_\ell < \pi,$$

where α_ℓ is the vertex angle of the crescent associated with ℓ. Here we are ignoring the degenerate (but important) special case of one leaf with associated angle π.

We are aiming at using the associated set of crescents as a domain for angle scaling mappings. But first we will address the following converse question: Given a finite or infinite discrete measured lamination $(\Lambda, \mu) = \{(\ell, \alpha_\ell)\}$, $0 < \alpha_\ell < \pi$, can we find an associated set of crescents with exactly the angles $\{\alpha_\ell\}$?

3. The ortho-condition

Suppose $(\Lambda, \mu) = \{(\ell, \alpha_\ell)\}$ is a finite measured lamination in $\mathbb{H}^2$. We will present a necessary and sufficient condition that there exists an associated set of nonoverlapping

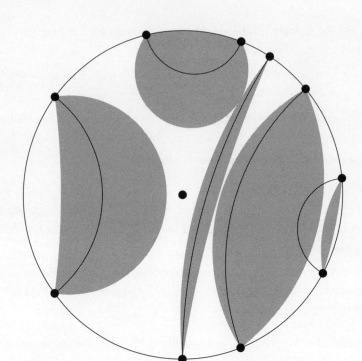

Figure 2: This picture shows five geodesics in the hyperbolic plane, and five associated crescents. The crescents have to be disjoint from each other and lie inside the unit circle. They do not need to be disjoint from the geodesics. A crescent is allowed to have an edge lying on the unit circle, as is the case for one of the crescents shown here. Note that the central gap containing the centre of the unit circle has three boundary pieces lying on the unit circle. That is, the gap is not separated, even though two crescents touch, because a crescent is open, by definition.

crescents such that the crescent with vertices at the endpoints of a leaf ℓ has vertex angle α_ℓ.

First we will examine the case that Λ is a ladder: By this term we mean that there is a geodesic γ transverse to and intersecting all the leaves of Λ. Choosing a direction along γ we can then order the leaves $\ell_1, \ldots, \ell_n$. Denote the associated angles by $\alpha_1, \ldots, \alpha_n$.

First assume that the leaves have no common endpoints. Find the hyperbolic distances $d_1, \ldots, d_{n-1}$ between successive leaves. If γ is orthogonal to all the leaves, then these distances are just the distances between the successive points of intersection of the leaves with γ.

Then with the data $\{\alpha_i, d_i\}$ construct the following model piecewise linear arc in say the unit disk model $\mathbb{D}$ of $\mathbb{H}^2$. Start with the ray $a_0 = (-1, 0]$ along the negative real axis. At $z = 0$ construct the arc a_1 of length d_1 with exterior bending angle α_1 with respect to a_0. Then construct a_2 of length d_2 with exterior bending angle α_2 with respect to a_1. And so on. The last arc a_n has exterior bending angle α_n with respect to a_{n-1} and will extend to a point on $\partial \mathbb{D}$. Our model arc is called the *ortho-curve* σ determined by the ladder.

If σ has no self-intersections, we say that the ladder (Λ, μ) satisfies the *ortho-condition*. If in addition the endpoints of σ on $\partial \mathbb{D}$ are distinct, we say the *strong ortho-condition* is satisfied. When the strong ortho-condition is satisfied, it will remain satisfied under all small enough changes in $\{\alpha_i\}$, or in the position of the leaves.

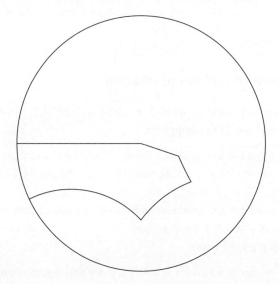

Figure 3: This ortho-curve is associated to a ladder (Λ, μ). Each vertex in the ortho-curve corresponds to a geodesic in Λ (or to several geodesics in Λ with common endpoints). The bending angles of the ortho-curve at a vertex is equal to the measure of the corresponding geodesic in Λ. The length of each finite side of the ortho-curve is equal to the distance between corresponding geodesics.

We must modify the construction if the leaves have common endpoints. Suppose the leaves $\ell_i, \ldots, \ell_{i+k}$ share the same endpoint $\zeta \in \partial \mathbb{D}$, but neither ℓ_{i-1} nor ℓ_{i+k+1} share ζ. Set $\beta = \sum_{j=i}^{i+k} \alpha_j$. In the construction replace the set of leaves $\{\ell_j : i \leq j \leq i+k\}$ by just ℓ_i and replace α_i by β. Apply this fix at all common fixed points until we return to the situation considered above. Then construct the ortho-curve.

Now return to a general finite positive measured lamination (Λ, μ). We say that it satisfies the *ortho-condition* (respectively, *strong ortho-condition*) if *every* ladder in Λ

satisfies the ortho-condition (respectively, strong ortho-condition).

Exactly the same analysis applies to infinite discrete laminations. Such a system arises as follows. Suppose $S = \mathbb{H}^2/G$ is a surface of finite area. Choose a simple geodesic $\gamma \subset S$ and a maximal area tubular neighborhood A about γ. The collection of lifts of A to $\mathbb{H}^2$ will be a system of crescents with the lifts of γ. The system will satisfy the ortho- but not the strong ortho-condition. In this situation, all the vertex angles are the same—see section 6 for an example.

Theorem 3.1 ([EMM02]). *Suppose* $(\Lambda, \mu) = \{(\ell, \alpha_\ell)\}$ *is a finite or infinite discrete positive measured lamination in* $\mathbb{H}^2$. *There is an associated system of nonoverlapping crescents if and only if it satisfies the ortho-condition. If* (Λ, μ) *is invariant under the action of a fuchsian group G, the associated crescents can be chosen to be invariant as well.*

4. Angle scaling on systems of crescents

In this section we will work in the disk model $\mathbb{D}$ of $\mathbb{H}^2$. Refer to Figure 2 as we describe the construction of the mappings.

Assume that $(\Lambda, \mu) = \{(\ell, \alpha_\ell)\}$ is a finite or infinite discrete positive measured lamination with an associated set of nonoverlapping crescents $\{C\}$ in $\mathbb{D}$. Reflect the set of crescents in $\partial \mathbb{D}$ to get a set of nonoverlapping crescents $\{C'\}$ in the exterior $\mathbb{D}^{ext}$. We may assume the measured laminations and the sets of crescents are invariant under G where G is a fuchsian group without elliptics—or simply $G = \{id\}$ (this will be the situation in the finite case).

Given (Λ, μ) as above, for each $t = u + iv \in \mathbb{C}$ we will define a continuous map $\mathbb{S}^2$ into or onto $\mathbb{S}^2$, $E_t(z) = E_t(z; \Lambda, \mu)$, as follows.

Start by choosing a crescent C_1, an edge e_1, its reflected partner C_1' in the exterior $\mathbb{D}^{ext}$, and the reflected edge e_1'. For simplicity we will assume that the region bounded by the closed curve $e_1 \cup e_1'$ contains no crescents.

On the boundary component $e_1 \cup e_1'$ of $C_1 \cup C_1'$, $E_t(z)$ is the identity. On $C_1 \cup C_1'$, E_t is angle scaling f_t. On the union of the other pair of boundary edges, $\varepsilon_1 \cup \varepsilon_1'$, $E_t(z)$ is the restriction of a Möbius transformation, say T_1. We will describe how to continue f_t to the other side of $\varepsilon_1 \cup \varepsilon_1'$.

Suppose the crescents $C_2, C_2', C_3, \ldots$ are adjacent on the right to the curve $\varepsilon_1 \cup \varepsilon_1'$ in the sense that no crescent separates it from any of these C_i. Let e_2^* denote the union of the boundary components of the $\{C_i, C_i'\}$, $i > 1$, which likewise are not separated from $\varepsilon_1 \cup \varepsilon_1'$. (We have to allow for tangencies between crescents.) Correspondingly

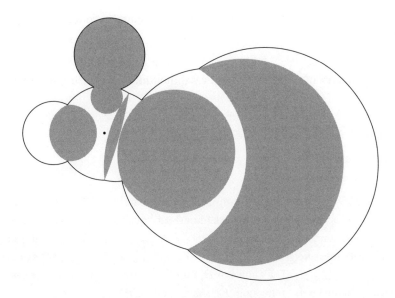

Figure 4: This is what we get by applying angle doubling to the crescents shown in Figure 2. Note that the image crescents may well have angles that are bigger than π. It is also possible (not shown) for one the image crescent to contain ∞, so that, in the plane, the image crescent surrounds the rest of the picture. We have marked the origin here and in Figure 2 with a black dot. The gap containing the origin is unchanged by angle doubling, though this may not be clear because the two pictures are drawn to different scales.

let D_1 be the union of those components (called *gaps*) of $\overline{\mathbb{D}} \cup \mathbb{D}^{ext} \setminus C_2 \cup C_2' \cup \cdots$ with the property that the boundary of each component is a union of arcs in $e_2^* \cup (\varepsilon_1 \cup \varepsilon_1')$.

For z in the closure $\overline{D}_1$ define $E_t(z) = T_1(z)$. In particular on the appropriate edge of each $C_i \cup C_i'$, $i > 1$, E_t is now well defined. In each $C_i \cup C_i'$ now define $E_t(z) = (f_t)_i(z)$, where $(f_t)_i$ is the angle scaling map in $C_i \cup C_i'$ whose boundary values are normalized to be T_1 on the appropriate edges.

Continuing in this pattern we end up with the following result.

Lemma 4.1. *Let* (Λ, μ) *be a finite or infinite discrete positive G-invariant measured lamination satisfying the ortho-condition. For each* $t = u + iv \in \mathbb{C}$, *the complex angle scaling maps on a set of associated crescents, interpolated in the gaps by Möbius transformations, fit together to give a continuous map, or G-equivariant continuous map,* $E_t(z; \Lambda, \mu) : \mathbb{S}^2 \to \mathbb{S}^2$.

The geometry of the map will in general be quite complicated as the images of the individual crescents may expand, contract, or fold over.

The proof in the case of an infinite lamination requires a convergence argument [EMM02].

4.1. The geometry of complex scaling: earthquaking and grafting/bending

The map $E_t(z; \Lambda, \mu)$ may be called again a *complex angle scaling map* or *complex earthquake*. It is often useful to regard it as a composition,

$$E_t = E_{iv} \circ E_u, \quad t = u + iv.$$

Here $E_u : \mathbb{S}^2 \to \mathbb{S}^2$ is a homeomorphism, its actions in $\mathbb{D}$ and $\mathbb{D}^{ext}$ being symmetric. In particular its action in $\mathbb{D}$ is a (smoothed out) earthquake sending the lamination Λ to a new lamination Λ_u; we write $E_u : (\Lambda, \mu) \to (\Lambda_u, \mu)$ and understand that each leaf of Λ_u takes its measure from its preimage. Equally E_u sends a crescent system for (Λ, μ) to one for (Λ_u, μ). The values of E_u on $\partial\mathbb{D}$ are exactly the boundary values of a classical earthquake. By reflection, E_u acts as well in $\mathbb{D}^{ext}$.

In contrast the action of E_{iv} is a (smoothed out) grafting on the crescents associated with (Λ_u, μ). When $v > 0$ each crescent in $\mathbb{D}$ is expanded $\alpha \mapsto (1+v)\alpha$ while each crescent in $\mathbb{D}^{ext}$ is contracted, collapsed, or folded over according to the sign of $(1-v)\alpha$.

The action of E_{iv} can also be described as *bending*. We will describe this operation on $\mathbb{D}$; there is a parallel action in $\mathbb{D}^{ext}$.

Thurston taught us that given a measured lamination (Λ, μ), there is an isometry $\mathbb{H}^2 \to \mathbb{H}^3$ taking $\mathbb{D}$ onto a pleated surface. Each leaf $\ell \subset \Lambda$ is sent a geodesic, and each component of $\mathbb{D} \setminus \Lambda$ is sent to a geodesic polygon. The exterior bending angles are given by μ (for complete details see [EM87]). We could equally well construct, instead of the mapping E_{iv} into $\mathbb{S}^2$, a mapping E_{iv}^* onto a (normalized) pleated surface in $\mathbb{H}^3$. From this point of view, associated with each *bending line* $\ell \subset E_{iv}^*(\mathbb{D})$ is the exterior bending angle $v\alpha_\ell$; the two flat pieces that border ℓ have exterior bending angle $v\alpha_\ell$.

The most interesting special case is *angle doubling*, the case that $t = i$.

We summarize our results (concentrating on the case $\operatorname{Im}(t) > -1$) as follows.

Theorem 4.2 ([EMM02]). *Suppose (Λ, μ) is a finite or infinite discrete G-invariant measured lamination that satisfies the ortho-condition. There exists a G-equivariant complex angle scaling mapping $E_t(z) = E_t(z; \Lambda, \mu)$ for every $t \in \mathbb{C}$. It has the following properties.*

(i) *For* $\{-1 < Im(t) < +1\}$, $E_t : \mathbb{S}^2 \to \mathbb{S}^2$ *is a K_t-quasiconformal homeomorphism. Furthermore $\{E_t(z)\}$ is a holomorphic motion of $\mathbb{S}^2$.*

(ii) *For* $\{-1 < Im(t)\}$, $E_t : \mathbb{D} \to \mathbb{S}^2$ *is a K_t-quasiregular developing mapping. For each $z \in \overline{\mathbb{D}}$, $E_t(z)$ is defined and holomorphic in t.*

(iii) *When $t = i$, the angle doubling map $E_i : \mathbb{D} \to \Omega_i = E_i(\mathbb{D})$ is a 2-quasiconformal homeomorphism. It has a continuous extension to $\overline{\mathbb{D}}$.*

   If $\partial \Omega_i$ is not a Jordan curve, there exists a ladder $\{\ell_j\} \subset \Lambda$ which has a common orthogonal and which does not satisfy the strong ortho-condition. Conversely if there is such a ladder, Ω_i is not a Jordan region.

A quasiregular developing mapping is locally quasiconformal but not in general a global homeomorphism. The quasiconformal factor from Equation(2.1) is

$$K_t = \frac{|t+2i| + |t|}{|t+2i| - |t|}, \quad Im(t) > -1.$$

5. Simply connected regions and their domes

Suppose $\Omega \subset \mathbb{C}$ is a simply connected region, but not $\mathbb{C}$ itself. The hyperbolic convex hull in $\mathbb{H}^3$ of the closed set $V = \mathbb{S}^2 \setminus \Omega$ is the intersection of all hyperbolic halfspaces whose closures on $\partial \mathbb{H}^3$ contain V. The convex hull has a single boundary component in $\mathbb{H}^3$. The boundary lies over Ω as the dome lies over the floor in a domed stadium. Accordingly we will denote the convex hull boundary component by Dome(Ω).

The dome is the union of bending lines and flat pieces. The bending lines form a closed set of geodesics which have their endpoints in $\partial \Omega$. In the present exposition the bending lines will form a finite or infinite discrete set. The complement is the union of flat pieces, in general an ideal hyperbolic polygon. The bending angle at an isolated bending line is the exterior angle formed by the two flat pieces with the line as a common edge. Thus the limiting case of bending angle zero means the two flat pieces are not bent with respect to each other at all.

The key to analyzing the dome is Thurston's theorem that the hyperbolic metric in $\mathbb{H}^3$ restricts to the dome to give a path metric, which is then called the hyperbolic metric of the dome. In terms of this metric and the hyperbolic metric, say in $\mathbb{D}$, there is an isometry

$$\iota : \text{Dome}(\Omega) \to \mathbb{D}$$

that carries the set of bending lines with their bending measure to a measured lamination in $\mathbb{D}$. For the full details see [EM87].

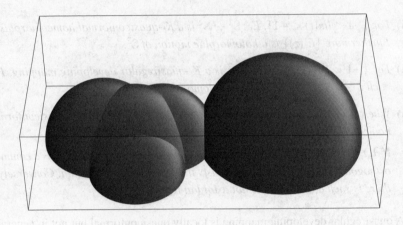

Figure 5: A dome

Sullivan stated the following theorem in [Sul81b] with half a page indicating how one might set about proving it. A complete (but very complicated) proof was later given in [EM87]. The theorem says that the hyperbolic metrics on Ω and Dome(Ω) are of uniformly bounded Teichmüller distance apart.

Theorem 5.1 ([Sul81b, EM87]). *There exists a universal constant $K > 1$ with the following property. For any simply connected $\Omega \in \mathbb{C}$, $\Omega \neq \mathbb{C}$, there exists a K-quasi-conformal mapping $\Phi : \text{Dome}(\Omega) \to \Omega$ that extends continuously to be the identity on the common boundary $\partial\Omega$. If Ω is invariant under a discrete group Γ of Möbius transformations, then Φ is equivariant under Γ.*

The boundary values pin down the homotopy class of Φ.

It is natural to ask: What is the least value K for which Sullivan's theorem holds? The simple example of the plane slit along the positive real axis shows that it cannot be that $K < 2$. The best result is by Bishop [Bis01], who showed that $K \leq 7.82$.

Conjecture 5.2. $K = 2$.

There are actually two possible constants: the universal K as stated, but also a constant $K_{eq} \geq K$ which is an upper bound for the maximal dilatation of maps Φ with the additional property that Φ must be equivariant whenever Ω is preserved by a discrete group Γ such that Ω/Γ has finite hyperbolic area. It is the latter constant that Thurston seems to have in mind because of its possible application to hyperbolic 3-manifolds.

Conjecture 5.3 ([Thu86b]). $K_{eq} = 2$.

To verify either of these conjectures it must be shown that for *any* simply connected region with the required properties, the maximal dilatation of *some* appropriate quasiconformal mapping does not exceed two.

Teichmüller space theory implies that $K \leq K_{eq}^3$ [Leh87, Thm. 4.7].

Actually *both* of these conjectures are false.

Later in section 6 we will outline a proof that $K_{eq} > 2$. The counterexamples are regular sets of groups in an open set of the Teichmüller space of once-punctured tori quasifuchsian groups.

More recently it has been shown that $K > 2$ without any requirement of equivariance. The counterexample is the complement of an infinite logarithmic spiral [EMk]. For the example the maximal dilatation of the extremal quasiconformal map is approximately 2.1.

On the other hand, there are many regions such that the corresponding constant satisfies $K = 2$. A large class of such regions can be constructed by using complex angle scaling. Moreover, $K = 2$ for all regions "sufficiently close" to a round disk (Theorem 5.3) and for all euclidean convex regions [EMM02].

To suggest the relationship between the image region $\mathbb{D} \to \Omega$ obtained by angle scaling and $\text{Dome}(\Omega)$ observe the following. Suppose $C \subset \mathbb{D}$ is a crescent of angle $0 < \alpha < \pi$. Under angle scaling C becomes a crescent C' of angle $(1 + v)\alpha$. If $0 < (1 + v)\alpha < 2\pi$, C' is embedded. $\text{Dome}(\Omega)$ consists of two flat pieces with exterior bending angle $v\alpha$. When $v = 1$ (angle doubling) the bending angle is α.

The following statement augments Theorem 4.3.

Theorem 5.4 ([EMM02]). *Suppose* $(\Lambda, \mu) \subset \mathbb{D}$ *is a finite or infinite discrete positive measured lamination invariant under a fuchsian group* G, *that satisfies the ortho-condition. Set* $t = i$ *(angle doubling). The map* $E_i : \mathbb{D} \to \Omega = E_i(\mathbb{D})$ *has the following properties:*

(i) *E_i is 2-quasiconformal and equivariant with respect to* G.

(ii) *E_i has a continuous extension to* $\overline{\mathbb{D}}$.

(iii) *The isometry* $\iota^{-1} : \mathbb{D} \to \text{Dome}(\Omega)$ *takes* $(\Lambda, \mu) \subset \mathbb{D}$ *to the bending measure of* $\text{Dome}(\Omega)$ *and has a continuous extension to* $\partial\mathbb{D}$. *Moreover if* E_i *and* ι *are normalized in the same way, the map*

$$\iota^{-1} \circ E_i^{-1} : \Omega \to \mathbb{D} \to \text{Dome}(\Omega)$$

is 2-quasiconformal and has a continuous extension to $\partial\Omega$ *where it is the identity.*

We remark that if instead we take $t \in \mathbb{C}$ with $\mathrm{Im}(t) = +1$ while $\mathrm{Re}t \neq 0$, E_t can again be interpreted as angle doubling. Namely, it is angle doubling on the lamination $(\Lambda_u, \mu) \in \mathbb{H}^2$ which results from applying the earthquake $E_u : \mathbb{D} \to \mathbb{D}$ to (Λ, μ), $t = u + iv$. Thus property (ii) holds for E_t and (i), (iii) hold for E_i with respect to (Λ_u, μ).

5.1. Nearly round domains

We digress to raise the following issue. Given a general positive Borel measured lamination $(\Lambda, \mu) \subset \mathbb{D}$, it is a hard question to determine whether or not it is the bending measure of the dome of a simply connected region Ω. Of course in the discrete case, we know the answer is yes if it satisfies the ortho-condition. But a much more general result is possible if we stick to regions Ω which are "close" to round circles.

The precise statement is in terms of the norm

$$\|\mu\| = \|(\Lambda, \mu)\| = \sup_\gamma \mu(\gamma) \in [0, \infty]$$

as γ varies over all transverse geodesic intervals γ of unit length. In particular, $\|\mu\| = 0$ if and only if Ω is a round disk. One may therefore use the norm as a measure of roundness of Ω.

Theorem 5.5 ([EMM01]). *Suppose (Λ, μ) is a positive Borel measured lamination. There exists a constant*

$$0 < c \le 2\arcsin(\tanh(1/2)) \approx 0.96$$

with the following property: if $\|\mu\| < c$, then (Λ, μ) is the bending measure of Dome(Ω) *for a simply connected region Ω, which is a quasidisk. There exists a 2-quasiconformal map $\Phi : \Omega \to$ Dome(Ω) which extends continuously to the identity on $\partial\Omega$. Moreover, if Ω is invariant under a discrete group Γ, then Φ is equivariant under Γ.*

It is known that $c = 0.73$ will work, and that any constant which is greater than $2\arcsin(\tanh(1/2))$ will not work. It is conjectured that the latter value is the correct upper bound for c.

In the other direction, it is known that if a pleated surface is embedded, its bending measure necessarily satisfies $\|\mu\| < 4.88$, [Bri98], [Bri02]. Actually the estimate in [Bri02] is stated to hold in the presence of a group action, but according to Bridgeman (personal communication), a small modification of its proof shows that it holds independently of any group structure.

6. Earthquake disks and $K_{eq} > 2$

We will present the examples that show $K_{eq} > 2$. This also gives the opportunity of further illuminating the method of angle scaling.

The starting point is a square, once punctured torus $T = \mathbb{D}/G$. That T is square means that there are simple closed geodesics $\mathbf{a}, \mathbf{b} \subset T$ which are of equal length and orthogonal at their point of intersection. Let A be a maximal collar about $\mathbf{a}$. Lift $\mathbf{a}$ and A to $\mathbb{D}$. The lifts $\{\mathbf{a}^*\}$ of $\mathbf{a}$ are arranged in ladders each with a common orthogonal which is a lift of $\mathbf{b}$. The lifts $\{A^*\}$ of A form an associated system of crescents, all with the same vertex angle which we will designate as α. The crescents satisfy the ortho-condition but not the strong condition. Let (Λ, μ) denote the associated measured lamination.

Now bring in the angle scaling map

$$E_t(z) = E_t(z; \Lambda, \mu), \quad t = u + iv.$$

We know that $E_u : \mathbb{D} \to \mathbb{D}$ is an earthquake. Denote the length of $\mathbf{a}$ by L. Replacing the projection $E_{u*} : T \to T$ by $E_{(u+L/\alpha)_*}$ is akin to replacing E_{u*} by its composition with a Dehn twist about $\mathbf{a}$.

Consider the following three sets in the parameter t-space.

$$S = \{t \in \mathbb{C} \mid -1 < \operatorname{Im}(t) < 1\}.$$

This strip is the set for which E_t extends to a quasiconformal mapping of $\mathbb{S}^2$.

$$\mathfrak{T}_0 = \{t \mid E_t : \mathbb{D} \to \Omega_t \text{ is injective}\}_0 \supset S.$$

Here the subscript 0 indicates that we must take the component containing $t = 0$, as the indicated set is not connected. Since all bending angles are less than π, for $t \in \mathfrak{T}_0$, $|\operatorname{Im}(t)| < \pi/\alpha$. Like S, $\mathfrak{T}_0$ is symmetric about $\mathbb{R}$.

$$\mathfrak{T} = \mathfrak{T}_0 \cup \{t \mid \operatorname{Im}(t) > 0\}.$$

The set $\mathfrak{T}$ is called an *earthquake disk*. It is the following theorem that shows it is simply connected.

Theorem 6.1 (McMullen [McM98]). $\mathfrak{T}$ *is biholomorphically equivalent to the Teich-müller space of* T.

358   D. B. A. Epstein, A. Marden V. Markovic

In this realization, Teichmüller space is parameterized by earthquakes on Λ then graftings on the images $E_u(\Lambda)$.

For $t \in S$, E_t is quasiconformal in $\mathbb{D}$. As $t \in \mathfrak{T}$ goes beyond S, E_t becomes instead a developing mapping; it sends $\mathbb{D}$ to a simply connected region spread over $\mathbb{S}^2$, and sends the initial fuchsian group G to a group Γ acting on the image so that the quotient surface is a quasiconformally equivalent to T. These surfaces fill out Teichmüller space. However Γ itself will in general not be discrete.

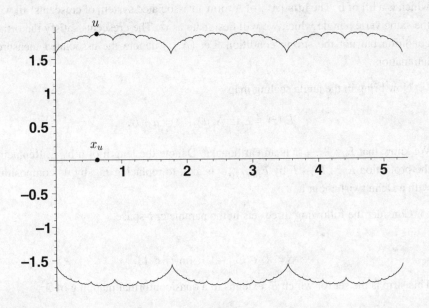

Figure 6: This figure from [EMM01] is the trace $A = 2\sqrt{2}$ slice of quasifuchsian space with basepoint the square torus with $\mathbf{a}, \mathbf{b}$ of length $\mathrm{arccosh}\,3$. On this torus, the vertex angle of the maximal collar symmetrical about $\mathbf{a}$ is $\alpha = \pi/2$. The horizontal strip bounded by the upper and lower curves shown represents $\mathfrak{T}_0$. The union of this strip and the upper halfplane is $\mathfrak{T}$. However the scaling indicated in the figure is not according to our conventions here. Instead the top and bottom curves should cut the imaginary axis in the points $\pm i$ and the periodicity of the strip $\mathfrak{T}_0$ is $2(\mathrm{arccosh}\,3)/\pi$. Thus S is bounded by the horizontal lines that go through the cusps $2n(\mathrm{arccosh}\,3)/\pi \pm i$, $-\infty < n < \infty$, on the upper and lower curves. The point $u = x_u + iy_u$ is a highest point on the upper curve. The halfplane $H \supset \mathfrak{T}$ is bounded by the horizontal line through the lowest points $2n(\mathrm{arccosh}\,3)/\pi - y_u i$ of the lower curve. This figure was drawn by David Wright.

Properties:

- $S, \mathfrak{T}_o, \mathfrak{T}$ are invariant under the translations $t \mapsto t + nL/\alpha$,

- $\{t = i\} \in \partial \mathfrak{T}_o \cap \partial S$ and for all small $\varepsilon > 0$, $\{t = \pm\varepsilon + i\} \in \mathfrak{T}_o$.

- $E_i : \mathbb{D} \rightarrow E_i(\mathbb{D}) = \Omega_i$ is 2-quasiconformal,

- $E_i : \mathbb{D}^{ext} \rightarrow \Omega_i' = \mathbb{S}^2 \setminus \overline{\Omega_i} \neq \emptyset$ (Angle collapse on exterior crescents),

- $\partial \Omega_i$ is not a Jordan curve,

- $\Omega_i \cup \Omega_i'$ is the ordinary set of Γ, where $E_{i*} : G \rightarrow \Gamma$ is the induced isomorphism,

- Ω_i'/Γ is the 3-punctured sphere.

The point $t = i$ is a minimal point on the top curve because after any nonperiodic earthquake, the geodesics **a**, **b** are no longer orthogonal, allowing larger crescent angles. This in turn permits larger values of $\mathrm{Im}(t)$ appear in $\mathfrak{T}_o$.

Set

$$\chi = \sup\{\mathrm{Im}(t) \mid t \in \mathfrak{T}_o\} < \pi/\alpha,$$

$$-\chi = \inf\{\mathrm{Im}(t) \mid t \in \mathfrak{T}_o\},$$

and introduce the halfplane

$$H = \{t \in \mathbb{C} \mid \mathrm{Im}(t) > -\chi\} \supset \mathfrak{T}.$$

The hyperbolic distances satisfy

$$\rho_H(z, w) < \rho_{\mathfrak{T}}(z, w), \quad z, w \in \mathfrak{T}.$$

Here we have made essential use of the fact that $\mathfrak{T} \neq H$. In particular,

$$\log 2 = \rho_H(u, u + i\chi) < \rho_{\mathfrak{T}}(u, u + i\chi).$$

Choose $u + iv \in \mathfrak{T}_o$ such that $\rho_{\mathfrak{T}}(u, u + iv) > \log 2$.

If K^* denotes the maximal dilatation of the extremal quasiconformal map between the Teichmüller points $u, u + iv$ then since Teichmüller distance coincides with hyperbolic distance in $\mathfrak{T}$,

$$\log K^* > \log 2, \text{ therefore, } K^* > 2.$$

What does this have to do with the relation between the floor and the dome? Using the point $t = u + iv \in \mathfrak{T}_o$ we know that

- The earthquake E_u sends the lamination $\Lambda \to \Lambda_u$, the initial fuchsian group G to a fuchsian deformation G_u where the surface $\mathbb{D}/G_u$ corresponds to the Teichmüller point u (the marking is given by the isomorphism induced by E_u).

- $E_{iv}(z; \Lambda_u, \mu) : \mathbb{D} \to \Omega_t$ induces an isomorphism $G_u \to \Gamma_t$. Ω_t is a Jordan region and Ω_t/Γ_t corresponds to the Teichmüller point $t = u + iv$.

- A Riemann map $\Phi : \Omega_t \to \mathbb{D}$ sends Γ_t to a fuchsian group G_t where $\mathbb{D}/G_t \equiv \Omega_t/\Gamma_t$.

- The isometry $\iota : \text{Dome}(\Omega_t) \to \mathbb{D}$ chosen so that $\iota \circ E_{iv}|_{\partial\mathbb{D}} = \text{id}$ takes Γ_t back to G_u.

Now we can complete the proof. Suppose $f : \text{Dome}(\Omega_t) \to \Omega_t$ is a K-quasiconformal, Γ_t-equivariant mapping which extends continuously to the identity on $\partial\Omega_t$. Then

$$\Phi \circ f \circ \iota^{-1} : \mathbb{D} \to \mathbb{D}$$

is K-quasiconformal. It induces an isomorphism

$$G_u \xrightarrow{\iota^{-1}} \Gamma_t \xrightarrow{f} \Gamma_t \xrightarrow{\Phi} G_t.$$

That is

$$u \equiv \mathbb{D}/G_u \mapsto \mathbb{D}/G_t \equiv u + iv.$$

We conclude that

$$\log K \geq d_{\mathfrak{T}}(u, u + iv) > \log 2.$$

Since $K > 2$ for any f associated with Ω_t, in particular for the extremal f, it follows that we necessarily have $K_{eq} > 2$.

References

[Bis01]   C. J. Bishop (2001). An explicit constant for Sullivan's convex hull theorem, *to appear Proc. 2002 Ahlfors-Bers Colloq., Contemp. Math., Amer. Math. Soc.*.

[Bri98]   M. Bridgeman (1998). Average bending of convex pleated planes in hyperbolic 3-space, *Invent. Math.* **132**, 381–391.

[Bri02]   M. Bridgeman (2002). Bounds on the average bending of the convex hull boundary of a Kleinian group, *to appear Michigan Math. J.*.

[EM87]   D.B.A. Epstein and A. Marden (1987). Convex hulls in hyperbolic space, a theorem of Sullivan, and measured pleated surfaces. In *Analytical and Geometrical Aspects of Hyperbolic Space (LMS Lecture Notes* **111***)*, Cambridge University Press.

[EMk]   D.B.A. Epstein and V. Markovic. The logarithmic spiral: a counterexample to the $K = 2$ conjecture, *to appear Ann. Math.*.

[EMM01]   D.B.A. Epstein, A. Marden and V. Markovic. Quasiconformal homeomorphisms and the convex hull boundary, *to appear Ann. Math.*.

[EMM02]   D.B.A. Epstein, A. Marden and V. Markovic, *to appear*

[Leh87]   O. Lehto (1987). *Univalent Functions*, Springer-Verlag.

[McM98]   C. McMullen (1998). Complex earthquakes and Teichmüller theory, *J. Amer. Math. Soc.* **11**, 283–320.

[Str76]   K. Strebel (1976). On the existence of extremal Teichmüller mappings, *Jour. d'Anal. Math.* **30**, 441–447.

[Sul81b] D.P. Sullivan (1981). *Travaux de Thurston sur les Groupes Quasi-fuchsiens et les Variétés Hyperboliques de Dimension 3 Fibrés sur* $\mathbb{S}^2$ *(Lec. Notes Math.* **842***)*, Springer-Verlag.

[Thu86b] W.P. Thurston (1986). Hyperbolic structures on 3-manifolds II: Surface groups and 3-manifolds which fiber over the circle, *arXiv:math.GT/9801045*.

David Epstein

Mathematics Institute
Warwick University
Coventry, CV4 7AL
England

dbae@maths.warwick.ac.uk

Albert Marden

School of Mathematics
University of Minnesota
Minneapolis, MN 55455
USA

am@math.umn.edu

Vladimir Markovic

Mathematics Institute
Warwick University
Coventry, CV4 7AL
England

markovic@maths.warwick.ac.uk

AMS Classification: 30C75, 30F40, 30F45, 30F60

Keywords: complex angle scaling, complex earthquakes, bending, convex hull, Sullivan's theorem, $K = 2$ conjecture

Kleinian Groups and Hyperbolic 3-Manifolds   Y. Komori, V. Markovic & C. Series (Eds.)
Lond. Math. Soc. Lec. Notes **299**, 363–384   Cambridge Univ. Press, 2003

Schwarz's lemma and the Kobayashi and Carathéodory pseudometrics on complex Banach manifolds

Clifford J. Earle, Lawrence A. Harris,
John H. Hubbard and Sudeb Mitra

Abstract

We discuss the Carathéodory and Kobayashi pseudometrics and their infinitesimal forms on complex Banach manifolds. Our discussion includes a very elementary treatment of the Kobayashi pseudometric as an integral of its infinitesimal form. We also prove new distortion theorems for the Carathéodory pseudometric under holomorphic maps from the open unit disk to a complex Banach manifold.

1. Introduction

The geometry of the hyperbolic plane has played an important role in complex function theory ever since the time of Poincaré, who brilliantly exploited the fact that the open unit disk $\mathbb{D}$ in the complex plane $\mathbb{C}$ carries a hyperbolic metric (commonly called the Poincaré metric) that is preserved by all holomorphic automorphisms of $\mathbb{D}$. This metric therefore descends to a hyperbolic metric on any Riemann surface whose universal covering surface is holomorphically isomorphic to $\mathbb{D}$, so (by the uniformization theorem) almost all Riemann surfaces carry natural complete hyperbolic metrics.

A further link between function theory and the hyperbolic geometry of the unit disk is provided by the Schwarz–Pick lemma. It states that a holomorphic map of $\mathbb{D}$ into itself does not increase the hyperbolic lengths of tangent vectors or the Poincaré distances between points. This result also descends to Riemann surfaces whose universal covering surface is holomorphically isomorphic to $\mathbb{D}$. If X and Y are such surfaces with their natural hyperbolic metrics then a holomorphic map from X to Y does not increase the hyperbolic lengths of tangent vectors or the hyperbolic distances between points.

No higher dimensional complex manifold has the privileged position that $\mathbb{D}$ occupies in the one-dimensional case, so the study of natural metrics on such manifolds continues to be an active area of research. In particular, systems of pseudometrics on complex spaces that satisfy the Schwarz–Pick lemma have been much studied. See for example the papers [Har79], [Kob67], [PSh89], [Roy88], and [Ven89] and the books [Din89], [FV80], [JP93], and [Kob98]. In [Roy88] Royden calls such systems hyperbolic metrics, which partly explains the presence of this paper in this volume.

The Carathéodory and Kobayashi pseudometrics are the most widely studied of these systems, and they are our focus of attention. In section 3 we combine results of Harris [Har79] with an idea of Royden [Roy88] to give a very elementary proof that the Kobayashi pseudometric can be defined by appropriately integrating an associated infinitesimal metric. Sharper results that apply to a more general class of pseudometrics can be found in Venturini's incisive paper [Ven89], which uses less elementary methods and places the results of [Har79] in a broader setting.

In section 4 we shall define both the Carathéodory pseudometric and an associated inner pseudometric, and in section 5 we shall obtain sharp new forms of the Schwarz–Pick lemma for holomorphic maps from $\mathbb{D}$ to X equipped with either of these pseudometrics. These results describe in quantitative terms how close a holomorphic map from $\mathbb{D}$ to X comes to being a complex Carathéodory geodesic. They are inspired by and derived from results in the beautiful paper [BMi] of Beardon and Minda about holomorphic maps of $\mathbb{D}$ into itself. We thank David Minda for sending us a preprint of [BMi] and for an additional private communication that led us to formulate and prove Theorem 5.5.

2. Schwarz–Pick systems

General systems of pseudometrics with the Schwarz–Pick property were first studied systematically in [Har79] by Harris, who coined the term *Schwarz–Pick system* for them. These systems can be studied on various classes of complex spaces. We choose to use the class of complex manifolds modelled on complex Banach spaces of positive, possibly infinite, dimension. All our manifolds are assumed to be connected Hausdorff spaces.

If X and Y are complex Banach manifolds, we shall denote the set of all holomorphic maps of X into Y by $\mathcal{O}(X,Y)$.

Definition 2.1. (see Harris [Har79]) A Schwarz–Pick system is a functor, denoted by $X \mapsto d_X$, that assigns to each complex Banach manifold X a pseudometric d_X so that the following conditions hold:

(a) The pseudometric assigned to $\mathbb{D}$ is the Poincaré metric

$$d_{\mathbb{D}}(z_1, z_2) = \tanh^{-1} \left| \frac{z_1 - z_2}{1 - z_1 \overline{z_2}} \right| \quad \text{if } z_1 \in \mathbb{D} \text{ and } z_2 \in \mathbb{D}. \tag{2.1}$$

(b) If X and Y are complex Banach manifolds then

$$d_Y(f(x_1), f(x_2)) \leq d_X(x_1, x_2) \quad \text{if } x_1 \in X, x_2 \in X \text{ and } f \in \mathscr{O}(X, Y). \qquad (2.2)$$

Remark 2.2. Because of conditions (a) and (b) the sets $\mathscr{O}(\mathbb{D}, X)$ and $\mathscr{O}(X, \mathbb{D})$ provide upper and lower bounds for d_X. These upper and lower bounds lead to the definitions of the Kobayashi and Carathéodory pseudometrics, which we shall study in the remainder of this paper.

3. The Kobayashi pseudometric and its infinitesimal form

3.1. The classical definition

In this paper $d_{\mathbb{D}}$ will always be the Poincaré metric (2.1) on the unit disk $\mathbb{D}$.

Definition 3.1. A Schwarz–Pick pseudometric on the complex Banach manifold X is a pseudometric d such that

$$d(f(z), f(w)) \leq d_{\mathbb{D}}(z, w) \quad \text{for all } z \text{ and } w \text{ in } \mathbb{D} \text{ and } f \text{ in } \mathscr{O}(\mathbb{D}, X). \qquad (3.1)$$

If $X \mapsto d_X$ is a Schwarz–Pick system, then d_X is obviously a Schwarz–Pick pseudometric on X for every complex Banach manifold X.

Definition 3.2. The Kobayashi pseudometric K_X is the largest Schwarz–Pick pseudometric on the complex Banach manifold X.

As Kobayashi observed (see for example [Kob67] or [Kob98]), K_X is easily described in terms of the function $\delta_X : X \times X \to [0, \infty]$ defined by

$$\delta_X(x, x') = \inf\{d_{\mathbb{D}}(0, z) : x = f(0) \text{ and } x' = f(z) \text{ for some } f \in \mathscr{O}(\mathbb{D}, X)\}$$

for all x and x' in X. (As usual the infimum of the empty set is ∞.)

In fact (3.1), the definition of δ_X, and the triangle inequality imply that any Schwarz–Pick pseudometric d on X satisfies

$$d(x, x') \leq \inf\left\{ \sum_{j=1}^{n} \delta_X(x_{j-1}, x_j) \right\} \quad \text{for all } x \text{ and } x' \text{ in } X, \qquad (3.2)$$

where the infimum is taken over all positive integers n and all $(n+1)$-tuples of points $x_0, \ldots, x_n$ in X such that $x_0 = x$ and $x_n = x'$.

The infimum on the right side of the inequality (3.2) defines a function on $X \times X$ that is obviously a Schwarz–Pick pseudometric on X, so (3.2) implies that

$$K_X(x,x') = \inf \left\{ \sum_{j=1}^{n} \delta_X(x_{j-1},x_j) \right\} \qquad \text{for all } x \text{ and } x' \text{ in } X. \tag{3.3}$$

(The infimum is of course taken over the same set as in (3.2) above.)

Equation (3.3) is Kobayashi's definition of the pseudometric K_X. It follows readily from (3.3) and the Schwarz–Pick lemma that the functor assigning the Kobayashi pseudometric K_X to each complex Banach manifold X is a Schwarz–Pick system. A slightly stronger property of this functor will follow from the arc length description of K_X that we shall explain in the remainder of section 2.

3.2. The infinitesimal Kobayashi pseudometric and its integrated form

Every complex Banach manifold X has an infinitesimal Kobayashi pseudometric k_X, first introduced (in the finite dimensional case) by Kobayashi in [Kob67]. Since k_X is a function on the tangent bundle $T(X)$ of X, we shall briefly review some properties of tangent bundles.

For the moment let X be a C^1 manifold modelled on a real Banach space V. For each x in X the tangent space to X at x will be denoted by $T_x(X)$. The tangent bundle $T(X)$ of X consists of the ordered pairs (x,v) such that $x \in X$ and $v \in T_x(X)$ (see [Lan62]).

If X is an open set in V with the C^1 structure induced by the inclusion map, then each $T_x(X)$, x in X, is naturally identified with V, and $T(X) = X \times V$.

If X and Y are C^1 manifolds and x is a point of X, every C^1 map $f: X \to Y$ induces a linear map $f_*(x)$ from $T_x(X)$ to $T_{f(x)}(Y)$ (see [Lan62]). If X and Y are subregions of Banach spaces V and W and the tangent spaces $T_x(X)$ and $T_{f(x)}(Y)$ are identified with V and W in the natural way, then $f_*(x)$ is the usual Fréchet derivative of f at x.

The tangent bundle $T(X)$ has a natural C^0 manifold structure modelled on $V \times V$. A convenient atlas for $T(X)$ consists of the charts $T(\varphi)$ defined by the formula

$$T(\varphi)(x,v) = (\varphi(x), \varphi_*(x)v), \qquad (x,v) \in T(U),$$

where U is an open set in X, $T(U)$ is the open subset $\{(x,v) \in T(X) : x \in U\}$ of $T(X)$, and φ is a chart on X with domain U. The image of $T(U)$ under $T(\varphi)$ is the open set $\varphi(U) \times V$ in $V \times V$.

If X is a complex Banach manifold modelled on a complex Banach space V, then each tangent space $T_x(X)$ has a unique complex Banach space structure such that the map $\varphi_*(x)$ from $T_x(X)$ to V is a $\mathbb{C}$-linear isomorphism whenever φ is a (holomorphic) chart defined in some neighborhood of x. Furthermore, $T(X)$ has a unique complex Banach manifold structure such that the map $T(\varphi)$ from $T(U)$ to $\varphi(U) \times V$ is biholomorphic for every (holomorphic) chart φ on X with domain U (see [Dou66]).

Now we are ready for Kobayashi's definition of k_X.

Definition 3.3. The infinitesimal Kobayashi pseudometric on the complex Banach manifold X is the function k_X on $T(X)$ defined by the formula

$$k_X(x,v) = \inf\{|z| : x = f(0) \text{ and } v = f_*(0)z \text{ for some } f \in \mathscr{O}(\mathbb{D},X)\}. \tag{3.4}$$

Obviously $k_X(x,v) \geq 0$ and $k_X(x,cv) = |c|k_X(x,v)$ for all complex numbers c. The following Schwarz–Pick property is also an immediate consequence of the definition (see [Roy88] or Theorem 1.2.6 in [NO90]).

Proposition 3.4. *If X and Y are complex Banach manifolds and $f \in \mathscr{O}(X,Y)$, then*

$$k_Y(f(x), f_*(x)v) \leq k_X(x,v) \qquad \text{for all } (x,v) \in T(X).$$

In particular, if f is biholomorphic then $k_Y(f(x), f_*(x)v) = k_X(x,v)$.

Corollary 3.5. $k_{\mathbb{D}}(w,z) = \dfrac{|z|}{1-|w|^2}$ *for all (w,z) in $\mathbb{D} \times \mathbb{C}$.*

Proof. Definition 3.3 and Schwarz's lemma imply that $k_{\mathbb{D}}(0,z) = |z|$ for all complex numbers z. To prove the formula for $k_{\mathbb{D}}(w,z)$, apply Proposition 3.4 with $X = Y = \mathbb{D}$ and $f(\zeta) = (\zeta - w)/(1 - \zeta\overline{w})$, ζ in $\mathbb{D}$. $\qquad\square$

We shall use the function k_X to measure the lengths of piecewise C^1 curves in X. As usual, if the curve $\gamma \colon [a,b] \to X$ is differentiable at t in $[a,b]$ the symbol $\gamma'(t)$ denotes the tangent vector $\gamma_*(t)1$ to X at $\gamma(t)$. If γ is piecewise C^1, it is natural to define the Kobayashi length of γ by integrating the function $k_X(\gamma(t), \gamma'(t))$ over the parameter interval of γ. That function is upper semicontinuous when X is either a domain in a complex Banach space (see [Har79] or [Din89]) or a finite dimensional complex manifold (see [Roy71] and [Roy74] or [NO90]), but the case of infinite dimensional complex manifolds is harder to deal with. In [Roy88] Royden evades that difficulty by using the upper Riemann integral. Venturini [Ven89] gets more refined results by using upper and lower Lebesgue integrals. We shall follow Royden's example, as it allows the very elementary arguments that we shall now present.

The required upper Riemann integrals exist because the function k_X is locally bounded on $T(X)$. To prove this we use special charts on X. By definition, a *standard chart at x* in X is a biholomorphic map φ of an open neighborhood of x onto the open unit ball of V with $\varphi(x) = 0$.

Lemma 3.6. *If φ is a standard chart at the point x in X, then*

$$k_X(y,v) \le 2\|\varphi_*(y)v\|$$

for all (y,v) in $T(X)$ such that y is in the domain of φ and $\|\varphi(y)\| \le 1/2$.

Proof. If $v = 0$ the inequality is trivial. If $v \ne 0$ we derive it from (3.4) by setting

$$f(z) = \varphi^{-1}\left(\varphi(y) + z\frac{\varphi_*(y)v}{2\|\varphi_*(y)v\|}\right), \qquad z \in \mathbb{D},$$

so that $f \in \mathcal{O}(\mathbb{D},X)$, $f(0) = y$, and $f_*(0)c = \dfrac{c}{2\|\varphi_*(y)v\|}v$ for all c in $\mathbb{C}$. $\qquad\square$

Corollary 3.7. *The function k_X is locally bounded in $T(X)$.*

Proof. Let U be the domain of the standard chart φ in Lemma 3.6. Since holomorphic maps are C^1, the function $(y,v) \mapsto 2\|\varphi_*(y)v\|$ is locally bounded in the open set $T(U) = \{(y,v) \in T(X) : y \in U\}$. $\qquad\square$

Following Royden [Roy88], we can now define the arc length $L_X(\gamma)$ of a piecewise C^1 curve $\gamma \colon [a,b] \to X$ in X to be the upper Riemann integral

$$L_X(\gamma) = \overline{\int_a^b} k_X(\gamma(t), \gamma'(t))\, dt, \tag{3.5}$$

and the distance $\rho_X(x,y)$ to be the infimum of the lengths of all piecewise C^1 curves joining x to y in X. The resulting pseudometric ρ_X on X is the integrated form of the Kobayashi infinitesimal pseudometric k_X. By Proposition 3.4,

$$\rho_Y(f(x_1), f(x_2)) \le \rho_X(x_1, x_2) \qquad \text{for all } x_1 \text{ and } x_2 \text{ in } X \tag{3.6}$$

whenever $f \in \mathcal{O}(X,Y)$. In fact even more is true. If $f \in \mathcal{O}(X,Y)$ and γ is a piecewise C^1 curve in X, then $L_Y(f \circ \gamma) \le L_X(\gamma)$.

Remark 3.8. The upper Riemann integral of an upper semicontinuous function equals its Lebesgue integral, so we can use a Lebesgue integral in (3.5) if X is finite dimensional or a region in a complex Banach space, but that is an unnecessary luxury.

Remark 3.9. By Corollary 3.5, $\rho_{\mathbb{D}}$ is the Poincaré metric $d_{\mathbb{D}}$ on $\mathbb{D}$. Therefore, by (3.6), the functor that assigns ρ_X to each complex Banach manifold X is a Schwarz–Pick system. In particular ρ_X is a Schwarz–Pick pseudometric on X, so $\rho_X(x,y) \le K_X(x,y)$ for all x and y in X. In the next subsection we shall use methods of Harris [Har79] to prove that the pseudometrics ρ_X and K_X are in fact equal.

3.3. Upper bounds for Schwarz–Pick pseudometrics

We begin with a simple estimate.

Lemma 3.10. *Let d be a Schwarz–Pick pseudometric on X and let φ be a standard chart at the point x_0 in X. If $0 < r < 1/3$ there is a constant $C(r)$ such that*

$$d(x,y) \leq C(r)\|\varphi(x) - \varphi(y)\|$$

for all x and y in the domain of φ such that $\|\varphi(x)\| \leq r$ and $\|\varphi(y)\| \leq r$.

Proof. There is nothing to prove if $x = y$. If x, y and r satisfy the stated conditions and $x \neq y$, define f in $\mathscr{O}(\mathbb{D}, X)$ by

$$f(z) = \varphi^{-1}\left(\varphi(x) + (1-r)z\frac{\varphi(y) - \varphi(x)}{\|\varphi(y) - \varphi(x)\|}\right), \qquad z \in \mathbb{D}.$$

Our hypothesis implies that $z_0 = \|\varphi(y) - \varphi(x)\|/(1-r)$ belongs to $\mathbb{D}$. Since $f(z_0) = y$ and $f(0) = x$, inequality (3.1) gives

$$d(x,y) \leq d_{\mathbb{D}}(0, \|\varphi(y) - \varphi(x)\|/(1-r)).$$

Since $d_{\mathbb{D}}(0,s)/s$ is an increasing function of s in the interval $(0,1)$ the required inequality holds with $C(r) = d_{\mathbb{D}}(0, 2r/(1-r))/2r$. $\qquad\square$

The crucial step is the following result from [Har79] (see pp. 368 and 371).

Lemma 3.11 (Harris). *Let $\gamma \colon [a,b] \to X$ be a C^1 curve in X. If d is a Schwarz–Pick metric on X then*

$$\limsup_{t \to s} \frac{d(\gamma(t), \gamma(s))}{|t - s|} \leq k_X(\gamma(s), \gamma'(s)) \qquad \text{for all } s \in [a,b].$$

Proof. Choose s in $[a,b]$ and a standard chart φ at $\gamma(s)$. Let c be a complex number such that there is f in $\mathscr{O}(\mathbb{D}, X)$ with $f(0) = \gamma(s)$ and $f_*(0)c = \gamma'(s)$.

Since the curves $t \mapsto \varphi(\gamma(t))$ and $t \mapsto \varphi(f(c(t-s)))$ are tangent at $t = s$,

$$\varphi(\gamma(t)) - \varphi(f(c(t-s))) = o(t-s)$$

as t approaches s in $[a,b]$. Therefore, by Lemma 3.10 and inequality (3.1),

$$d(\gamma(t), \gamma(s)) \leq d(f(c(t-s)), f(0)) + o(t-s) \leq d_{\mathbb{D}}(c(t-s), 0) + o(t-s)$$

and

$$\limsup_{t \to s} \frac{d(\gamma(t), \gamma(s))}{|t - s|} \leq \limsup_{t \to s} \frac{d_{\mathbb{D}}(c(t-s), 0)}{|t - s|} = |c|.$$

Since $k_X(\gamma(s), \gamma'(s))$ is the infimum of all such complex numbers $|c|$, the lemma is proved. $\qquad\square$

We also need a simple fact from integration theory. (See pp. 369 and 370 of [Har79] for a proof that uses only the Riemann integrability of the function h.)

Lemma 3.12. *Let h be a positive continuous function on the closed interval $[a,b]$ and let ρ be a pseudometric on $[a,b]$. If*

$$\limsup_{t\to s} \frac{\rho(s,t)}{|s-t|} \le h(s) \qquad \text{for all } s \in [a,b], \tag{3.7}$$

then $\rho(a,b) \le \int_a^b h(t)\, dt$.

Proof. Set $J = \rho(a,b)/\int_a^b h$. If $a \le x < c < y \le b$ and $\rho(x,y) \ge J \int_x^y h$ then, by the triangle inequality, either $\rho(x,c) \ge J \int_x^c h$ or $\rho(c,y) \ge J \int_c^y h$.

We can therefore inductively define sequences $\{x_n\}$ and $\{y_n\}$ in $[a,b]$ such that $x_1 = a$, $y_1 = b$, $x_n \le x_{n+1} < y_{n+1} \le y_n$, $(y_{n+1} - x_{n+1}) = \frac{1}{2}(y_n - x_n)$, and

$$\rho(x_n, y_n) \ge J \int_{x_n}^{y_n} h(t)\, dt$$

for every n. Let $s = \lim_{n\to\infty} x_n = \lim_{n\to\infty} y_n$. For each n we can choose t_n equal to one of the points x_n or y_n so that $t_n \ne s$ and

$$\frac{\rho(s,t_n)}{|s-t_n|} \ge J \left| \frac{\int_s^{t_n} h(t)\, dt}{s - t_n} \right|.$$

As $n \to \infty$ the right side of this inequality converges to $Jh(s)$. By hypothesis $h(s) > 0$ and the lim sup of the left side is at most $h(s)$, so $J \le 1$. $\qquad\square$

The following theorem, which follows readily from Lemmas 3.11 and 3.12, is a special case of Theorem 3.1 of Venturini [Ven89]. Venturini's treatment of arc length uses both a more refined theory of integration and a wider class of admissible curves.

Theorem 3.13. *If d is a Schwarz–Pick metric on X, then $d(x,y) \le \rho_X(x,y)$ for all x and y in X.*

Proof. We follow the proof of Proposition 14 in [Har79]. It suffices to prove that $d(\gamma(a), \gamma(b)) \le L_X(\gamma)$ for all C^1 curves $\gamma \colon [a,b] \to X$. By definition

$$L_X(\gamma) = \overline{\int}_a^b f(t)\, dt,$$

with $f(t) = k_X(\gamma(t), \gamma'(t))$ for t in $[a,b]$.

Let $\varepsilon > 0$ be given. By the definition of upper Riemann sums and integrals there is a continuous function h on $[a,b]$ such that $h(t) > f(t)(\geq 0)$ for all t in $[a,b]$ and

$$\int_a^b h(t)\,dt < L_X(\gamma) + \varepsilon. \tag{3.8}$$

Consider the pseudometric $\rho(s,t) = d(\gamma(s),\gamma(t))$ on $[a,b]$. By Lemma 3.11, ρ and h satisfy (3.7), so Lemma 3.12 and the inequality (3.8) give

$$d(\gamma(a),\gamma(b)) = \rho(a,b) \leq \int_a^b h(t)\,dt < L_X(\gamma) + \varepsilon.$$

Since ε is arbitrary the proof is complete. $\qquad\qquad\qquad\qquad\qquad\qquad\Box$

Corollary 3.14. *The pseudometric ρ_X is the Kobayashi pseudometric on X.*

This is obvious since Theorem 3.13 identifies ρ_X as the largest Schwarz–Pick pseudometric on X. We conclude that K_X is the integrated form of the infinitesimal pseudometric k_X.

4. The Carathéodory pseudometric and its infinitesimal form

4.1. The definitions

Let x and y be points of the complex Banach manifold X and let v be a tangent vector to X at x. Since the Kobayashi pseudometrics form a Schwarz–Pick system, Definition 2.1 implies that

$$d_{\mathbb{D}}(f(x),f(y)) \leq K_X(x,y) \qquad \text{for all } f \text{ in } \mathscr{O}(X,\mathbb{D}). \tag{4.1}$$

Similarly, Proposition 3.4 and Corollary 3.5 give

$$|f_*(x)v| \leq \frac{|f_*(x)v|}{1 - |f(x)|^2} \leq k_X(x,v) \qquad \text{for all } f \text{ in } \mathscr{O}(X,\mathbb{D}). \tag{4.2}$$

Therefore the numbers

$$C_X(x,y) = \sup\{d_{\mathbb{D}}(f(x),f(y)) : f \in \mathscr{O}(X,\mathbb{D})\} \tag{4.3}$$

and

$$c_X(x,v) = \sup\{|f_*(x)v| : f \in \mathscr{O}(X,\mathbb{D})\} \tag{4.4}$$

are finite and are bounded by $K_X(x,y)$ and $k_X(x,v)$ respectively.

By definition, $C_X(x, y)$ is the Carathéodory pseudo-distance between x and y, and $c_X(x, v)$ is the Carathéodory length of v. The functions C_X and c_X are called the Carathéodory pseudometric and the infinitesimal Carathéodory pseudometric respectively.

When X is the unit disk $\mathbb{D}$, we can take f in $\mathcal{O}(\mathbb{D}, \mathbb{D})$ to be the identity map and obtain

$$d_{\mathbb{D}}(z, w) \leq C_{\mathbb{D}}(z, w) \leq K_{\mathbb{D}}(z, w) = d_{\mathbb{D}}(z, w) \qquad \text{for all } z \text{ and } w \text{ in } \mathbb{D}$$

from (4.3) and (4.1).

Similarly, given any w in $\mathbb{D}$, we use the map $\zeta \mapsto (\zeta - w)/(1 - \bar{\zeta} w)$ in $\mathcal{O}(\mathbb{D}, \mathbb{D})$ to obtain

$$\frac{|z|}{1 - |w|^2} \leq c_{\Delta}(w, z) \leq k_{\mathbb{D}}(w, z) = \frac{|z|}{1 - |w|^2} \qquad \text{for all } z \text{ in } \mathbb{C}$$

from (4.4), (4.2), and Corollary 3.5. These observations verify the well-known fact that the functions $C_{\mathbb{D}}$ and $c_{\mathbb{D}}$ are respectively the Poincaré metric and the infinitesimal Poincaré metric on $\mathbb{D}$.

It is evident from the definition (4.3) that the functor assigning the Carathéodory pseudometric C_X to each complex Banach manifold X is a Schwarz–Pick system. This Carathéodory functor is the first Schwarz–Pick system, having been introduced by Carathéodory in [Car26] for domains in $\mathbb{C}^2$.

4.2. The derivative of C_X

It is well known (see for example [Din89], [Har79], [JP93], or [Ven89]) that the function c_X is continuous on $T(X)$ and is the derivative of C_X in the sense that

$$\lim_{t \to 0} \frac{C_X(\gamma(0), \gamma(t))}{|t|} = c_X(\gamma(0), \gamma'(0)) \tag{4.5}$$

whenever $\gamma: (-\varepsilon, \varepsilon) \to X$ is a C^1 curve in X. Since the cited references provide detailed proofs of (4.5) only when X is an open subset of V, we shall prove (4.5) here by extending the treatment of C_X and c_X in [EH70] to the manifold case. Our proof will show that (4.5) also holds when t is a complex variable and $\gamma(t)$ is a holomorphic map of a neighborhood of 0 into X (see Corollary 4.5 below).

We need two well-known lemmas.

Lemma 4.1. *Let $H^{\infty}(X)$ be the Banach space of bounded holomorphic functions on X, and let $H^{\infty}(X)^*$ be its dual space. The map $\phi: X \to H^{\infty}(X)^*$ defined by*

$$\phi(x)(f) = f(x), \qquad x \in X \text{ and } f \in H^{\infty}(X),$$

is holomorphic.

Proof. Fix x_0 in X and a neighborhood U of x_0 that is biholomorphically equivalent to a bounded open set in V. It follows readily from Cauchy's estimates that the formula $\phi_U(x)(f) = f(x)$, x in U and f in $H^\infty(U)$, defines a holomorphic embedding of U in $H^\infty(U)^*$. (See [EH70] for details.) The restriction of ϕ to U is the composition of ϕ_U with the bounded linear map from $H^\infty(U)^*$ to $H^\infty(X)^*$ that takes ℓ in $H^\infty(U)^*$ to the linear functional $f \mapsto \ell(f|U)$ on $H^\infty(X)$, so ϕ is holomorphic in U. $\quad\square$

Corollary 4.2. *If x is a point of X and v is a tangent vector to X at x, then the Carathéodory length $c_X(x,v)$ of v equals the norm $\|\phi_*(x)v\|$ of the linear functional $\phi_*(x)v$ on $H^\infty(X)$. In particular, c_X is a continuous function on $T(X)$.*

Proof. Let $\gamma(t)$ be a C^1 curve in X, defined in a neighborhood of 0, with $\gamma(0) = x$ and $\gamma'(0) = v$. For each f in $H^\infty(X)$ we have

$$(\phi_*(x)v)(f) = \lim_{t \to 0} \frac{\phi(\gamma(t))(f) - \phi(\gamma(0))(f)}{t} = \lim_{t \to 0} \frac{f(\gamma(t)) - f(x)}{t} = f_*(x)v.$$

Therefore $\phi_*(x)v$ is the linear functional that takes f in $H^\infty(X)$ to $f_*(x)v$. By (4.4), the norm of that functional is $c_X(x,v)$. $\quad\square$

The statements and proofs of the following lemma and its first corollary are implicit in Lewittes's paper [Lew66], particularly in the reasoning on its final page.

Lemma 4.3. *If a and b are points in $\mathbb{D}$, then $|a - b| \leq 2\tanh \dfrac{d_\mathbb{D}(a,b)}{2}$, with equality if and only if $b = \pm a$.*

Proof. If $a = b$ there is nothing to prove.

If $a \neq b$ choose a positive number $r < 1$ and a conformal map g of $\mathbb{D}$ onto itself so that $g(r) = a$ and $g(-r) = b$. Then $d_\mathbb{D}(a,b) = d_\mathbb{D}(r,-r) = 2d_\mathbb{D}(r,0)$, so

$$2\tanh \frac{d_\mathbb{D}(a,b)}{2} = 2\tanh d_\mathbb{D}(r,0) = 2r.$$

It therefore suffices to prove that $|g(r) - g(-r)| \leq 2r$ with equality if and only if $g(0) = 0$. That is easily done by writing g in the form

$$g(z) = e^{i\theta}(z - \alpha)/(1 - \overline{\alpha}z)$$

with $|\alpha| < 1$. $\quad\square$

Corollary 4.4. $\|\phi(x) - \phi(y)\| = 2\tanh \dfrac{c_X(x,y)}{2}$ *for all x and y in X,*

Proof. Let x and y be given. For any f in $H^\infty(X)$ with norm less than one, let g_f be a conformal map of $\mathbb{D}$ onto itself that satisfies $g_f(f(x)) = -g_f(f(y))$. Lemma 4.3 and the definition of ϕ give

$$|(\phi(x) - \phi(y))(f)| \leq 2\tanh\frac{d_{\mathbb{D}}(f(x), f(y))}{2} = |(\phi(x) - \phi(y))(g_f \circ f)|.$$

The corollary follows by taking suprema over f.   □

We have already established the convention that if a curve γ in X is differentiable at some point t in its parameter interval, then $\gamma'(t)$ is the tangent vector $\gamma_*(t)1$ to X at $\gamma(t)$. It will be convenient from now on to use the same convention when t is a complex variable and γ is a holomorphic map from a neighborhood of t to X.

Corollary 4.5. *If $t \mapsto \gamma(t)$ is either a C^1 map of an open interval $(-\varepsilon, \varepsilon)$ into X or a holomorphic map of an open disk $\{t \in \mathbb{C} : |t| < \varepsilon\}$ into X, then (4.5) holds.*

Proof. Under either hypothesis on γ, Corollary 4.4 and Lemma 4.1 imply that

$$\lim_{t \to 0} \frac{C_X(\gamma(0), \gamma(t))}{|t|} = \lim_{t \to 0} \frac{2}{|t|} \tanh^{-1} \frac{\|\phi(\gamma(0)) - \phi(\gamma(t))\|}{2} = \|(\phi \circ \gamma)'(0)\|,$$

which equals $c_X(\gamma(0), \gamma'(0))$ by Corollary 4.2.   □

4.3. The Carathéodory–Reiffen pseudometric

In general $C_X(x, y)$ cannot be defined as the infimum of lengths of curves joining x to y. In fact there are bounded domains of holomorphy X in $\mathbb{C}^2$ such that not all open C_X balls are connected (see [JP93]). For that reason the pseudometric $\widetilde{C}_X$ generated by the infinitesimal Carathéodory metric c_X has independent interest.

Since it was first systematically studied (in the finite dimensional case) by Reiffen in [Ref65], $\widetilde{C}_X$ is called the Carathéodory–Reiffen pseudometric on X. It is defined in the obvious way. The Carathéodory length of a piecewise C^1 curve $\gamma \colon [a, b] \to X$ in X is

$$\widetilde{L}_X(\gamma) = \int_a^b c_X(\gamma(t), \gamma'(t))\, dt \tag{4.6}$$

and the distance $\widetilde{C}_X(x, y)$ is the infimum of the lengths of all piecewise C^1 curves joining x to y. Observe that the integrand in (4.6) is piecewise continuous.

Since $c_{\mathbb{D}}$ is the infinitesimal Poincaré metric on $\mathbb{D}$, its integrated form $\widetilde{C}_{\mathbb{D}}$ is the Poincaré metric. In addition, it follows readily from the definitions (4.4) and (4.6) that $\widetilde{L}_Y(f \circ \gamma) \leq \widetilde{L}_X(\gamma)$ for every piecewise C^1 curve in X and $f \in \mathcal{O}(X, Y)$. Therefore the functor assigning $\widetilde{C}_X$ to each complex Banach manifold X is a Schwarz–Pick system. In particular, if x and y are points in X, then

$$d_{\mathbb{D}}(f(x), f(y)) \leq \widetilde{C}_X(x, y) \qquad \text{for all } f \text{ in } \mathscr{O}(X, \mathbb{D}).$$

Definition (4.3) therefore implies that

$$C_X(x, y) \leq \widetilde{C}_X(x, y) \qquad \text{for all } x \text{ and } y \text{ in } X. \tag{4.7}$$

The relationship between $\widetilde{C}_X$ and C_X is explored more fully in [Din89], [Har79], and [JP93].

5. Distortion theorems for complex non-geodesics

5.1. Complex geodesics

Since $X \mapsto \widetilde{C}_X$ is a Schwarz–Pick system, $\widetilde{C}_X$ is a Schwarz–Pick pseudometric on X for every complex Banach manifold X. Therefore $\widetilde{C}_X \leq K_X$ for every X. Combining that inequality with (4.7) we obtain

$$C_X(f(z), f(w)) \leq \widetilde{C}_X(f(z), f(w)) \leq K_X(f(z), f(w)) \leq d_{\mathbb{D}}(z, w) \tag{5.1}$$

whenever X is a complex Banach manifold, $f \in \mathscr{O}(\mathbb{D}, X)$, and z and w are points of $\mathbb{D}$.

Following Vesentini [Ves81], we call f in $\mathscr{O}(\mathbb{D}, X)$ a complex geodesic (more precisely a complex C_X-geodesic) if there is a pair of distinct points z and w in $\mathbb{D}$ with

$$C_X(f(z), f(w)) = d_{\mathbb{D}}(z, w), \tag{5.2}$$

so that none of the inequalities in (5.1) is strict. By a theorem of Vesentini (see Proposition 3.3 in [Ves81]), if f is a complex geodesic then in fact (5.2) holds for all z and w in $\mathbb{D}$ (see also [Din89], [JP93], and [Ves82]).

In other words, if the inequality $C_X(f(z), f(w)) < d_{\mathbb{D}}(z, w)$ holds for some pair of points then it must hold for all pairs of distinct points in $\mathbb{D}$. Our results in this section study quantitatively how the distortion of distance at one pair of points influences the degree of distortion at another.

5.2. The Beardon–Minda quotient

The prototypes for our theorems can be found in the paper [BMi], where Beardon and Minda make an elegant systematic study of the situation when the target manifold

$X = \mathbb{D}$. Their results indicate that the distortion of distance is appropriately measured by the quotient

$$Q_X(f,a,b) = \frac{\tanh C_X(f(a),f(b))}{\tanh d_\mathbb{D}(a,b)}, \tag{5.3}$$

where $f \in \mathscr{O}(\mathbb{D},X)$ and a and b are distinct points of $\mathbb{D}$. We call $Q_X(f,a,b)$ the Beardon–Minda quotient because for f in $\mathscr{O}(\mathbb{D},\mathbb{D})$ the number $Q_\mathbb{D}(f,a,b)$ is the absolute value of the "hyperbolic difference quotient" $f^*(a,b)$ on which Beardon and Minda base their study (see section 2 of [BMi]).

It is obvious that $0 \leq Q_X(f,a,b) \leq 1$, and $Q_X(f,a,b) = 1$ if and only if f is a complex geodesic. The size of Q_X provides a quantitative measure of how close f is to being a complex geodesic.

Our next result is inspired by and follows readily from Theorem 3.1 of [BMi]. Its proof provides a new proof (without use of normal family arguments) of Vesentini's theorem.

Theorem 5.1. *If X is a complex Banach manifold, $f \in \mathscr{O}(\mathbb{D},X)$, and equation (5.2) fails for some pair of points in $\mathbb{D}$, then $Q_X(f,a,b) < 1$ whenever the points a and b in $\mathbb{D}$ are distinct. Moreover, if a, b, and c are points in $\mathbb{D}$ and neither b nor c equals a, then*

$$d_\mathbb{D}(Q_X(f,a,b),Q_X(f,a,c)) \leq d_\mathbb{D}(b,c). \tag{5.4}$$

5.3. Proof of Theorem 5.1

Assume first that $X = \mathbb{D}$. In this case $C_X = d_\mathbb{D}$ and Theorem 5.1 reduces to a weak version of Theorem 3.1 in [BMi]. For the reader's convenience we include its proof, imitating the proof in [BMi].

By pre- and post-composing f with appropriate Poincaré isometries, we may assume that $a = f(a) = 0$. Then $Q_\mathbb{D}(f,0,z) = |f(z)|/|z|$ for all nonzero z in $\mathbb{D}$, so the conclusion of Theorem 5.1 reduces to the inequalities $|f(b)| < |b|$ for all nonzero b in $\mathbb{D}$ and

$$d_\mathbb{D}\left(\frac{|f(b)|}{|b|}, \frac{|f(c)|}{|c|}\right) \leq d_\mathbb{D}(b,c) \text{ for all nonzero } b \text{ and } c \text{ in } \mathbb{D}.$$

Since $f(0) = 0$ and f is not a rotation, the first inequality follows immediately from Schwarz's lemma. To obtain the second, define g in $\mathscr{O}(\mathbb{D},\mathbb{D})$ by the formula

$$g(0) = f'(0) \quad \text{and} \quad g(z) = f(z)/z \text{ if } 0 < |z| < 1,$$

apply the Schwarz-Pick lemma to g, and observe that $d_\mathbb{D}(|\zeta|,|\zeta'|) \leq d_\mathbb{D}(\zeta,\zeta')$ for all ζ and ζ' in $\mathbb{D}$.

Now we consider the general case. Given f in $\mathcal{O}(\mathbb{D},X)$ we choose any pair of distinct points a and b in $\mathbb{D}$. Suppose $Q_X(f,a,b) < 1$.

Given any g in $\mathcal{O}(X,\mathbb{D})$, set $h = g \circ f$. Then $Q_{\mathbb{D}}(h,a,b) < 1$, so (by what we already proved) for all c in $\mathbb{D} \setminus \{a\}$ we have $Q_{\mathbb{D}}(h,a,c) < 1$ and

$$d_{\mathbb{D}}(Q_{\mathbb{D}}(h,a,b),Q_{\mathbb{D}}(h,a,c)) \le d_{\mathbb{D}}(b,c). \tag{5.5}$$

Set $k = \tanh d_{\mathbb{D}}(b,c)$, $r = Q_{\mathbb{D}}(h,a,b)$, and $\widehat{r} = Q_X(f,a,b)$. By hypothesis these three numbers all lie in the half-open interval $[0,1)$. In terms of k and r, (5.5) becomes the double inequality

$$\frac{r-k}{1-kr} \le Q_{\mathbb{D}}(h,a,c) \le \frac{r+k}{1+kr}. \tag{5.6}$$

Recall that $h = g \circ f$ for some g in $\mathcal{O}(X,\mathbb{D})$. Taking suprema in (5.6) over all such g, we obtain

$$\frac{\widehat{r}-k}{1-k\widehat{r}} \le Q_X(f,a,c) \le \frac{\widehat{r}+k}{1+k\widehat{r}}. \tag{5.7}$$

Since $0 \le \widehat{r} < 1$, (5.7) implies the inequalities $Q_X(f,a,c) < 1$ and (5.4).

We have proved that if $Q_X(f,a,b) < 1$ for some a and b in $\mathbb{D}$ then for all c in $\mathbb{D} \setminus \{a\}$ we have both $Q_X(f,a,c) < 1$ and the inequality (5.4). Theorem 5.1 follows readily. $\square$

5.4. Two corollaries of Theorem 5.1

In [BMi] Beardon and Minda derive many consequences of their Theorem 3.1. Some of them require only the inequality (5.5) and can therefore be generalized to our situation. We shall concentrate on results that involve the distortion of the infinitesimal Carathéodory pseudometric at points of $\mathbb{D}$. We measure that distortion by the Carathéodory norm of the derivative of the map f in $\mathcal{O}(\mathbb{D},X)$, which is defined as follows.

Definition 5.2. Let X be a complex Banach manifold and let $f'(a) = f_*(a)1$ be the derivative of the map f in $\mathcal{O}(\mathbb{D},X)$ at the point a in $\mathbb{D}$. The Carathéodory norm of $f'(a)$ is the ratio of the Carathéodory length of the tangent vector 1 to $\mathbb{D}$ at a and the Carathéodory length of its image in the tangent space to X at $f(a)$.

Since it equals the limit of the Beardon–Minda quotient $Q_X(f,a,c)$ as c approaches a (see (5.10)), we shall denote the Carathéodory norm of $f'(a)$ by the symbol $q_X(f,a)$. Explicitly,

$$q_X(f,a) = \frac{c_X(f(a),f'(a))}{c_{\mathbb{D}}(a,1)} = (1-|a|^2)c_X(f(a),f'(a)). \tag{5.8}$$

378   C. J. Earle, L. A. Harris, J. H. Hubbard & S. Mitra

For f in $\mathscr{O}(\mathbb{D},\mathbb{D})$ the number $q_{\mathbb{D}}(f,a)$ is the absolute value of the "hyperbolic derivative" $f^h(a)$ of Beardon and Minda (see section 2 of [BMi]).

Our next result quantifies Vesentini's theorem (Proposition 3.2 in [Ves81]) that if f in $\mathscr{O}(\mathbb{D},X)$ is not a complex geodesic then it shortens the Carathéodory lengths of all nonzero tangent vectors. For f in $\mathscr{O}(\mathbb{D},\mathbb{D})$ it is a weak special case of Theorem 3.1 in Beardon-Minda [BMi].

Corollary 5.3. *If X is a complex Banach manifold and f in $\mathscr{O}(\mathbb{D},X)$ is not a complex geodesic, then $q_X(f,a) < 1$ for all a in $\mathbb{D}$ and*

$$d_{\mathbb{D}}(Q_X(f,a,b),q_X(f,a)) \le d_{\mathbb{D}}(a,b) \tag{5.9}$$

whenever a and b are distinct points of $\mathbb{D}$.

Proof. Let a and b be distinct points of $\mathbb{D}$. As $Q_X(f,a,b) < 1$ by hypothesis, (5.4) holds for any c in $\mathbb{D} \setminus \{a\}$. We shall obtain (5.9) as the limiting case of (5.4) when $c \to a$ while a and b are held fixed.

As our first step, we apply Corollary 4.5 to the holomorphic maps $z \mapsto a + z$ and $z \mapsto f(a+z)$ from $\{z \in \mathbb{C} : |z| < 1 - |a|\}$ to $\mathbb{D}$ and X, obtaining the equations

$$\lim_{z \to a} \frac{C_X(f(a),f(z))}{|z-a|} = c_X(f(a),f'(a))$$

and

$$\lim_{z \to a} \frac{d_{\mathbb{D}}(a,z)}{|z-a|} = c_{\mathbb{D}}(a,1) \left(= \frac{1}{1-|a|^2}\right).$$

Using these equations, (5.3), and Definition 5.2, we obtain

$$\lim_{c \to a} Q_X(f,a,c) = \lim_{c \to a} \frac{C_X(f(a),f(c))}{d_{\mathbb{D}}(a,c)} = \frac{c_X(f(a),f'(a))}{c_{\mathbb{D}}(a,1)} = q_X(f,a). \tag{5.10}$$

As $d_{\mathbb{D}}(Q_X(f,a,b),Q_X(f,a,c)) \le d_{\mathbb{D}}(b,c) \le d_{\mathbb{D}}(a,b) + d_{\mathbb{D}}(c,a)$ for all c in $\mathbb{D} \setminus \{a\}$, it follows readily from (5.10) that $q_X(f,a)$ belongs to $\mathbb{D}$ and satisfies (5.9). $\square$

For f in $\mathscr{O}(\mathbb{D},\mathbb{D})$ our next corollary is a restatement of Corollary 3.7 of Beardon and Minda [BMi]. The general case has essentially the same proof.

Corollary 5.4. *If X is a complex Banach manifold and f in $\mathscr{O}(\mathbb{D},X)$ is not a complex geodesic, then*

$$d_{\mathbb{D}}(q_X(f,a),q_X(f,b)) \le 2d_{\mathbb{D}}(a,b) \qquad \text{for all a and b in $\mathbb{D}$.} \tag{5.11}$$

Proof. If $a = b$ there is nothing to prove. If $a \ne b$ then Corollary 5.3 implies that the number

$$d_{\mathbb{D}}(Q_X(f,a,b),q_X(f,a)) + d_{\mathbb{D}}(Q_X(f,a,b),q_X(f,b))$$

is bounded by $2d_{\mathbb{D}}(a,b)$, so (5.11) follows from the triangle inequality. $\square$

5.5. The $\widetilde{C}_X$ version of Corollary 5.3

We do not know whether Theorem 5.1 remains valid when C_X is replaced by $\widetilde{C}_X$ in its conclusion, but Corollary 5.3 does, as we shall now prove. Our method of proof is to use the inequality (5.11) to bound the Carathéodory length of appropriate curves in X. It was suggested to us in a private communication from David Minda about the classical case when $X = \mathbb{D}$.

Theorem 5.5. *If X is a complex Banach manifold and f in $\mathcal{O}(\mathbb{D}, X)$ is not a complex C_X-geodesic, then $\widetilde{C}_X(f(a), f(b)) < d_{\mathbb{D}}(a, b)$ and*

$$d_{\mathbb{D}}\left(\frac{\tanh \widetilde{C}_X(f(a), f(b))}{\tanh d_{\mathbb{D}}(a, b)}, q_X(f, a)\right) \leq d_{\mathbb{D}}(a, b) \tag{5.12}$$

whenever the points a and b in $\mathbb{D}$ are distinct.

Proof. Let a and b be distinct points of $\mathbb{D}$. Set $k = \tanh d_{\mathbb{D}}(a, b)$ and $r = q_X(f, a)$, so that (5.12) can be written as the double inequality

$$k\frac{r-k}{1-rk} \leq \tanh \widetilde{C}_X(f(a), f(b)) \leq k\frac{r+k}{1+rk}. \tag{5.13}$$

By hypothesis, $0 < k < 1$ and (by Corollary 5.3) $0 \leq r < 1$. Therefore the right side of (5.13) is less than k, so (5.13) implies both $\widetilde{C}_X(f(a), f(b)) < d_{\mathbb{D}}(a, b)$ and (5.12). We shall prove (5.13).

In terms of r and k, the known inequality (5.9) can be written in the form

$$k\frac{r-k}{1-rk} \leq \tanh C_X(f(a), f(b)) \leq k\frac{r+k}{1+rk}. \tag{5.14}$$

Since $C_X(f(a), f(b)) \leq \widetilde{C}_X(f(a), f(b))$, the left side of (5.14) implies the left side of (5.13) but the right side of (5.13) requires proof.

For the proof we may pre-compose f with a Poincaré isometry so that $a = 0$ and $b > 0$. Then $b = \tanh d_{\mathbb{D}}(0, b) = k$ and the curve $\gamma(t) = f(t)$, $0 \leq t \leq k$, joins $f(a)$ to $f(b)$. Therefore

$$\widetilde{C}_X(f(a), f(b)) \leq \widetilde{L}_X(\gamma) = \int_0^k c_X(f(t), f'(t)) \, dt = \int_0^k \frac{q_X(f, t)}{1 - t^2} \, dt,$$

and it suffices to prove the inequality

$$\int_0^k \frac{q_X(f, t)}{1 - t^2} \, dt \leq \tanh^{-1}\left(k\frac{r+k}{1+rk}\right). \tag{5.15}$$

Following a suggestion of David Minda, we use the identity

$$\tanh^{-1}\left(k\frac{r+k}{1+rk}\right) = \int_0^k \frac{r(1+t^2)+2t}{(1-t^2)(1+2rt+t^2)}\,dt. \tag{5.16}$$

Comparison of the integrands in (5.15) and (5.16) shows that it suffices to prove that

$$q_X(f,t) \le \frac{r(1+t^2)+2t}{1+2rt+t^2} \qquad \text{if } 0<t<1. \tag{5.17}$$

If $X=\mathbb{D}$ and $f(0)=0$, (5.17) is contained in the inequality (6.1) in Beardon and Minda [BMi]. For the general case, we imitate their proof, using our Corollary 5.4 as follows. If $0<t<1$, then

$$2d_{\mathbb{D}}(0,t) = d_{\mathbb{D}}(-t,t) = d_{\mathbb{D}}\left(0,\frac{2t}{1+t^2}\right)$$

and (by definition of r)

$$d_{\mathbb{D}}(q_X(f,0),q_X(f,t)) = d_{\mathbb{D}}(r,q_X(f,t)) = d_{\mathbb{D}}\left(0,\frac{q_X(f,t)-r}{1-rq_X(f,t)}\right).$$

Therefore, by Corollary 5.4, $d_{\mathbb{D}}\left(0,\dfrac{q_X(f,t)-r}{1-rq_X(f,t)}\right) \le d_{\mathbb{D}}\left(0,\dfrac{2t}{1+t^2}\right)$, so

$$\left|\frac{q_X(f,t)-r}{1-rq_X(f,t)}\right| \le \frac{2t}{1+t^2} \qquad \text{if } 0<t<1.$$

The required inequality (5.17) follows readily.   □

The qualitative part of Theorem 5.5 provides the following strengthening of Proposition 3.3 in [Ves81].

Corollary 5.6. *If X is a complex Banach manifold, $f \in \mathcal{O}(\mathbb{D},X)$, and the equation $\widetilde{C}_X(f(a),f(b)) = d_{\mathbb{D}}(a,b)$ holds for some pair of distinct points a and b in $\mathbb{D}$, then f is a complex C_X-geodesic.*

Remark 5.7. In [BMi] Beardon and Minda show that the inequalities (5.4), (5.9), (5.11), and (5.12) are already sharp when $X=\mathbb{D}$. For example, equality occurs in all four of them if $f(z)=z^2$, $a=0$, and c in (5.4) is a positive multiple of b. The cases of equality are fully analysed in [BMi].

Remark 5.8. As Theorems 5.1 and 5.5 illustrate, C_X and $\widetilde{C}_X$ are on an equal footing in the theory of complex geodesics. If f in $\mathcal{O}(\mathbb{D},X)$ preserves the distance between two distinct points in either pseudometric, then it preserves the distance between any two points in both pseudometrics.

The Kobayashi pseudometric, however, stands apart. The methods we have used in this section do not apply to it, and simple examples show that the stated results cannot be extended to the Kobayashi pseudometric without additional hypotheses. For instance, holomorphic universal covering maps are known (see [Din89] or [Kob98]) to be local isometries in the Kobayashi pseudometric, so if X is any hyperbolic Riemann surface other than $\mathbb{D}$ none of the theorems and corollaries in this section holds true when C_X or $\tilde{C}_X$ is replaced by K_X.

One might ask whether K_X can be substituted for C_X or $\tilde{C}_X$ in these results if X is required to be simply connected, but there are difficulties even for contractible domains of holomorphy in $\mathbb{C}^n$, $n > 1$. For example, let

$$X = \{(z,w) \in \mathbb{C}^2 : |z| < 1, |w| < 1, \text{ and } |zw| < a^2\},$$

where $0 < a < \frac{1}{2}$. Vigué [Vig85] remarks that the map $\zeta \mapsto (a\zeta, a\zeta)$ from $\mathbb{D}$ to X preserves the Kobayashi lengths of the tangent vectors at $\zeta = 0$ but decreases the Kobayashi distances from 0. Thus the K_X version of Corollary 5.3 fails for holomorphic maps of $\mathbb{D}$ into this domain X.

But Theorem 5.1 could still hold, for Venturini uses the same map in [Ven89] to show that the infinitesimal Kobayashi metric is not the derivative of K_X at the origin in X. The limiting argument by which we obtained Corollary 5.3 from Theorem 5.1 therefore fails in this case, and the question whether some form of Theorem 5.1 holds for K_X in this domain appears to be still open.

Remark 5.9. Samuel Krushkal's preprint [Kru01] contains the remarkable result that the Carathéodory and Teichmüller metrics on the universal Teichmüller space are equal. Using that result we can use the Teichmüller metric in the formulas for Q_X and q_X when X is the universal Teichmüller space. The inequalities (5.4), (5.9), (5.11) then become remarkable strengthenings of the classical principle of Teichmüller contraction (see [Ear02]).

References

[BMi]   A.F. Beardon and D. Minda (in press). A multi-point Schwarz–Pick lemma, *J. Anal. Math.*.

[Car26] C. Carathéodory (1926). Über das Schwarzsche Lemma bei analytischen Funktionen von zwei komplexen Veränderlichen, *Math. Ann.* **97**, 76–98.

[Din89]   S. Dineen (1989). *The Schwarz Lemma*, Oxford Mathematical Monographs, Clarendon Press.

[Dou66]   A. Douady (1966). Le problème des modules pour les sous-espaces analytiques compacts d'un espace analytique donné, *Ann. Inst. Fourier (Grenoble)* **16**, 1–95.

[Ear02]   C. J. Earle (2002). Schwarz's lemma and Teichmüller contraction. In *Complex Manifolds and Hyperbolic Geometry, Contemp. Math.* **311**, Amer. Math. Soc., 79–85.

[EH70]   C. J. Earle and R. S. Hamilton (1970). A fixed point theorem for holomorphic mappings. In *Global Analysis, Proc. Sympos. Pure Math.* **XVI**, Amer. Math. Soc., 61–65.

[FV80]   T. Franzoni and E. Vesentini (1980). Holomorphic Maps and Invariant Distances, *North-Holland Mathematical Studies* **40**, North Holland.

[Har79]   L. A. Harris (1979). Schwarz–Pick systems of pseudometrics for domains in normed linear spaces. In *Advances in Holomorphy (North-Holland Mathematical Studies* **34***)*, North Holland, 345–406.

[JP93]   M. Jarnicki and P. Pflug (1993). Invariant Distances and Metrics in Complex Analysis, *de Gruyter Expositions in Mathematics* **9**, Walter de Gruyter.

[Kob67]   S. Kobayashi (1967). Invariant distances on complex manifolds and holomorphic mappings, *J. Math. Soc. Japan* **19**, 460–480.

[Kob98]   S. Kobayashi (1998). Hyperbolic Complex Spaces, *Grundlehren der Mathematischen Wissenschaften* **318**, Springer-Verlag.

[Kru01]   S. L. Krushkal (2001). Complex geometry of the universal Teichmüller space, Grunsky coefficients and plurisubharmonic functions, *preprint*.

[Lan62]   S. Lang (1962). *Introduction to Differentiable Manifolds*, Wiley Interscience.

[Lew66]   J. Lewittes (1966). A note on parts and hyperbolic geometry, *Proc. Amer. Math. Soc.* **17**, 1087–1090.

[NO90]   J. Noguchi and T. Ochiai (1990). Geometric Function Theory in Several Complex Variables, *Translations of Mathematical Monographs* **80**, Amer. Math. Soc..

[PSh89] E. A. Poletskii and B. V. Shabat (1989). Invariant metrics. In *Several Complex Variables III, Encyclopædia of Mathematical Sciences* **9**, Springer-Verlag, 63–111.

[Ref65] H.-J. Reiffen (1965). Die Carathéodory Distanz und ihre zugehörige Differentialmetrik, *Math. Ann.* **161**, 315–324.

[Roy71] H. L. Royden (1971). Remarks on the Kobayashi metric. In *Several Complex Variables II, Maryland 1970 (Lec. Notes Math.* **185***)*, Springer-Verlag, 125–137.

[Roy74] H. L. Royden (1974). The extension of regular holomorphic maps, *Proc. Amer. Math. Soc.* **43**, 306–310.

[Roy88] H. L. Royden (1988). Hyperbolicity in complex analysis, *Ann. Acad. Sci. Fenn.* **13**, 387–400.

[Ven89] S. Venturini (1989). Pseudodistances and pseudometrics on real and complex manifolds, *Ann. Mat. Pura Appl.* **154**, 385–402.

[Ves81] E. Vesentini (1981). Complex geodesics, *Compositio Math.* **44**, 375–394.

[Ves82] E. Vesentini (1982). Complex geodesics and holomorphic mappings, *Sympos. Math.* **26**, 211–230.

[Vig85] J.-P. Vigué (1985/86). Sur la caractérisation des automorphismes analytiques d'un domaine borné, *Portugal. Math.* **43**, 439–453.

Clifford J. Earle

Department of Mathematics
Cornell University
Ithaca, NY 14853-4201
USA

cliff@polygon.math.cornell.edu

Lawrence A. Harris

Department of Mathematics
University of Kentucky
Lexington, KY 40506–0027
USA

larry@ms.uky.edu

John H. Hubbard

Department of Mathematics
Cornell University
Ithaca, NY 14853-4201
USA

hubbard@math.cornell.edu

Sudeb Mitra[1]

Department of Mathematics
University of Connecticut
Storrs, CT 06269-3009
USA

mitra@math.uconn.edu

AMS Classification: 32F45, 46G20

Keywords: Schwarz–Pick lemma, Kobayashi pseudometric, Carathéodory pseudo-metric

[1] New address: Department of Mathematics, Queens College of CUNY, Flushing, NY 11367-1597, USA; email: smitra@qc1.qc.edu